Molecular Nano Dynamics

*Edited by
Hiroshi Fukumura,
Masahiro Irie,
Yasuhiro Iwasawa,
Hiroshi Masuhara,
and Kohei Uosaki*

Related Titles

Matta, C. F. (ed.)

Quantum Biochemistry

2010
ISBN: 978-3-527-32322-7

Meyer, H.-D., Gatti, F., Worth, G. A. (eds.)

Multidimensional Quantum Dynamics

MCTDH Theory and Applications

2009
ISBN: 978-3-527-32018-9

Reiher, M., Wolf, A.

Relativistic Quantum Chemistry

The Fundamental Theory of Molecular Science

2009
ISBN: 978-3-527-31292-4

Höltje, H.-D., Sippl, W., Rognan, D., Folkers, G.

Molecular Modeling

Basic Principles and Applications
Third, Revised and Expanded Edition

2008
ISBN: 978-3-527-31568-0

Matta, C. F., Boyd, R. J. (eds.)

The Quantum Theory of Atoms in Molecules

From Solid State to DNA and Drug Design

2007
ISBN: 978-3-527-30748-7

Rode, B. M., Hofer, T., Kugler, M.

The Basics of Theoretical and Computational Chemistry

2007
ISBN: 978-3-527-31773-8

Molecular Nano Dynamics

Volume I: Spectroscopic Methods and Nanostructures

Edited by
Hiroshi Fukumura, Masahiro Irie, Yasuhiro Iwasawa,
Hiroshi Masuhara, and Kohei Uosaki

WILEY-VCH Verlag GmbH & Co. KGaA

The Editors

Prof. Dr. Hiroshi Fukumura
Tohoku University
Graduate School of Science
6-3 Aoba Aramaki, Aoba-ku
Sendai 980-8578
Japan

Prof. Dr. Masahiro Irie
Rikkyo University
Department of Chemistry
Nishi-Ikebukuro 3-34-1
Toshima-ku
Tokyo 171-8501
Japan

Prof. Dr. Yasuhiro Iwasawa
University of Electro-Communications
Department of Applied Physics and Chemistry
1-5-1 Chofu
Tokyo 182-8585

and

Emeritus Professor
University of Tokyo
7-3-1 Hongo, Bunkyo-ku
Tokyo 113-0033
Japan

Dr. Hiroshi Masuhara
Nara Institute of Science and Technology
Graduate School of Material Science
8916-5 Takayama, Ikoma
Nara, 630-0192
Japan

and

National Chiao Tung University
Department of Applied Chemistry and
Institute of Molecular Science
1001 Ta Hsueh Road
Hsinchu 30010
Taiwan

Prof. Dr. Kohei Uosaki
Hokkaido University
Graduate School of Science
N 10, W 8 , Kita-ku
Sapporo 060-0810
Japan

■ All books published by Wiley-VCH are carefully produced. Nevertheless, authors, editors, and publisher do not warrant the information contained in these books, including this book, to be free of errors. Readers are advised to keep in mind that statements, data, illustrations, procedural details or other items may inadvertently be inaccurate.

Library of Congress Card No.: applied for

British Library Cataloguing-in-Publication Data
A catalogue record for this book is available from the British Library.

Bibliographic information published by the Deutsche Nationalbibliothek
The Deutsche Nationalbibliothek lists this publication in the Deutsche Nationalbibliografie; detailed bibliographic data are available on the Internet at http://dnb.d-nb.de.

© 2009 WILEY-VCH Verlag GmbH & Co. KGaA, Weinheim

All rights reserved (including those of translation into other languages). No part of this book may be reproduced in any form – by photoprinting, microfilm, or any other means – nor transmitted or translated into a machine language without written permission from the publishers. Registered names, trademarks, etc. used in this book, even when not specifically marked as such, are not to be considered unprotected by law.

Cover Design Adam-Design, Weinheim
Typesetting Thomson Digital, Noida, India
Printing and Binding betz-druck GmbH, Darmstadt

Printed in the Federal Republic of Germany
Printed on acid-free paper

ISBN: 978-3-527-32017-2

Contents to Volume 1

Contents to Volume 2 *XIII*
Preface *XVII*
About the Editors *XIX*
List of Contributors for Both Volumes *XXIII*

Part One Spectroscopic Methods for Nano Interfaces *1*

1 Raman and Fluorescence Spectroscopy Coupled with Scanning Tunneling Microscopy *3*
Noriko Nishizawa Horimoto and Hiroshi Fukumura
1.1 Introduction *3*
1.2 Outline of STM Combined with Optical Spectroscopy *4*
1.2.1 Raman Spectroscopy *4*
1.2.2 Fluorescence Spectroscopy *6*
1.3 Theoretical Approaches *6*
1.4 Experimental Approaches *8*
1.4.1 STM Combined with Raman Spectroscopy *8*
1.4.2 STM Combined With Fluorescence Spectroscopy *12*
1.5 Future Prospects *13*
References *16*

2 Vibrational Nanospectroscopy for Biomolecules and Nanomaterials *19*
Yasushi Inouye, Atsushi Taguchi, and Taro Ichimura
2.1 Introduction *19*
2.2 Surface Plasmon Polaritons *20*
2.3 Near-Field Optical Microscopy Using a Metallic Nano-Tip *22*
2.4 Tip-Enhanced Near-Field Raman Spectroscopy and Imaging *24*
2.4.1 Raman Spectroscopy *25*
2.4.2 Near-Field Nano-Raman Microscopy *25*
2.4.3 Tip-Enhanced Near-Field Raman Spectroscopy and Imaging *26*

Molecular Nano Dynamics, Volume I: Spectroscopic Methods and Nanostructures
Edited by H. Fukumura, M. Irie, Y. Iwasawa, H. Masuhara, and K. Uosaki
Copyright © 2009 WILEY-VCH Verlag GmbH & Co. KGaA, Weinheim
ISBN: 978-3-527-32017-2

2.5	Tip Effect on Near-Field Raman Scattering	30
2.6	Conclusion	36
	References	36

3	**Near-Field Optical Imaging of Localized Plasmon Resonances in Metal Nanoparticles**	**39**
	Hiromi Okamoto and Kohei Imura	
3.1	Introduction	39
3.2	Near-Field Spectroscopic Method	40
3.3	Fundamental Spectroscopic Characteristics of Gold Nanoparticles	42
3.4	Wavefunction Images of Plasmon Modes of Gold Nanorod — Near-Field Transmission Method	42
3.5	Ultrafast Time-Resolved Near-Field Imaging of Gold Nanorods	45
3.6	Near-Field Two-Photon Excitation Images of Gold Nanorods	47
3.7	Enhanced Optical Fields in Spherical Nanoparticle Assemblies and Surface Enhanced Raman Scattering	48
3.8	Concluding Remarks	51
	References	52

4	**Structure and Dynamics of a Confined Polymer Chain Studied by Spatially and Temporally Resolved Fluorescence Techniques**	**55**
	Hiroyuki Aoki	
4.1	Introduction	55
4.2	Conformation of a Confined Polymer Chain	56
4.2.1	Polymer Ultra-Thin Film	56
4.2.2	Near-Field Optical Microscopy	56
4.2.3	Structure of a Single Polymer Chain	58
4.3	Dynamics of a Confined Polymer Chain	61
4.3.1	Polymer Brush	61
4.3.2	Fluorescence Depolarization Method	61
4.3.3	Dynamics of a Polymer Brush	63
4.4	Summary	67
	References	68

5	**Real Time Monitoring of Molecular Structure at Solid/Liquid Interfaces by Non-Linear Spectroscopy**	**71**
	Hidenori Noguchi, Katsuyoshi Ikeda, and Kohei Uosaki	
5.1	Introduction	71
5.2	Sum Frequency Generation Spectroscopy	72
5.2.1	Brief Description of SFG	72
5.2.2	Origin of SFG Process	73
5.2.3	SFG Spectroscopy	74
5.2.4	Experimental Arrangement for SFG Measurements	77

5.2.4.1	Laser and Detection Systems	77
5.2.4.2	Spectroscopic Cells	78
5.3	SFG Study of the Potential-Dependent Structure of Water at a Pt Electrode/Electrolyte Solution Interface	80
5.3.1	Introduction	80
5.3.2	Results and Discussion	80
5.3.3	Conclusions	83
5.4	Photoinduced Surface Dynamics of CO Adsorbed on a Platinum Electrode	84
5.4.1	Introduction	84
5.4.2	Results and Discussion	85
5.4.3	Conclusions	88
5.5	Interfacial Water Structure at Polyvinyl Alcohol (PVA) Gel/Quartz Interfaces Investigated by SFG Spectroscopy	89
5.5.1	Introduction	89
5.5.2	Results and Discussions	90
5.5.3	Conclusions	92
5.6	Hyper-Raman Spectroscopy	94
5.6.1	Selection Rules for Hyper-Raman Scattering	94
5.6.2	Enhancement of Hyper-Raman Scattering Intensity	94
5.6.3	Conclusion	96
5.7	General Conclusion	96
	References	97
6	**Fourth-Order Coherent Raman Scattering at Buried Interfaces**	**103**
	Hiroshi Onishi	
6.1	Why Buried Interfaces?	103
6.2	Optical Transitions	104
6.3	Experimental Scheme	106
6.4	Application to a Liquid Surface	107
6.5	Application to a Liquid/Liquid Interface	108
6.6	Applications to Solid Surfaces	109
6.7	Frequency Domain Detection	112
6.8	Concluding Remarks	113
	References	113
7	**Dynamic Analysis Using Photon Force Measurement**	**117**
	Hideki Fujiwara and Keiji Sasaki	
7.1	Introduction	117
7.1.1	Weak Force Measurements	117
7.1.2	Potential Analysis Method Using Photon Force Measurement	118
7.2	Measurement of the Hydrodynamic Interaction Force Acting between Two Trapped Particles Using the Potential Analysis Method	121
7.2.1	Two-Beam Photon Force Measurement System	121

7.2.2	Potential Analysis Method for Hydrodynamic Force Measurement	122
7.2.3	Trapping Potential Analysis	124
7.3	Kinetic Potential Analysis	125
7.4	Summary	129
	References	130

8 Construction of Micro-Spectroscopic Systems and their Application to the Detection of Molecular Dynamics in a Small Domain 133

Syoji Ito, Hirohisa Matsuda, Takashi Sugiyama, Naoki Toitani, Yutaka Nagasawa, and Hiroshi Miyasaka

8.1	Introduction	133
8.2	Development of a Near-Infrared 35 fs Laser Microscope and its Application to Higher Order Multiphoton Excitation	133
8.2.1	Confocal Microscope with a Chromium: Forsterite Ultrafast Laser as an Excitation Source	134
8.2.2	Detection of Higher Order Multiphoton Fluorescence from Organic Crystals	135
8.2.3	Multiphoton Fluorescence Imaging with the Near-Infrared 35 fs Laser Microscope	137
8.3	Application of Fluorescence Correlation Spectroscopy to the Measurement of Local Temperature at a Small Area in Solution	139
8.3.1	Experimental System of FCS	139
8.3.2	The Principle of the Method of Measurement of Local Temperature Using FCS	140
8.3.3	Measurement of Local Temperature for Several Organic Solvents	141
8.3.4	Summary	146
8.4	Relaxation Dynamics of Non-Emissive State for Water-Soluble CdTe Quantum Dots Measured by Using FCS	147
8.4.1	Samples and Analysis of Experimental Data Obtained with FCS	147
8.4.2	Non-Emissive Relaxation Dynamics in CdTe Quantum dots	148
8.5	Summary	150
	References	152

9 Nonlinear Optical Properties and Single Particle Spectroscopy of CdTe Quantum Dots 155

Lingyun Pan, Yoichi Kobayashi, and Naoto Tamai

9.1	Introduction	155
9.2	Nonlinear Optical Properties of CdTe QDs	156
9.3	Optical Trapping of CdTe QDs Probed by Nonlinear Optical Properties	158
9.4	Single Particle Spectroscopy of CdTe QDs	162
9.5	Summary	166
	References	167

Part Two Nanostructure Characteristics and Dynamics *171*

10 Morphosynthesis in Polymeric Systems Using Photochemical Reactions *173*
Hideyuki Nakanishi, Tomohisa Norisuye, and Qui Tran-Cong-Miyata
10.1 Introduction *173*
10.2 Morphosynthesis of Polymeric Systems by Using Light *174*
10.2.1 Significance of Photochemical Reactions *174*
10.2.2 Polymer Mixtures Used in this Study *175*
10.2.3 Polymers with Spatially Graded Morphologies Designed from Photo-Induced Interpenetrating Polymer Networks (IPNs) *175*
10.2.4 Designing Polymers with an Arbitrary Distribution of Characteristic Length Scales by the Computer-Assisted Irradiation (CAI) Method *177*
10.2.5 Reversible Phase Separation Driven by Photodimerization of Anthracene: A Novel Method for Processing and Recycling Polymer Blends *181*
10.3 Concluding Remarks *184*
References *185*

11 Self-Organization of Materials Into Microscale Patterns by Using Dissipative Structures *187*
Olaf Karthaus
11.1 Self-Organization and Self-Assembly *187*
11.2 Dissipative Structures *189*
11.3 Dynamics and Pattern Formation in Evaporating Polymer Solutions *191*
11.4 Applications of Dewetted Structures in Organic Photonics and Electronics *196*
11.5 Summary *198*
References *198*

12 Formation of Nanosize Morphology of Dye-Doped Copolymer Films and Evaluation of Organic Dye Nanocrystals Using a Laser *203*
Akira Itaya, Shinjiro Machida, and Sadahiro Masuo
12.1 Introduction *203*
12.2 Position-Selective Arrangement of Nanosize Polymer Microspheres Onto a PS-b-P4VP Diblock Copolymer Film with Nanoscale Sea–island Microphase Structure *205*
12.3 Nanoscale Morphological Change of PS-b-P4VP Block Copolymer Films Induced by Site-Selective Doping of a Photoactive Chromophore *208*
12.3.1 Nanoscale Surface Morphology of PS-b-P4VP Block Copolymer Films *208*

12.3.2	Nanoscale Surface Morphological Change of PS-b-P4VP Block Copolymer Films Induced by Site-Selective Doping of a Photoactive Chromophore *208*	
12.4	Site-Selective Modification of the Nanoscale Surface Morphology of Dye-Doped Copolymer Films Using Dopant-Induced Laser Ablation *211*	
12.5	Photon Antibunching Behavior of Organic Dye Nanocrystals on a Transparent Polymer Film *217*	
	References *221*	

13 Molecular Segregation at Periodic Metal Nano-Architectures on a Solid Surface *225*
Hideki Nabika and Kei Murakoshi

13.1	Molecular Manipulation in Nano-Space *225*	
13.1.1	Lipid Bilayer and its Fluidic Nature *225*	
13.1.2	Controlling Molecular Diffusion in the Fluidic Lipid Bilayer *227*	
13.1.3	Self-Spreading of a Lipid Bilayer or Monolayer *229*	
13.1.4	Controlling the Self-Spreading Dynamics *230*	
13.1.5	Molecular Manipulation on the Self-Spreading Lipid Bilayer *233*	
13.2	Summary *235*	
	References *236*	

14 Microspectroscopic Study of Self-Organization in Oscillatory Electrodeposition *239*
Shuji Nakanishi

14.1	Introduction *239*	
14.2	Dynamic Self-Organization in Electrochemical Reaction Systems *240*	
14.3	Oscillatory Electrodeposition *241*	
14.3.1	Formation of a Layered Nanostructure of Cu–Sn Alloy *242*	
14.3.2	Layered Nanostructures of Iron-Group Alloys *246*	
14.3.3	Layered Nanostructure of Cu/Cu_2O *247*	
14.3.4	Nanostructured Metal Filaments *250*	
14.4	Raman Microspectroscopy Study of Oscillatory Electrodeposition of Au at an Air/Liquid Interface *252*	
14.5	Summary *255*	
	References *256*	

15 Construction of Nanostructures by use of Magnetic Fields and Spin Chemistry in Solid/Liquid Interfaces *259*
Hiroaki Yonemura

15.1	Introduction *259*	
15.2	Construction of Nanostructures by the use of Magnetic Fields *260*	
15.2.1	Magnetic Orientation and Organization of SWNTs or their Composite Materials Using Polymer Wrapping *260*	

15.2.2	Effects of Magnetic Processing on the Morphological, Electrochemical, and Photoelectrochemical Properties of Electrodes Modified with C_{60}-Phenothiazine Nanoclusters *264*	
15.2.3	Effects of Magnetic Processing on the Luminescence Properties of Monolayer Films with Mn^{2+}-Doped ZnS Nanoparticles *268*	
15.3	Spin Chemistry at Solid/Liquid Interfaces *270*	
15.3.1	Magnetic Field Effects on the Dynamics of the Radical Pair in a C_{60} Clusters–Phenothiazine System *270*	
15.3.2	Magnetic Field Effects on Photoelectrochemical Reactions of Electrodes Modified with the C_{60} Nanocluster-Phenothiazine System *272*	
15.4	Summary *274*	
	References *274*	

16 **Controlling Surface Wetting by Electrochemical Reactions of Monolayers and Applications for Droplet Manipulation** *279*
Ryo Yamada

16.1	Introduction *279*	
16.1.1	Self-Assembled Monolayers *279*	
16.1.2	Preparation of Gradient Surfaces *280*	
16.1.3	Spontaneous Motion of a Droplet on Wetting Gradients *281*	
16.1.4	Surface Switching *282*	
16.2	Ratchet Motion of a Droplet *284*	
16.2.1	Ratchet Motion of a Droplet on Asymmetric Electrodes *284*	
16.2.2	Ratchet Motion of a Droplet Caused by Dynamic Motions of the Wetting Boundary *285*	
16.3	Conclusion *289*	
	References *291*	

17 **Photoluminescence of CdSe Quantum Dots: Shifting, Enhancement and Blinking** *293*
Vasudevanpillai Biju and Mitsuru Ishikawa

17.1	Introduction *293*	
17.2	Synthesis of CdSe Quantum Dots *295*	
17.2.1	Synthesis of CdSe Quantum Dots in Organic Phases *295*	
17.2.1.1	Synthesis of CdSe Quantum Dots from Dimethyl Cadmium *295*	
17.2.1.2	Synthesis of CdSe Quantum Dots from Cadmium Sources Other Than Dimethyl Cadmium *296*	
17.2.2	Synthesis of Water-Soluble Quantum Dots *296*	
17.3	Bandgap Structure and Photoluminescence of CdSe Quantum Dots *298*	
17.4	Photoluminescence Spectral Shifts *299*	
17.4.1	Physical Effects on Spectral Shifts *300*	
17.4.2	Chemical Effects on Spectral Shifts *301*	

17.5	Enhancement of Photoluminescence in CdSe Quantum Dots	*303*
17.6	On and Off Luminescence Blinking in Single Quantum Dots	*306*
17.6.1	Power-Law Statistics of On and Off Time Distributions	*308*
17.6.2	Modified Blinking *308*	
17.7	Conclusions *312*	
	References *312*	

Contents to Volume 2

Part Three Active Surfaces *315*

18 The Genesis and Principle of Catalysis at Oxide Surfaces: Surface-Mediated Dynamic Aspects of Catalytic Dehydration and Dehydrogenation on $TiO_2(110)$ by STM and DFT *317*
    Yohei Uemura, Toshiaki Taniike, Takehiko Sasaki, Mizuki Tada, and Yasuhiro Iwasawa

19 Nuclear Wavepacket Dynamics at Surfaces *337*
    Kazuya Watanabe

20 Theoretical Aspects of Charge Transfer/Transport at Interfaces and Reaction Dynamics *357*
    Hisao Nakamura and Koichi Yamashita

21 Dynamic Behavior of Active Ag Species in NOx Reduction on Ag/Al_2O_3 *401*
    Atsushi Satsuma and Ken-ichi Shimizu

22 Dynamic Structural Change of Pd Induced by Interaction with Zeolites Studied by Means of Dispersive and Quick XAFS *427*
    Kazu Okumura

Part Four Single Crystals *441*

23 Morphology Changes of Photochromic Single Crystals *443*
    Seiya Kobatake and Masahiro Irie

24 Direct Observation of Change in Crystal Structures During Solid-State Reactions of 1,3-Diene Compounds *459*
    Akikazu Matsumoto

Molecular Nano Dynamics, Volume I: Spectroscopic Methods and Nanostructures
Edited by H. Fukumura, M. Irie, Y. Iwasawa, H. Masuhara, and K. Uosaki
Copyright © 2009 WILEY-VCH Verlag GmbH & Co. KGaA, Weinheim
ISBN: 978-3-527-32017-2

| 25 | Reaction Dynamics Studies on Crystalline-State Photochromism of Rhodium Dithionite Complexes *487*
Hidetaka Nakai and Kiyoshi Isobe |
|---|---|
| 26 | Dynamics in Organic Inclusion Crystals of Steroids and Primary Ammonium Salts *505*
Mikiji Miyata, Norimitsu Tohnai, and Ichiro Hisaki |
| 27 | Morphology Changes of Organic Crystals by Single-Crystal-to-Single-Crystal Photocyclization *527*
Hideko Koshima |

Part Five   Single Biocells *545*

| 28 | Femtosecond Laser Tsunami Processing and Light Scattering Spectroscopic Imaging of Single Animal Cells *547*
Hiroshi Masuhara, Yoichiroh Hosokawa, Takayuki Uwada, Guillaume Louit, and Tsuyoshi Asahi |
|---|---|
| 29 | Super-Resolution Infrared Microspectroscopy for Single Cells *571*
Makoto Sakai, Keiichi Inoue, and Masaaki Fujii *571* |
| 30 | Three-Dimensional High-Resolution Microspectroscopic Study of Environment-Sensitive Photosynthetic Membranes *589*
Shigeichi Kumazaki, Makotoh Hasegawa, Mohammad Ghoneim, Takahiko Yoshida, Masahide Terazima, Takashi Shiina, and Isamu Ikegami |
| 31 | Fluorescence Lifetime Imaging Study on Living Cells with Particular Regard to Electric Field Effects and pH Dependence *607*
Nobuhiro Ohta and Takakazu Nakabayashi |
| 32 | Multidimensional Fluorescence Imaging for Non-Invasive Tracking of Cell Responses *623*
Ryosuke Nakamura and Yasuo Kanematsu |
| 33 | Fluorescence Correlation Spectroscopy on Molecular Diffusion Inside and Outside a Single Living Cell *645*
Kiminori Ushida and Masataka Kinjo |
| 34 | Spectroscopy and Photoreactions of Gold Nanorods in Living Cells and Organisms *669*
Yasuro Niidome and Takuro Niidome |

**35 Dynamic Motion of Single Cells and its Relation to
Cellular Properties** *689*
*Hideki Matsune, Daisuke Sakurai, Akitomo Hirukawa,
Sakae Takenaka, and Masahiro Kishida*

Index *703*

Preface

Over the past two decades, studies of chemical reaction dynamics have shifted from ideal systems of isolated molecules in the gas phase, of molecular clusters in jet beams, on ultra-clean surfaces, in homogeneous and in dilute molecular solutions, and in bulk crystals, towards nanosystems of supramolecules, colloids, and ultra-small materials, following the contemporary trends in nanoscience and nanotechnology. The preparation, characterization, and functionalization of supramolecules, molecular assemblies, nanoparticles, nanodots, nanocrystals, nanotubes, nanowires, and so on, have been conducted extensively, and their chemical reactions and dynamic processes are now being elucidated. The systematic investigation of molecular nanosystems gives us a platform from which we can understand the nature of the dynamic behavior and chemical reactions occurring in complex systems such as molecular devices, catalysts, living cells, and so on. Thus we have conducted the KAKENHI (The Grant-in-Aid for Scientific Research) Project on Priority Area "Molecular Nanodynamics" (Project Leader: Hiroshi Masuhara) for the period from 2004 April to 2007 March, involving 86 laboratories in Japan.

For the investigation of such complex systems new methodologies which enable us to analyze dynamics and mechanisms in terms of space and time are indispensable. Methods for simultaneous direct dynamic measurements in both time and real space domains needed to be devised and applied. Spectroscopy with novel space-resolution and ultrafast spectroscopy with high sensitivity have been developed, the manipulation and fabrication of single molecules, nanoparticles, and single living cells have been realized, molecules and nanoparticles for probing chemical reactions spectroscopically and by imaging have been synthesized, new catalyses for cleaning air and new reactions have been found, and the way in which a reaction in a single molecular crystal leads to its morphological change has been elucidated under the umbrella of this research program. The recent development of these new methods and the advances in understanding chemical reaction dynamics in nanosystems are summarized in the present two volumes.

The presented results are based on our activities over three years, including 1146 published papers and 1112 presentations at international conferences. We hope readers will understand the present status and new movement in Molecular

Molecular Nano Dynamics, Volume I: Spectroscopic Methods and Nanostructures
Edited by H. Fukumura, M. Irie, Y. Iwasawa, H. Masuhara, and K. Uosaki
Copyright © 2009 WILEY-VCH Verlag GmbH & Co. KGaA, Weinheim
ISBN: 978-3-527-32017-2

Nano Dynamics and its relevant research fields. The editors thank the contributors and the Ministry of Education, Culture, Sport, Science, and Technology (MEXT), Japan for their support of the project. We would also like to thank our publishers for their constant support.

Sendai, Tokyo, Nara
and Sapporo
August 2009

Hiroshi Fukumura
Masahiro Irie
Yasuhiro Iwasawa
Hiroshi Masuhara
Kohei Uosaki

About the Editors

Hiroshi Fukumura received his M.Sc and Ph.D. degrees from Tohoku University, Japan. He studied biocompatibility of polymers in the Government Industrial Research Institute of Osaka from 1983 to 1988. He became an assistant professor at Kyoto Institute of Technology in 1988, and then moved to the Department of Applied Physics, Osaka University in 1991, where he worked on the mechanism of laser ablation and laser molecular implantation. Since 1998, he is a professor in the Department of Chemistry at Tohoku University. He received the Award of the Japanese Photochemistry Association in 2000, and the Award for Creative Work from The Chemical Society Japan in 2005. His main research interest is the physical chemistry of organic molecules including polymeric materials studied with various kinds of time-resolved techniques and scanning probe microscopes.

Masahiro Irie received his B.S. and M.S. degrees from Kyoto University and his Ph.D. in radiation chemistry from Osaka University. He joined Hokkaido University as a research associate in 1968 and started his research on photochemistry. In 1973 he moved to Osaka University and developed various types of photoresponsive polymers. In 1988 he was appointed Professor at Kyushu University. In the middle of the 1980's he invented a new class of photochromic molecules – diarylethenes - which undergo thermally irreversible and fatigue resistant photochromic reactions. He is currently interested in developing single-crystalline photochromism of the diarylethene derivatives.

Molecular Nano Dynamics, Volume I: Spectroscopic Methods and Nanostructures
Edited by H. Fukumura, M. Irie, Y. Iwasawa, H. Masuhara, and K. Uosaki
Copyright © 2009 WILEY-VCH Verlag GmbH & Co. KGaA, Weinheim
ISBN: 978-3-527-32017-2

Yasuhiro Iwasawa received his B.S., M.S. and Ph.D. degrees in chemistry from The University of Tokyo. His main research interests come under the general term "Catalytic Chemistry" and "Surface Chemistry", but more specifically, catalyst surface design, new catalytic materials, reaction mechanism, in situ characterization, oxide surfaces by SPM, time-resolved XAFS, etc. His honors include the Progress Award for Young Chemists in The Chemical Society of Japan (1979), The Japan IBM Science Award (1990), Inoue Prize for Science (1996), Catalysis Society of Japan Award (1999), The Surface Science Society of Japan Award (2000), Medal with Purple Ribbon (2003), and The Chemical Society of Japan Award (2004). The research reported by Yasuhiro Iwasawa represents a pioneering integration of modern surface science and organometallic chemistry into surface chemistry and catalysis in an atomic/ molecular scale. Iwasawa is a leader in the creation of the new filed of catalysis and surface chemistry at oxide surfaces by XAFS and SPM techniques.

Hiroshi Masuhara received his B.S. and M.S. degrees from Tohoku University and Ph.D. from Osaka University. He started his research in photochemistry and was the first to use nanosecond laser spectroscopy in Japan. He studied electronic states, electron transfer, ionic photodissociation of molecular complexes, polymers, films, and powders by developing various time-resolved absorption, fluorescence, reflection, and grating spectroscopies until the mid 1990s. The Masuhara Group combined microscope with laser and created a new field on Microchemistry, which has now developed to Laser Nano Chemistry. After retiring from Osaka University he shifted to Hamano Foundation and is now extending his exploratory research on femtosecond laser crystallization and laser trapping crystallization in National Chiao Tung University in Taiwan and Nara Institute of Science and Technology. He is a foreign member of Royal Flemish Academy of Belgium for Science and the Arts and his honors include The Purple Ribbon Medal, Doctor Honoris Causa de Ecole Normale Superier de Cachan, Porter Medal, the Chemical Society of Japan Award, Osaka Science Prize, and Moet Hennessy Louis Vuitton International Prize "Science for Art" Excellence de Da Vinci.

Kohei Uosaki received his B.Eng. and M.Eng. degrees from Osaka University and his Ph.D. in Physical Chemistry from Flinders University of South Australia. He was a Research Chemist at Mitsubishi Petrochemical Co. Ltd. From 1971 to 1978 and a Research Officer at Inorganic Chemistry Laboratory, Oxford University, U.K. between 1978 and 1980 before joining Hokkaido University in 1980 as Assistant Professor in the Department of Chemistry. He was promoted to Associate Professor in 1981 and Professor in 1990. He is also a Principal Investigator of International Center for Materials Nanoarchitectonics (MANA) Satellite, National Institute for Materials Science (NIMS) since 2008. His scientific interests include photoelectrochemistry of semiconductor electrodes, surface electrochemistry of single crystalline metal electrodes, electrocatalysis, modification of solid surfaces by molecular layers, and non-linear optical spectroscopy at interfaces.

List of Contributors for Both Volumes

Hiroyuki Aoki
Kyoto University
Department of Polymer Chemistry
Katsura, Nishikyo
Kyoto 615-8510
Japan

Tsuyoshi Asahi
Osaka University
Department of Applied Physics
Suita 565-0871
Japan

Vasudevanpillai Biju
National Institute of Advanced
Industrial Science and Technology
(AIST)
Health Technology Research Center
Nano-bioanalysis Team
2217-14 Hayashi-cho, Takamatsu
Kagawa 761-0395
Japan

Masaaki Fujii
Tokyo Institute of Technology
Chemical Resources Laboratory
4259 Nagatsuta-cho, Midori-ku
Yokohama 226-8503
Japan

Hideki Fujiwara
Hokkaido University
Research Institute for Electronic
Science
Kita-12, Nishi-6, Sapporo
Hokkaido 060-0812
Japan

Hiroshi Fukumura
Tohoku University
Graduate School of Science
6-3 Aramaki Aoba
Sendai 980-8578
Japan

Mohammad Ghoneim
Kyoto University
Graduate School of Science
Department of Chemistry
Kyoto 606-8502
Japan

Makotoh Hasegawa
Kyoto University
Graduate School of Science
Department of Chemistry
Kyoto 606-8502
Japan

Molecular Nano Dynamics, Volume I: Spectroscopic Methods and Nanostructures
Edited by H. Fukumura, M. Irie, Y. Iwasawa, H. Masuhara, and K. Uosaki
Copyright © 2009 WILEY-VCH Verlag GmbH & Co. KGaA, Weinheim
ISBN: 978-3-527-32017-2

Akitomo Hirukawa
Kyushu University
Faculty of Engineering
Department of Chemical Engineering
Moto-Oka, Nishi Ku
Fukuoka 819-0395
Japan

Ichiro Hisaki
Osaka University
Graduate School of Engineering
2-1 Yamadaoka, Suita
Osaka 565-0871
Japan

Noriko Nishizawa Horimoto
Tohoku University
Graduate School of Science
6-3 Aramaki Aoba
Sendai 980-8578
Japan

Yoichiroh Hosokawa
Nara Institute of Science and Technology
Graduate School of Materials Science
Takayama 8916-5
Ikoma 630-0192
Japan

Taro Ichimura
Osaka University
Graduate School of Frontier Biosciences
& Graduate School of Engineering
Suita, Osaka
Japan

Katsuyoshi Ikeda
Hokkaido University
Graduate School of Science
Division of Chemistry
Sapporo 060-0810
Japan

Isamu Ikegami
Kyoto University
Graduate School of Science
Department of Chemistry
Kyoto 606-8502
Japan

Kohei Imura
The Graduate University for Advanced Studies
Institute for Molecular Science
Myodaiji, Okazaki
Aichi 444-8585
Japan

Keiichi Inoue
Tokyo Institute of Technology
Chemical Resources Laboratory
4259 Nagatsuta-cho, Midori-ku
Yokohama 226-8503
Japan

Yasushi Inouye
Osaka University
Graduate School of Frontier Biosciences
& Graduate School of Engineering
Suita, Osaka
Japan

Masahiro Irie
Rikkyo University
Department of Chemistry
Nishi-Ikebukuro 3-34-1, Toshima-ku
Tokyo 171-8501
Japan

Mitsuru Ishikawa
National Institute of Advanced Industrial Science and Technology (AIST)
Health Technology Research Center
Nano-bioanalysis Team
2217-14 Hayashi-cho, Takamatsu
Kagawa 761-0395
Japan

Kiyoshi Isobe
Kanazawa University
Graduate School of Natural Science and Technology
Department of Chemistry
Kakuma-machi
Kanazawa 920-1192
Japan

Akira Itaya
Kyoto Institute of Technology
Department of Polymer Science and Engineering
Matsugasaki, Sakyo-ku
Kyoto 606-8585
Japan

Syoji Ito
Osaka University
Graduate School of Engineering Science
Center for Quantum Science and Technology under Extreme Conditions
Division of Frontier Materials Science
Toyonaka
Osaka 560-8531
Japan

Yasuhiro Iwasawa
The University of Tokyo
Graduate School of Science
Department of Chemistry
Hongo, Bunkyo-ku
Tokyo 113-0033
Japan

Yasuo Kanematsu
Osaka University
Center for Advanced Science and Innovation
Venture Business Laboratory
JST-CREST
Suita
Osaka 565-0871
Japan

Olaf Karthaus
Chitose Institute of Science and Technology
758-65 Bibi, Chitose
Hokkaido 066-8655
Japan

Masataka Kinjo
Riken
Hirosawa 2-1, Wako
Saitama 351-0198
Japan

Masahiro Kishida
Kyushu University
Faculty of Engineering
Department of Chemical Engineering
Moto-Oka, Nishi Ku
Fukuoka 819-0395
Japan

Yoichi Kobayashi
Kwansei Gakuin University
School of Science and Technology
Department of Chemistry
2-1 Gakuen
Sanda 669-1337
Japan

Seiya Kobatake
Osaka City University
Graduate School of Engineering
Department of Applied Chemistry
Sugimoto 3-3-138, Sumiyoshi-ku
Osaka 558-8585
Japan

Hideko Koshima
Ehime University
Graduate School of Science and Engineering
Department of Materials Science and Biotechnology
Matsuyama 790-8577
Japan

Shigeichi Kumazaki
Kyoto University
Graduate School of Science
Department of Chemistry
Kyoto 606-8502
Japan

Guillaume Louit
Osaka University
Department of Applied Physics
Suita 565-0871
Japan

Shinjiro Machida
Kyoto Institute of Technology
Department of Polymer Science and
Engineering
Matsugasaki, Sakyo-ku
Kyoto 606-8585
Japan

Hiroshi Masuhara
Nara Institute of Science and
Technology
Graduate School of Materials Science
Takayama 8916-5
Ikoma 630-0192
Japan

and

Osaka University
Department of Applied Physics
Suita 565-0871
Japan

Sadahiro Masuo
Kyoto Institute of Technology
Department of Polymer Science and
Engineering
Matsugasaki, Sakyo-ku
Kyoto 606-8585
Japan

Hirohisa Matsuda
Osaka University
Graduate School of Engineering Science
Center for Quantum Science and
Technology under Extreme Conditions
Division of Frontier Materials Science
Toyonaka
Osaka 560-8531
Japan

Akikazu Matsumoto
Osaka City University
Graduate School of Engineering
Department of Applied Chemistry
3-3-138 Sugimoto, Sumiyoshi-ku
Osaka 558-8585
Japan

Hideki Matsune
Kyushu University
Faculty of Engineering
Department of Chemical Engineering
Moto-Oka, Nishi Ku
Fukuoka 819-0395
Japan

Hiroshi Miyasaka
Osaka University
Graduate School of Engineering Science
Center for Quantum Science and
Technology under Extreme Conditions
Division of Frontier Materials Science
Toyonaka
Osaka 560-8531
Japan

Mikiji Miyata
Osaka University
Graduate School of Engineering
2-1 Yamadaoka, Suita
Osaka 565-0871
Japan

Kei Murakoshi
Hokkaido University
Graduate School of Science
Department of Chemistry
Sapporo
Hokkaido 060-0810
Japan

Takakazu Nakabayashi
Hokkaido University
Research Institute for Electronic
Science (RIES)
Sapporo 001-0020
Japan

Hideki Nabika
Hokkaido University
Graduate School of Science
Department of Chemistry
Sapporo
Hokkaido 060-0810
Japan

Yutaka Nagasawa
Osaka University
Graduate School of Engineering Science
Center for Quantum Science and
Technology under Extreme Conditions
Division of Frontier Materials Science
Toyonaka
Osaka 560-8531
Japan

Hidetaka Nakai
Kanazawa University
Graduate School of Natural Science and
Technology
Department of Chemistry
Kakuma-machi
Kanazawa 920-1192
Japan

Hisao Nakamura
The University of Tokyo
Graduate School of Engineering
Department of Chemical System
Engineering
Tokyo 113-8656
Japan

Ryosuke Nakamura
Osaka University
Center for Advanced Science and
Innovation
Venture Business Laboratory
JST-CREST
Suita
Osaka 565-0871
Japan

Hideyuki Nakanishi
Graduate School of Science and
Technology
Kyoto Institute of Technology
Department of Macromolecular Science
and Engineering
Matsugasaki
Kyoto 606-8585
Japan

Shuji Nakanishi
Osaka University
Graduate School of Engineering Science
Division of Chemistry
Toyonaka
Osaka 560-8531
Japan

Takuro Niidome
Kyushu University
Department of Applied Chemistry
744 Moto-Oka, Nishi Ku
Fukuoka 819-0395
Japan

Yasuro Niidome
Kyushu University
Department of Applied Chemistry
744 Moto-Oka, Nishi Ku
Fukuoka 819-0395
Japan

Hidenori Noguchi
Hokkaido University
Graduate School of Science
Division of Chemistry
Sapporo 060-0810
Japan

Tomohisa Norisuye
Graduate School of Science and Technology
Kyoto Institute of Technology
Department of Macromolecular Science and Engineering
Matsugasaki
Kyoto 606-8585
Japan

Nobuhiro Ohta
Hokkaido University
Research Institute for Electronic Science (RIES)
Sapporo 001-0020
Japan

Hiromi Okamoto
The Graduate University for Advanced Studies
Institute for Molecular Science
Myodaiji, Okazaki
Aichi 444-8585
Japan

Kazu Okumura
Tottori University
Faculty of Engineering
Department of Materials Science
Koyama-cho, Minami
Tottori 680-8552
Japan

Hiroshi Onishi
Kobe University
Faculty of Science
Department of Chemistry
Rokko-dai, Nada, Kobe
Hyogo 657-8501
Japan

Lingyun Pan
Kwansei Gakuin University
School of Science and Technology
Department of Chemistry
2-1 Gakuen
Sanda 669-1337
Japan

Makoto Sakai
Tokyo Institute of Technology
Chemical Resources Laboratory
4259 Nagatsuta-cho, Midori-ku
Yokohama 226-8503
Japan

Daisuke Sakurai
Kyushu University
Faculty of Engineering
Department of Chemical Engineering
Moto-Oka, Nishi Ku
Fukuoka 819-0395
Japan

Keiji Sasaki
Hokkaido University
Research Institute for Electronic Science
Kita-12, Nishi-6, Sapporo
Hokkaido 060-0812
Japan

Takehiko Sasaki
The University of Tokyo
Graduate School of Frontier Science
Department of Chemistry
Kashiwanoha, Kashiwa
Chiba 277-8561
Japan

Atsushi Satsuma
Nagoya University
Graduate School of Engineering
Department of Molecular Design and Engineering
Chikusa
Nagoya 464-8603
Japan

Takashi Shiina
Kyoto University
Graduate School of Science
Department of Chemistry
Kyoto 606-8502
Japan

Ken-ichi Shimizu
Nagoya University
Graduate School of Engineering
Department of Molecular Design and Engineering
Chikusa
Nagoya 464-8603
Japan

Takashi Sugiyama
Osaka University
Graduate School of Engineering Science
Center for Quantum Science and Technology under Extreme Conditions
Division of Frontier Materials Science
Toyonaka
Osaka 560-8531
Japan

Mizuki Tada
The University of Tokyo
Graduate School of Frontier Science
Department of Chemistry
Kashiwanoha, Kashiwa
Chiba 277-8561
Japan

Atsushi Taguchi
Osaka University
Graduate School of Frontier Biosciences
& Graduate School of Engineering
Suita, Osaka
Japan

Sakae Takenaka
Kyushu University
Faculty of Engineering
Department of Chemical Engineering
Moto-Oka, Nishi Ku
Fukuoka 819-0395
Japan

Naoto Tamai
Kwansei Gakuin University
School of Science and Technology
Department of Chemistry
2-1 Gakuen
Sanda 669-1337
Japan

Toshiaki Taniike
The University of Tokyo
Graduate School of Science
Department of Chemistry
Hongo, Bunkyo-ku
Tokyo 113-0033
Japan

Masahide Terazima
Kyoto University
Graduate School of Science
Department of Chemistry
Kyoto 606-8502
Japan

Norimitsu Tohnai
Osaka University
Graduate School of Engineering
2-1 Yamadaoka, Suita
Osaka 565-0871
Japan

Naoki Toitani
Osaka University
Graduate School of Engineering Science
Center for Quantum Science and Technology under Extreme Conditions
Division of Frontier Materials Science
Toyonaka
Osaka 560-8531
Japan

Qui Tran-Cong-Miyata
Graduate School of Science and Technology
Kyoto Institute of Technology
Department of Macromolecular Science and Engineering
Matsugasaki
Kyoto 606-8585
Japan

Yohei Uemura
The University of Tokyo
Graduate School of Science
Department of Chemistry
Hongo, Bunkyo-ku
Tokyo 113-0033
Japan

Kohei Uosaki
Hokkaido University
Graduate School of Science
Division of Chemistry
Sapporo 060-0810
Japan

Kiminori Ushida
Riken
Hirosawa 2-1, Wako
Saitama 351-0198
Japan

Takayuki Uwada
Nara Institute of Science and Technology
Graduate School of Materials Science
Takayama 8916-5
Ikoma 630-0192
Japan

and

National Chiao Tung University
Institute of Molecular Science
Department of Applied Chemistry
1001 Ta Hsueh Road
Hsinchu 30010
Taiwan

Kazuya Watanabe
Kyoto University
Graduate School of Science
Department of Chemistry
Kyoto 606-8502
Japan

and

PRESTO, JST
4-1-8 Honcho Kawaguchi
Saitama
Japan

Ryo Yamada
Osaka University
Graduate School of Engineering Science
Division of Materials Physics
Department of Materials Engineering Science
Toyonaka, Osaka
Japan

Hiroaki Yonemura
Kyushu University
Department of Applied Chemistry
6-10-1 Hakozaki, Higashi-ku
Fukuoka 812-8581
Japan

Takahiko Yoshida
Kyoto University
Graduate School of Science
Department of Chemistry
Kyoto 606-8502
Japan

Part One
Spectroscopic Methods for Nano Interfaces

1
Raman and Fluorescence Spectroscopy Coupled with Scanning Tunneling Microscopy

Noriko Nishizawa Horimoto and Hiroshi Fukumura

1.1
Introduction

Optical spectroscopy is a very powerful tool in many fields, especially in chemistry. Light irradiates molecules (input) and, after interaction with the molecules, light is collected (output). In Raman spectroscopy, light is inelastically scattered by molecules and, owing to molecular vibration excitation or de-excitation, light with higher or lower energy is observed. Raman scattering is emitted together with a vast amount of elastically scattered light known as Rayleigh scattering, which generally makes the measurement of Raman spectra difficult. Raman spectra are, however, very valuable because we can obtain directly the vibration frequencies of bonds contained in matter by using visible light. In fluorescence spectroscopy, light is absorbed by molecules and the molecules are excited to the electronically excited state, resulting in the emission of fluorescence when the molecules relax to the ground state. Fluorescence spectra and decay processes are generally sensitive to intermolecular interactions in excited states. By the use of such optical spectroscopy, chemical species can be identified and the average environment surrounding target molecules can be deduced.

Conventional optical spectroscopy has the drawback that the spatial resolution is limited because of the diffraction limit of light. The resolution of a conventional microscope is about half of the observation wavelength, which is several hundred nm in the case of visible light. A variety of attempts to overcome the diffraction limit realizing better spatial resolution have been performed to date. Scanning near-field optical microscopy (SNOM) is one such methods, which has a spatial resolution of about 50 nm [1]. In aperture SNOM, light is irradiated from a tapered optical fiber tip so that the spatial resolution is limited by the aperture. Aperture-less SNOM uses a sharp tip which scatters the evanescent field containing the information in the sub-wavelength range. In both types of SNOM, the tip is placed close to or in contact with a sample, and light signals are collected while the tip scans the sample surface.

Molecular Nano Dynamics, Volume I: Spectroscopic Methods and Nanostructures
Edited by H. Fukumura, M. Irie, Y. Iwasawa, H. Masuhara, and K. Uosaki
Copyright © 2009 WILEY-VCH Verlag GmbH & Co. KGaA, Weinheim
ISBN: 978-3-527-32017-2

The combination of atomic force microscopy (AFM) and Raman spectroscopy is another approach to attain high spatial resolution. AFM also employs a sharp tip close to a sample surface. When the tip is made of metal and light is irradiated onto the tip and surface, Raman scattering is largely enhanced. In this way, a spatial resolution of 15 nm is achieved [2].

However, there is demand for better spatial resolution – ultimately atomic scale resolution. If this is realized we will be able to distinguish individually molecules densely packed on a surface, or even identify a part of a single molecule. For example, it could be clarified which part of a large molecule is modified with which kind of functional group. This kind of instrument may become essential for the preparation and analysis of single molecular devices. Combining scanning tunneling microscopy (STM) with optical spectroscopy is considered to be a feasible approach. This is because, at present, STM has the highest resolution among the various types of scanning probe microscopy. STM was developed by Binnig et al. in 1982 [3], and now enables us to observe a sample surface with atomic-scale resolution. The principle of the method is based on the tunneling current between the sample and a sharp metal tip as a tunneling probe [4, 5]. Very recently, attempts to combine STM with Raman or fluorescence spectroscopy have emerged and these will be described in this chapter. Here we will focus on studies performed under ambient conditions, that is, at room temperature and in air. Therefore, inelastic electron tunneling STM spectroscopy (IETS) is omitted from this chapter. STM induced luminescence, (alternatively called photon scanning tunneling microscopy, PSTM) which observes the luminescence induced by the STM tunneling current, is also beyond the scope of this chapter.

1.2
Outline of STM Combined with Optical Spectroscopy

1.2.1
Raman Spectroscopy

STM-Raman spectroscopy utilizes the effect that Raman scattering is enhanced for a molecule in the vicinity of a metal nanostructure. This enhancement effect is generally called surface-enhanced Raman scattering (SERS). When a sharp scanning probe, such as a tunneling tip for STM, is used as a metal nanostructure to enhance Raman intensity, it is called "tip-enhanced Raman scattering" (TERS). The concept of STM combined with Raman spectroscopy is presented in Figure 1.1.

The mechanisms of surface enhanced Raman scattering have been well discussed in the early review papers [6, 7]. Briefly, Raman enhancement is considered to occur by two mechanisms: chemical enhancement and electromagnetic enhancement. The chemical enhancement is due to resonance Raman scattering based on charge transfer between a metal nanostructure and a molecule, and takes place only when the metal nanostructure is in contact with the molecule. In STM, the tip is close to the sample molecule but not close enough to induce charge transfer, so the enhancement by the chemical effect may be negligibly small. The electromagnetic enhancement

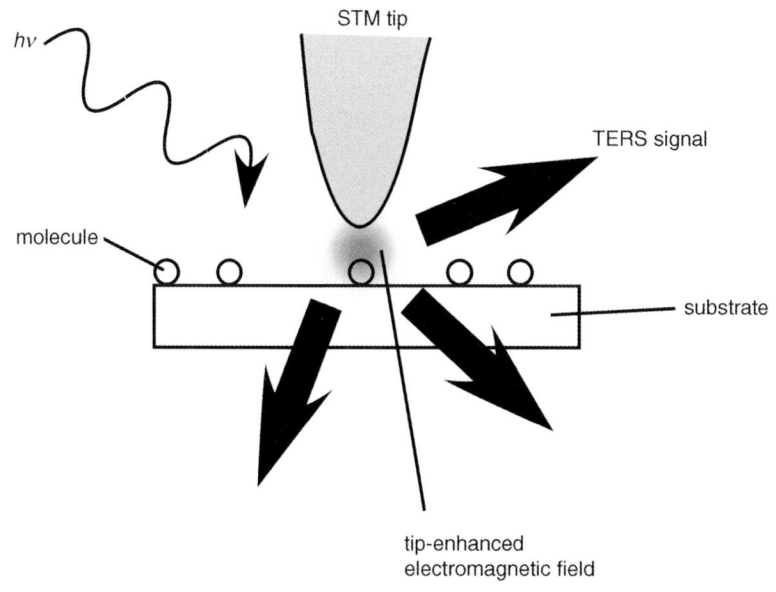

Figure 1.1 Concept of STM combined with Raman spectroscopy.

arises from the field enhancement in the vicinity of a metal nanostructure when it is illuminated by probe light. This is due to the excitation of localized surface plasmons in the metal nanostructure, which is a collective oscillation of the electron gas at the subsurface of the metal. The enhancement factor of Raman intensity is roughly proportional to M^4, where M is the factor indicating by how much the electric field is enhanced by the nanostructure, since both the incident light intensity and the scattered light intensity are enhanced [8, 9]. The enhancement factor is especially large when the nanostructure is made of silver, gold, or other metals containing free electrons. By changing the shape, size, or material of the metal nanostructure, and the surrounding medium, the wavelength dependence of the enhancement efficiency changes due to changes in plasmon resonance. The optical response of metal nanospheres can be described precisely by using the Mie theory, which is an analytical solution of Maxwell's equation. Its details and numerical calculation programs are fully and comprehensively presented by Bohren and Huffman [10]. For non-spherical particles, numerical approximation methods are generally required. The finite different time domain (FDTD) method [11], and the discrete dipole approximation (DDA) [12, 13] are commonly utilized. For a classical SERS substrate, which is a roughened silver electrode, the average enhancement factor is known to be about 10^6 [14]. For a selected single silver nanoparticle with a single molecule, the enhancement factor is reported to be 10^{14} [15]. A maximum enhancement factor of about 5×10^9 is reported for TERS [16]. It should be noted that the net enhancement factor by TERS is large but the observed signal increase is sometimes small because the spot having the intense electric field under a tunneling tip is very limited compared with the area illuminated by the probe light beam.

1.2.2
Fluorescence Spectroscopy

The effect of a tunneling tip on fluorescence is rather complicated compared with that on Raman scattering. The enhancement of the electric field is induced by approaching the metal tip to a fluorescent molecule; however, the quenching of fluorescence is also caused in the vicinity of or in contact with the surface of the metal. Generally, four major processes in fluorescent molecules are known to compete with the emissive decay: energy transfer, electron transfer, intersystem crossing, and internal conversion. Energy transfer can take place when an energy acceptor comes as close as about 5 nm to a fluorescent molecule, depending on the interaction between the excited state of the molecule and the ground state of the acceptor. Here the energy acceptor can be the nanostructured metal tip which shows an absorption spectrum in the visible region. The interaction is roughly estimated from the overlap of the molecular fluorescence and the acceptor absorption spectra. When the metal tip approaches to less than 1 nm from the fluorescent molecule, strong interactions like electron transfer and electron exchange can occur, which results in the fast decay of excited states. Intersystem crossing to triplet states is also known to be accelerated by approaching heavy atoms to fluorescent singlet states. This is caused by the enhancement of spin–orbit coupling, which requires the overlap of electron orbitals. Although internal conversion is rather intrinsic to molecular electronic states, it may be affected by approaching a metal tip if the vibronic coupling in a fluorescent molecule is modified through the molecular structure distortion. Thus, there are plural processes reducing fluorescence intensity, in addition to the electromagnetic field enhancement effect. Roughly speaking, the enhancement effect extends to 10 nm or more around the tunneling tip although the quenching effects become dominant only at a few nm from the tip, which results in a donut-like enhancement pattern for fluorescence.

1.3
Theoretical Approaches

Demming et al. calculated the field enhancement factor (FEF) for a silver tip and a flat gold surface by the boundary element method [17]. They demonstrated that the peak position and the magnitude of the FEF maximum depend on the geometry of the tip. For smaller tip radii, the FEF maximum increases and shifts to longer wavelength. The wavelength dependence of the FEF for a silver ellipsoid also behaves in this manner when its aspect ratio is changed. The spatial distribution of the FEF also becomes narrower and the peak intensity becomes stronger when the tip shape becomes sharper. The same research group calculated the FEF for a small spheroid shape as a model of tips made of gold, platinum, silver, p-doped silicon, and tungsten, by solving the Laplace equation analytically [18]. They calculated the dependence of FEF on the aspect ratio of the ellipsoid and wavelength, as shown in Figure 1.2. They also calculated the FEF space distribution for a tip with a flat shape by a boundary

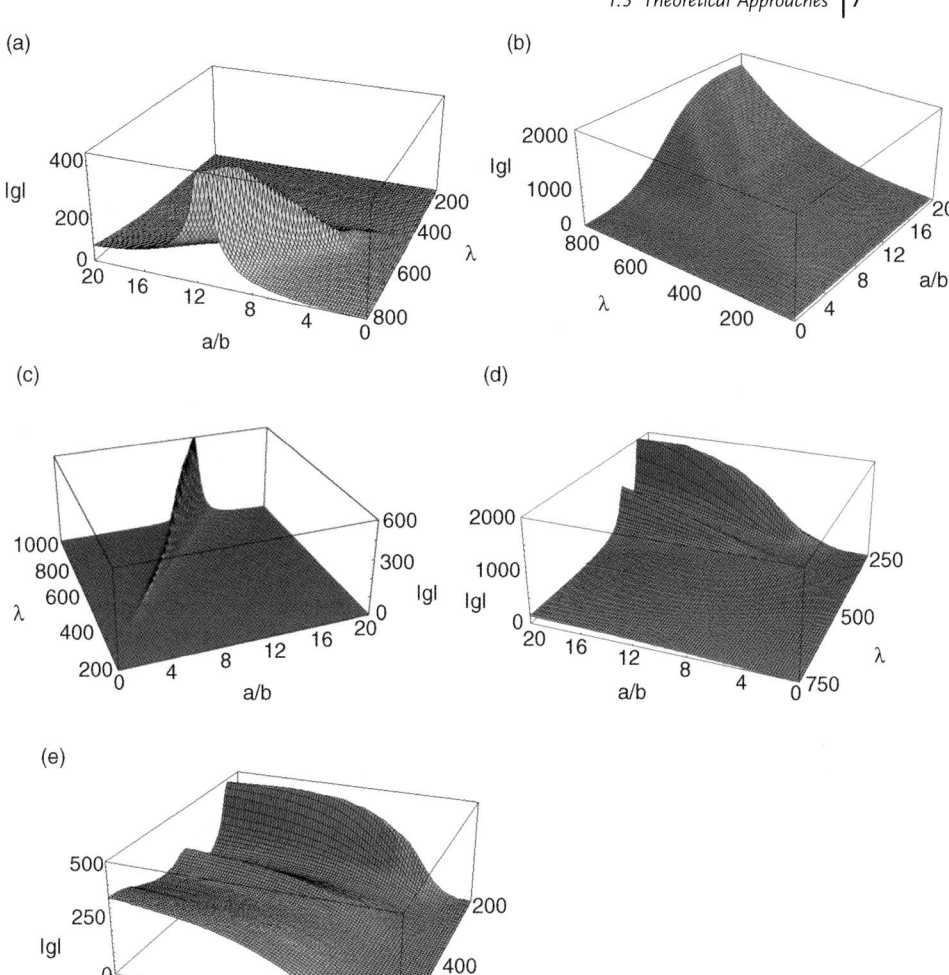

Figure 1.2 Modulus of the field enhancement factor versus the aspect ratio $a = b$ and wavelengths λ for SPM tips of different materials: (a) gold, (b) platinum, (c) silver, (d) p-doped silicon, (e) tungsten. Reprinted with permission from J. Jersch, *Applied Physics A*, 66, 29 (1998). Copyright 1998, Springer-Verlag.

element method. The intensity showed a double peak. These FEF calculation results were consistent with interesting experimental results: a flat Au surface was topographically changed to form a double structure by the illumination of the surface and a flat-shaped tip with laser light.

Mills calculated the enhancement of dynamic dipole moments for a dipole moment on a smooth silver or copper surface with a silver tip above it, which leads to fluorescence enhancement and TERS [19]. They used a sphere as a model of the tip and calculated the enhancement factors for various wavelengths. It was found that the enhancement factors are larger for a dipole moment oriented perpendicular to the surface than for one oriented parallel to the surface and are also larger when the tip–sample distance is small. Using the same condition and a similar method, Wu and Mills performed calculations of the enhancement factors for various wavelengths for several tip–sample distances [20]. They also calculated the enhancement factor for dipoles at different lateral distances from directly beneath the tip. They confirmed that the lateral resolution scales as $(RD)^{1/2}$, with R the radius of curvature of the tip, and D its distance from the substrate, as was first suggested by Rendell et al. [21].

Klein et al. used the same model as Demming et al. [17], and calculated the FEF for an Ag tip near an Au surface [22], which is similar to the experimental conditions of Pettinger et al. [23]. They showed that Raman signals of molecules below or in the near field of the tip can be enhanced to practically measurable levels, and the FEF is larger for smaller tip–sample gaps, being localized below the tip, as shown in Figure 1.3.

Downes et al. calculated the enhancement of the scattered light intensity for a tip over a substrate using finite element electromagnetic simulations [24]. A gold tip or silicon tip over a gold, mica, or silicon substrate in air or water was considered. They calculated the enhancement dependence on incident light polarization, wavelength, tip–substrate separation, Raman shift, distance beneath the tip apex, angle between the tip axis and incident radiation, tip radius, and lateral displacement from the tip apex. The enhancement exceeded 10^{12} or 10^8 for a 20 nm radius Au tip over an Au or mica substrate, respectively, for a tip–substrate separation of 1 nm. In both cases, the enhancement gradually decreased as the tip–substrate separation increased. They showed that with this enhancement, high-speed (10 ms integration time) mapping is possible considering the signal contrast against the far-field signal. The lateral resolution dependence on tip–substrate separation was also calculated, and the lateral resolution was better than 1 nm for a 10 or 20 nm radius Au tip over an Au substrate. The lateral resolution became worse as the tip–substrate separation increased.

1.4
Experimental Approaches

1.4.1
STM Combined with Raman Spectroscopy

Pettinger et al. observed a TERS spectrum of monolayer-thick brilliant cresyl blue (BCB) adsorbed on a smooth Au film surface by using a Ag tip, while no STM image of the adsorbed surface was shown [23]. The Raman intensity increased when the tip was in the tunneling position, meaning that the tip–surface distance was around 1 nm. They calculated the field enhancement factor by the method described by

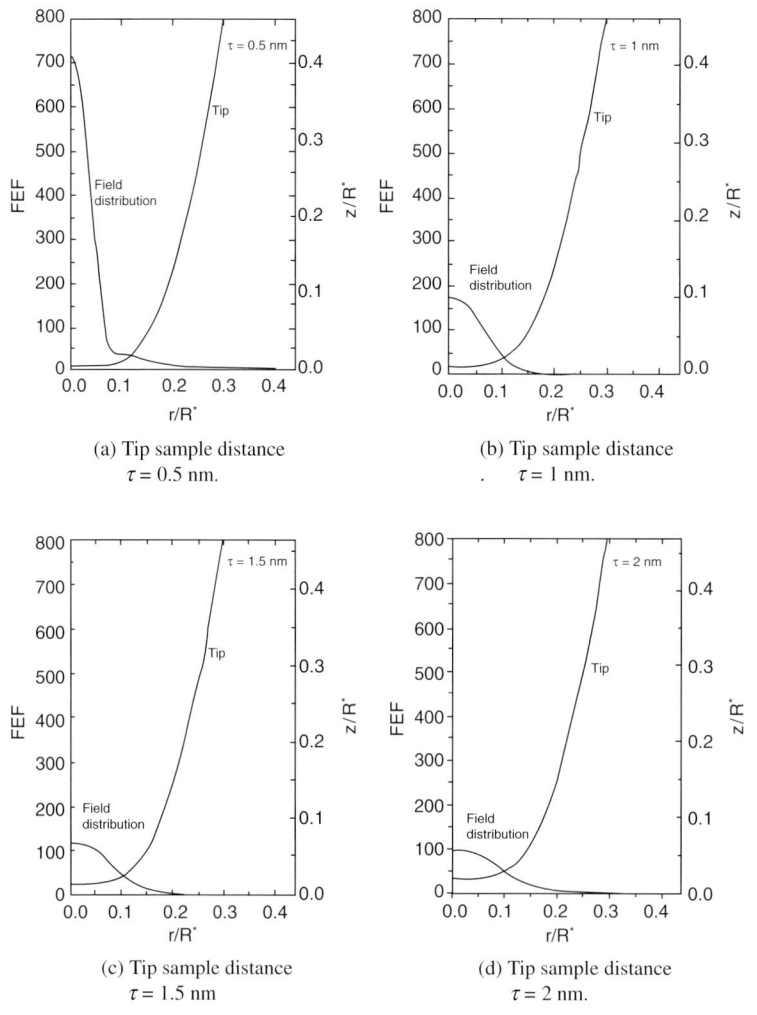

Figure 1.3 Field distributions along the Ag-tip surface and corresponding Ag-tip geometry. $z=0$ corresponds to the Au-substrate. r/R^* is the normalized radius from the point directly beneath the tip (R^* is the Rayleigh length $R^* = \lambda/2\pi$). Reprinted with permission from S. Klein, *Electrochemistry*, **71**, 114 (2003). Copyright 2003, The Electrochemical Society of Japan.

Demming et al. [17], and compared it with their experimental results. The observed increase in Raman intensity by a factor of 15 within the laser focus of about 2 μm, corresponds to a maximum local enhancement at the center of the tip apex by a factor of about 12 000 under the assumption that the tip radius is 100 nm. More precisely, about 20% of the Raman intensity is considered to originate from an area with a radius of 14 nm under the tip, which is roughly equivalent to 400 molecules out of 2×10^6 in the total focused area.

Pettinger et al. further reported the TERS spectra of CN⁻ and BCB on a smooth and rough Au substrate with a silver tip [25]. They observed a signal enhancement by a factor of 4 for CN⁻, and estimated the real enhancement to be at least three orders of magnitude, assuming the laser spot size to be 2 μm, and the tip radius at the apex to be 100 nm. They also observed TERS spectra of CN⁻ ions on SERS active Au surfaces [26], and compared the experimental result of the illumination from the bottom through a thin metal film with that of direct excitation from the top at an incident angle of 60°. It was found that subtraction of the SERS spectra from the TERS spectra shows narrow bandwidth spectra, suggesting that the TERS signal arises from a rather small amount of molecules.

Ren et al. reported a method to prepare a gold tip with a tip apex radius of 30 nm reproducibly [27]. They observed the TERS of a Malachite Green isothiocyanate (MGITC) monolayer on an Au(111) surface and obtained an enhancement factor of about 1.6×10^5, by using the relation, $q = I_{TERS}/I_{RRS} = g^4 a^2/R_{focus}^2$, where q is the net increase in the signal, I_{TERS} and I_{RRS} are the signal intensities for TERS and RRS (resonance Raman scattering), respectively; g^4 is the TERS enhancement (g is the field enhancement), a denotes the radius of the enhanced field, and R_{focus} the radius of the laser focus.

Pettinger et al. observed TERS with a sharp Au tip for MGITC dye on Au(111) with a side illumination [28]. They studied the bleaching of the dye and fitted the data by taking into account the radial varying intensity distribution of the field as a Gaussian profile instead of a Heaviside profile. For the former profile, the TERS radius is smaller by a factor of 1/2 than for the latter profile. They obtained a TERS enhancement factor of 6.25×10^6. The radius of the enhanced field R_{field} is about 50 nm, which results in a TERS radius of 25 nm, which is smaller than the radius of the tip apex (about 30 nm).

Domke et al. also studied the TERS of the MGITC dye on Au(111) in combination with the corresponding STM images of the probed surface region [29]. They estimated an enhancement factor of $10^6 - 10^7$, where five molecules should be present in the enhanced-field region, assuming an enhanced field with a radius of 20 nm, thus they claim that single-molecule Raman spectroscopy is possible. In this work, the origin of the background signal is considered to be mainly due to the adsorbate (or an adsorbate–metal complex) which most likely emits enhanced fluorescence, whereas, as a first approximation, the contribution from the substrate and contamination are negligibly small compared with resonant Raman scattering.

Wang et al. developed a method to prepare a sharp Au tip, as shown in Figure 1.4, and built an apparatus for Raman spectroscopy [30]. By using the original apparatus, they measured an STM image of a monolayer of 4, 4′-bipyridine (4BPY) on Au(111) as well as a TERS spectrum at the tip-approached condition. They found strong enhancement of the Raman spectrum compared to the tip-retracted condition. From the STM image, it can be seen that 4BPY is adsorbed flat on the surface, but the TERS signal seems to originate from 4BPY perpendicular to the surface.

Picardi et al. introduced a method to fabricate a sharp Au tip for STM by electrochemical etching [31]. The efficiency of TERS for a thin BCB dye layer using the etched sharp tip was then compared with that using an Au-coated AFM tip.

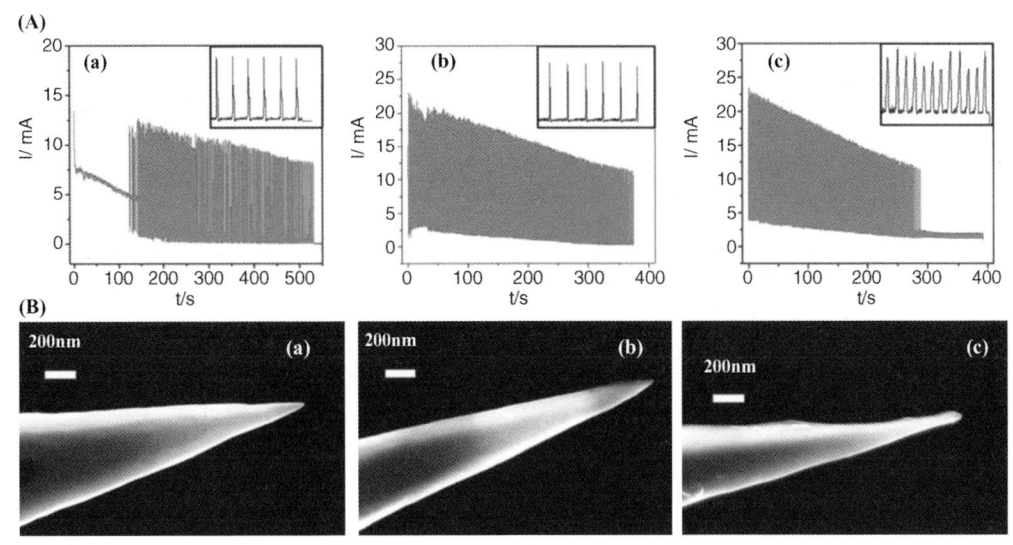

Figure 1.4 (A) Current–time curves for Au wires etched in a mixture of HCl:ethanol (1 : 1v/v) at different voltages. (B) SEM images of the etched Au tips. A: 2.1 V, B: 2.2 V, C: 2.4 V. Reprinted with permission from Xi Wang, *Applied Physics Letters*, 91, 101105 (2007). Copyright 2007, American Institute of Physics.

A higher enhancement factor and better reproducibility were obtained when the etched sharp tip was used. This is probably due to the intrinsically lower "optical quality" of the coated tips with respect to the massive metal ones as well as to the differences in tip shapes. They calculated the Raman signal enhancement factor (EF) using the equation $C = I_{tip\ engaged}/I_{tip\ withdrawn} = 1 + \text{EF}(\pi r_{tip}^2/A)$, where C is the ratio of the Raman scattered intensity with the tip engaged ($I_{tip\ engaged}$) and the tip withdrawn ($I_{tip\ withdrawn}$), r_{tip} is the tip radius, A is the laser focal area. They estimated the EF to be about 5×10^2 for coated AFM tips and 4×10^4 to 2×10^5 for etched STM tips. The same research group further studied TERS by changing the polarization of the incident laser light [32]. They measured TERS of Si(111) and BCB dye on Au(111). The experimental result with the former sample is in accord with a model reported by Ossikovski et al. [33]. For both samples, the TERS enhancement was larger when the incident light had an electric field component along the tip axis, while a non-negligible enhancement was also observed for the field component perpendicular to the tip axis.

We have observed the TERS of a single wall carbon nanotube (SWCNT) on highly oriented pyrolytic graphite (HOPG) [34]. The STM image and TERS image were obtained simultaneously, as shown in Figure 1.5. The TERS image was obtained by mapping the Raman signal intensity at the $1340\,\text{cm}^{-1}$ peak (D-band) of SWCNT. An aggregate of SWCNT was observed in the STM topographical image, and strong Raman signals were observed in the TERS image. Thus, we have succeeded in observing the simultaneous spectral mapping of TERS, although the position of the SWCNT in the STM image was shifted from the TERS image by several hundred nm.

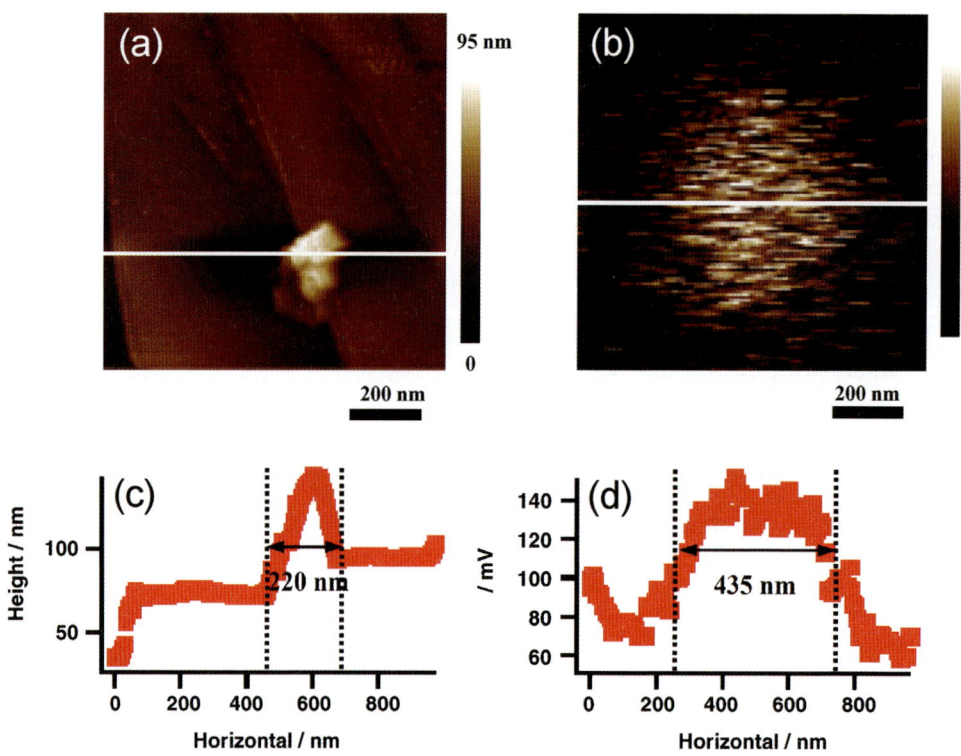

Figure 1.5 Simultaneously obtained STM (a) and TERS (b) image of SWCNT on HOPG. The TERS image was obtained for the 1340 cm^{-1} peak (D-band) of SWCNT.

This probably arises from the different position of tunneling (STM) and of field enhancement (TERS) owing to the tip shape and the height of the observation object.

1.4.2
STM Combined With Fluorescence Spectroscopy

Anger et al. measured the fluorescence rate of a single Nile Blue molecule as a function of its distance from a gold nanoparticle using a scanning probe technology [35]. The fluorescence rate shows a maximum around the molecule–nanoparticle distance (z) of 5 nm and decreases for smaller or larger z. This is due to the competition between the increased rate of excitation due to local field enhancement and the decrease in the quantum yield due to nonradiative energy transfer to the nanoparticle. They also calculated the quantum yield, excitation rate, and fluorescence rate by the multiple multipole method (MMA) and the dipole approximation. The curve of the quantum yield and fluorescence rate was reproduced only with the MMA calculation.

Kuhn et al. observed the fluorescence enhancement and fluorescence decay rate of a single terrylene molecule when a spherical gold nanoparticle was approached to the molecule, using scanning probe technology [36]. The dependence of the fluorescence intensity and fluorescence decay rate on the lateral and vertical displacement of the gold nanoparticle from the molecule was measured. The fluorescence intensity was enhanced 19 times with a simultaneous 22-fold shortening of the excited state lifetime when the gold nanoparticle was in the vicinity of the molecule. The decay rates c_r (radiative decay rate), c_{nr} (nonradiative decay rate), and c (total decay rate), were calculated neglecting the effect from the substrate following a formalism presented by Puri *et al.* [37], and were compared with the experimentally obtained c. At the closest molecule–nanoparticle separation achieved experimentally, the values $c_r = 11c_0$ and $c_{nr} = 11c_0$ were deduced, where c_0 is the unperturbed decay rate. The enhancement was larger when using an excitation wavelength near the gold nanoparticle plasmon resonance maximum wavelength than for a wavelength at the tail of the plasmon resonance, revealing the importance of the plasmon resonance in the excitation enhancement.

Nishitani *et al.* observed the tip-enhanced fluorescence of 8 nm thick meso-tetrakis (3,5-di-tertiarybutyl-phenyl)porphyrin (H2TBPP) films on ITO with an Ag tip [38]. They reported that the fluorescence was enhanced by the locally confined electromagnetic field in the vicinity of the tip. The enhancement factor is evaluated to be larger than 2000.

1.5
Future Prospects

We have reviewed the present situation of research on STM tip-enhanced Raman and fluorescence spectroscopy. There have been several papers showing the enhancement of Raman spectra when tips are in the approached condition. However, STM-TERS intensity mapping – scanning the tip to acquire a TERS intensity image and a STM topographical image simultaneously – seems to remain a difficult task. Very recently, Steidtner *et al.* have reported the TERS spectra and the TERS intensity mapping of a single BCB molecule on an Au(111) surface using a gold tip in ultrahigh vacuum [39]. Observation under ambient conditions is also expected for various systems.

We propose here a method to achieve single (or even sub-) molecule spectroscopy based on an additional principle to modify conventional TERS. The method utilizes the vibrational excitation of molecules by inelastic scattering of tunneling electrons, leading to further localization of the excited area. This can be applied to both fluorescence and Raman spectroscopy. For fluorescence spectroscopy, an area including the tip is illuminated with light having photon energies smaller than the absorption edge of the target molecules. When the molecule is observed by STM and electrons tunnel through the tip–molecule gap, some of the electrons are inelastically scattered and the molecule is vibrationally excited if the energy of the electron is higher than the energy of the molecular vibration. The vibrational excitation by tunneling electrons was first demonstrated by Stipe *et al.* for adsorbed single molecules under a high vacuum and at ultra-low temperature [40]. The

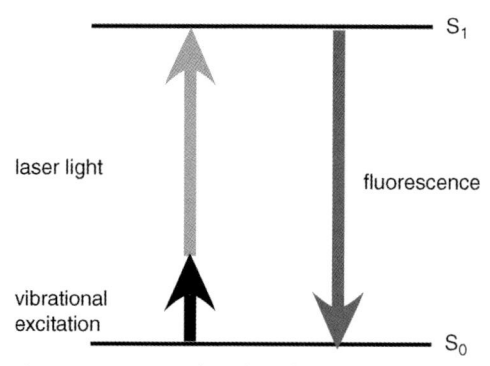

Figure 1.6 Concept of single molecule fluorescence observation using STM.

vibrationally excited molecule can be electronically excited by the absorption of light, resulting in the emission of fluorescence. The concept of this method is presented in Figure 1.6. Because STM has an atomic scale spatial resolution, only one molecule (or part of a molecule) among other surrounding molecules can be selectively excited. Therefore, the above mentioned method would enable us to perform single (sub-) molecule fluorescence spectroscopy, even if the molecules are densely packed.

The STM tunneling current and the light intensity necessary for this method can be estimated as follows. If both the tunneling current and the light intensity are continuous with time, the probability for simultaneous injection of electron and photon will be very low. Therefore, the tunneling bias of STM and the excitation light signal should be applied as pulses in order to induce vibrational excitation and absorption of a photon simultaneously. This is shown schematically in Figure 1.7.

First we consider the electronic excitation probability P_{elec} for a single molecule during a single laser pulse. When a molecule has the absorption coefficient ε_{abs}(dm^3 mol^{-1} cm^{-1}), its absorption cross section σ_{abs} is given by $3.81 \times 10^{-21}\varepsilon_{abs}$ cm^2 molecule^{-1}. Since the probability P_{elec} is proportional to the light intensity (photons s^{-1} cm^{-2}) under the objective lens, it is given by

$$P_{elec} = \frac{1.91 \times 10^{-5} \varepsilon_{abs} I_{ph} \lambda_{ph}}{T_{ph} \tau_{ph} S} \tag{1.1}$$

where I_{ph} (W) is the average laser power, T_{ph} (Hz) the laser repetition rate, λ_{ph} (nm) the laser wavelength, τ_{ph} (s) the laser pulse width, and S (cm^2) the illuminated area under the objective lens. For an allowed transition in typical organic molecules, ε_{abs} is larger than 10^4 dm^3 mol^{-1} cm^{-1}, and a conventional table-top ps laser can safely emit 1 µW green light (λ_{ph} = 500 nm) with a repetition rate of 10 MHz. Assuming that the illuminated area under the objective lens is 100 µm in diameter, we obtain $P_{elec} = 1.2 \times 10^5$. This means that the molecule under the microscope can be definitely excited during a short ps pulse.

The next step is to estimate how long a molecule can stay in its vibrational excited state during a single electric pulse. A conventional electronic circuit cannot generate a short pulse like 10 ps, so that the use of electric pulses longer than 100 ps is more

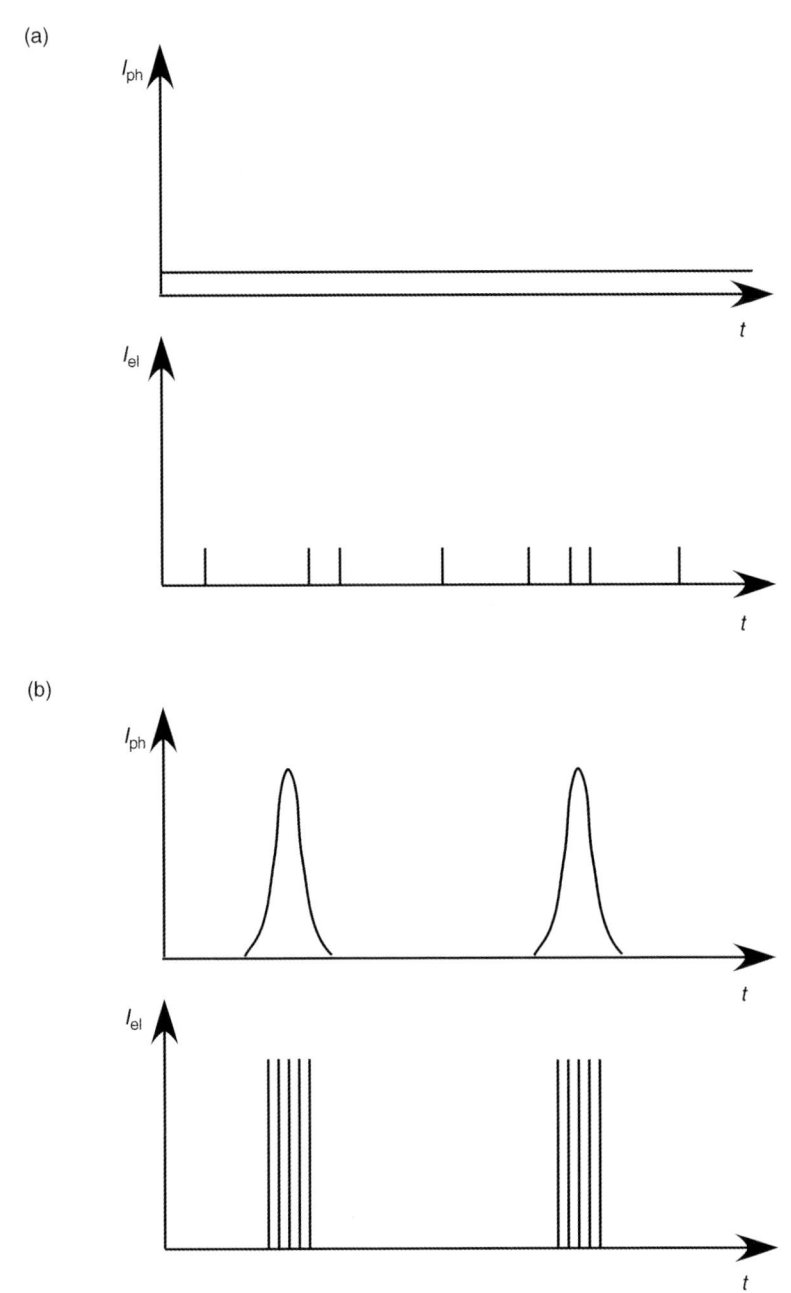

Figure 1.7 Schematics of simultaneous incoming probability of electrons and photons for (a) continuous mode and (b) pulsed mode.

realistic. The lifetime of vibrational excited states in organic molecules is generally in the range of 10 ps, which means that the molecule can relax immediately and be excited again during the electric pulses. The average dwelling time for the molecule in its vibrational excited states during a single electric pulse is given by

$$T_{dwell} = \frac{6.25 \times 10^9 \eta_{el} I_{el} \tau_{vib}}{T_{el}} \quad (1.2)$$

where η_{el} denotes the probability for vibrational excitation by a single electron, I_{el} (nA) average tunneling current, τ_{vib} (s) the lifetime of vibrational excited states, T_{el} (Hz) the tunneling bias repetition rate. Stipe et al. reported that the total conductance increase induced by inelastic electron tunneling for acetylene molecules is of the order of several percent [40]. Therefore, several percent of the electrons injected into the molecule are considered to induce vibrational excitation, although this is dependent on the molecule, its vibrational mode, and other surrounding conditions. By assuming $\eta_{el} = 1\%$, $I_{el} = 1$ nA, $\tau_{vib} = 10$ ps, and $T_{el} = 10$ MHz, we obtain $T_{dwell} = 62.5$ ps. If we can utilize electric pulses shorter than this time duration as well as synchronized with the light pulses, we would be able to further excite the molecule in its vibrationally excited state to the electronically excited states. Of course, the values of the above variables depend on each experimental system, and optimization of the values will be necessary.

For Raman spectroscopy, the experimental set-up is similar to that for fluorescence spectroscopy, but the observation wavelength should be in the range of anti-Stokes Raman spectra. When a molecule is excited vibrationally with pulsed tunneling currents, the molecule is expected to emit anti-Stokes scattered light when the probe light pulses are synchronized. In general, Raman scattering efficiency is lower than that of fluorescence, meaning that the anti-Stokes Raman measurement may be difficult. In any case, the key is how to increase the dwelling time for a molecule in its vibrationally excited states.

In summary, recent progress and future prospects in the research field of fluorescence and Raman spectroscopy combined with STM in order to achieve high spatial resolution spectroscopy have been reviewed. In the near future, single (sub-) molecule STM spectroscopy is expected to be applied to the nano-world of science and engineering.

References

1 Imura, K. and Okamoto, H. (2008) Development of novel near-field microspectroscopy and imaging of local excitations and wave functions of nanomaterials. *Bull. Chem. Soc. Jpn.*, **81**, 659–675.

2 Anderson, N., Hartschuh, A., Cronin, S. and Novotny, L. (2005) Nanoscale vibrational analysis of single-walled carbon nanotubes. *J. Am. Chem. Soc.*, **127**, 2533–2537.

3 Binnig, G., Rohrer, H., Gerber, Ch. and Weibel, W. (1982) Surface studies by scanning tunneling microscopy. *Phys. Rev. Lett.*, **49**, 57–61.

4 Bai, C. (2000) *Scanning Tunneling Microscopy and Its Applications*, (Springer

Series in Surface Sciences), 2nd Revised edn, Springer, New York.

5 Julian Chen, C. (2008) *Introduction to Scanning Tunneling Microscopy*, 2nd edn, (Monographs on the Physics and Chemistry of Materials) (64), Oxford University Press, Oxford.

6 Moskovits, M. (1985) Surface-enhanced spectroscopy. *Rev. Mod. Phys.*, **57**, 783–826.

7 Campion, A. and Kambhampati, P. (1998) Surface-enhanced Raman scattering. *Chem. Soc. Rev.*, **27**, 241–250.

8 Shen, C. K., Heinz, T. H., Ricard, D. and Shen, Y. R. (1983) Surface-enhanced second-harmonic generation and Raman scattering. *Phys. Rev. B*, **27**, 1965–1979.

9 Barber, P. W., Chang, R. K. and Massoudi, H. (1983) Surface-enhanced electric intensities on large silver spheroids. *Phys. Rev. Lett.*, **50**, 997–1000.

10 Bohren, C. F. and Huffman, D. R. (2004) *Absorption and Scattering of Light by Small Particles*, Wiley-VCH, Weinheim.

11 Kottmann, J. P., Martin, O. J. F., Smith, D. R. and Schultz, S. (2000) Spectral response of plasmon resonant nanoparticles with a non-regular shape. *Opt. Express*, **6**, 213–219.

12 Purcell, E. M. and Pennypacker, C. R. (1973) Scattering and absorption of light by non-spherical dielectric grains. *Astrophys. J.*, **186**, 705–714.

13 Draine, B. T. and Flatau, P. J. (1994) Discrete-dipole approximation for scattering calculations. *J. Opt. Soc. Am. A*, **11**, 1491–1499.

14 Jeanmaire, D. L. and Van Duyne, R. P. (1977) Surface Raman spectroelectrochemistry: Part I. Heterocyclic, aromatic, and aliphatic amines adsorbed on the anodized silver electrode. *J. Electroanal. Chem.*, **84**, 1–20.

15 Nie, S. and Emory, S. R. (1997) Probing single molecules and single nanoparticles by surface-enhanced Raman scattering. *Science*, **275**, 1102–1106.

16 Neacsu, C. C., Dreyer, J., Behr, N. and Raschke, M. B. (2006) Scanning-probe Raman spectroscopy with single-molecule sensitivity. *Phys. Rev. B*, **73**, 193406-1–193406-4.

17 Demming, F., Jersch, J., Dickmann, K. and Geshev, P. I. (1998) Calculation of the field enhancement on laser-illuminated scanning probe tips by the boundary element method. *Appl. Phys. B*, **66**, 593–598.

18 Jersch, J., Demming, F., Hildenhagen, L. J. and Dickmann, K. (1998) Field enhancement of optical radiation in the nearfield of scanning probe microscope tips. *Appl. Phys. A*, **66**, 29–34.

19 Mills, D. L. (2002) Theory of STM-induced enhancement of dynamic dipole moments on crystal surfaces. *Phys. Rev. B*, **65**, 125419-1–125419-11.

20 Wu, S. and Mills, D. L. (2002) STM-induced enhancement of dynamic dipole moments on crystal surfaces: theory of the lateral resolution. *Phys. Rev. B*, **65**, 205420-1–205420-7.

21 Rendell, R., Scalapino, D. and Mühlschlegel, B. (1978) Role of local plasmon modes in light emission from small-particle tunnel junctions. *Phys. Rev. Lett.*, **41**, 1746–1750.

22 Klein, S., Geshev, P., Witting, T., Dickmann, K. and Hietschold, M. (2003) Enhanced Raman scattering in the near field of a scanning tunneling tip – an approach to single molecule Raman spectroscopy. *Electrochemistry*, **71**, 114–116.

23 Pettinger, B., Picardi, G., Schuster, R. and Ertl, G. (2000) Surface enhanced Raman spectroscopy: towards single molecule spectroscopy. *Electrochemistry*, **68**, 942–949.

24 Downes, A., Salter, D. and Elfick, A. (2006) Finite element simulations of tip-enhanced Raman and fluorescence spectroscopy. *J. Phys. Chem. B*, **110**, 6692–6698.

25 Pettinger, B., Picardi, G., Schuster, R. and Ertl, G. (2002) Surface-enhanced and STM-Tip-enhanced Raman spectroscopy at metal surfaces. *Single Mol.*, **5–6**, 285–294.

26 Pettinger, B., Picardi, G., Schuster, R. and Ertl, G. (2003) Surface-enhanced and STM tip-enhanced Raman spectroscopy of CN⁻ ions at gold surfaces. *J. Electroanal. Chem.*, **554**, 293–299.

27 Ren, B., Picardi, G. and Pettinger, B. (2004) Preparation of gold tips suitable for tip-enhanced Raman spectroscopy and light emission by electrochemical etching. *Rev. Sci. Instrum.*, **75**, 837–841.

28 Pettinger, B., Ren, B., Picardi, G., Schuster, R. and Ertl, G. (2005) Tip-Enhanced Raman spectroscopy (TERS) of malachite green isothiocyanate at Au(111): bleaching behavior under the influence of high electromagnetic fields. *J. Raman Spectrosc.*, **36**, 541–550.

29 Domke, K. F., Zhang, D. and Pettinger, B. (2006) Toward Raman fingerprints of single dye molecules at atomically smooth Au(111). *J. Am. Chem. Soc.*, **128**, 14721–14727.

30 Wang, X., Liu, Z., Zhuang, M.-D., Zhang, H.-M., Wang, X., Xie, Z.-X., Wu, D.-Y., Ren, B. and Tian, Z.-Q. (2007) Tip-enhanced Raman spectroscopy for investigating adsorbed species on a single-crystal surface using electrochemically prepared Au tips. *Appl. Phys. Lett.*, **91**, 101105–101105-3.

31 Picardi, G., Nguyen, Q., Schreiber, J. and Ossikovski, R. (2007) Comparative study of atomic force mode and tunneling mode tip-enhanced Raman spectroscopy. *Eur. Phys. J. Appl. Phys.*, **40**, 197–201.

32 Picardi, G., Nguyen, Q., Ossikovski, R. and Schreiber, J. (2007) Polarization properties of oblique incidence scanning tunneling microscopy – tip-enhanced Raman spectroscopy. *Appl. Spectrosc.*, **61**, 1301–1305.

33 Ossikovski, R., Nguyen, Q. and Picardi, G. (2007) Simple model for the polarization effects in tip-enhanced Raman spectroscopy. *Phys. Rev. B*, **75**, 045412-1–045412-9.

34 Yoshidome, M. (2006) Study of molecular aggregates on solid surface using scanning tunneling microscopy and Raman spectroscopy, Ph.D. thesis, Tohoku University.

35 Anger, P., Bharadwaj, P. and Novotny, L. (2006) Enhancement and quenching of single-molecule fluorescence. *Phys. Rev. Lett.*, **96**, 113002-1–113002-4.

36 Kühn, S., Håkanson, U., Rogobete, L. and Sandoghdar, V. (2006) Enhancement of single-molecule fluorescence using a gold nanoparticle as an optical nanoantenna. *Phys. Rev. Lett.*, **97**, 017402-1–017402-4.

37 Das, P. C. and Puri, A. (2002) Energy flow and fluorescence near a small metal particle. *Phys. Rev. B*, **65**, 155416-1–155416-8.

38 Nishitani, R., Liu, H. W., Kasuya, A., Miyahira, H., Kawahara, T. and Iwasaki, H. (2007) STM Tip-enhanced photoluminescence from porphyrin film. *Surf. Sci.*, **601**, 3601–3604.

39 Steidtner, J. and Pettinger, B. (2008) Tip-enhanced Raman spectroscopy and microscopy on single dye molecules with 15 nm resolution. *Phys. Rev. Lett.*, **100**, 236101-1–236101-4.

40 Stipe, B. C., Rezaei, M. A. and Ho, W. (1998) Single-molecule vibrational spectroscopy and microscopy. *Science*, **280**, 1732–1735.

2
Vibrational Nanospectroscopy for Biomolecules and Nanomaterials

Yasushi Inouye, Atsushi Taguchi, and Taro Ichimura

2.1
Introduction

When metallic nanostructures are illuminated with white light, they give us a variety of colors by scattering the light. For example, old stained glass windows in a church show a distinguished and glorious scene because nanoparticles of noble metals, for example, gold, are present in the glass. Stained glass does not bleach, but keeps its beauty forever, as long as metallic particles exist in the glass. Why do the metallic nanoparticles provide various colors which are different from the color of the bulk metal?

In fact, their coloration is strongly related to the size of the nanoparticles. Plasmon polaritons, which are coupled quanta of the collective oscillation of free electrons in a metal and photons, are generated on/around the surface of the nanoparticles [1]. As the plasmon polaritons are resonant phenomena, the frequency of the resonance is very sensitive to the dielectric constant and the size. Hence, metallic nanoparticles produce a variety of colors according to their size and composition. Features of the plasmon polaritons correlated with resonant phenomenon are enhancement and confinement of light field/photons in the nano-dimensions. In particular, photons accompanied by plasmons are localized in the vicinity of the metallic nanostructures although the wavelength of the photon is of the order of several hundred nanometers. Hence we create a nano-light-source around a metallic nanostructure by virtue of the localization of plasmon polaritons. Furthermore, the field enhancement effect provides the nano-light-source capability for achieving highly sensitive optical measurement.

In 1994, we proposed that a metallic needle having a nano-tip at its apex be employed as a nano-light-source for microscopy attaining nanometric spatial resolution [2]. Later, we expanded the technique to Raman spectroscopy for molecular nano-identification, nano-analysis and nano-imaging. In this chapter, we give a brief introduction to local plasmons and microscopy using a metallic nano-needle to produce the local plasmons. Then, we describe the microscope that we built and show some experimental results of nano-Raman analysis and nano-imaging with the microscope.

2.2
Surface Plasmon Polaritons

Electrons in a metal move around freely as electron gas while the atomic nuclei are located at each lattice position. The free electrons are collectively oscillating due to Coulomb interaction between the electric charges. As the density of the electric charges moves back and forth in an oscillating electric field, such a phenomenon is called plasma oscillation and the quantum of the plasma oscillation is called a plasmon. In such a phenomenon, an electric field is induced owing to the collective oscillation of electric charges, thereby a magnetic field arises due to generation of the electric field. This means that an electromagnetic wave is accompanied by the plasmon. Reciprocally, plasma oscillation is induced because an electric field or light exerts a Coulomb force on free electrons. Hence, the plasma oscillation and the electromagnetic field are coupled with each other. The coupled quanta are called plasmon polaritons.

Plasmons generated on a metallic surface are a longitudinal wave propagating on the surface as a transverse magnetic wave [3]. Such plasmons are called surface plasmons and can arise on a metallic thin film deposited on a glass substrate [4]. The metallic film works as a waveguide on which the plasmons propagate. Here, the wavenumber of the plasmon has to be matched with that of the photons in order for coupling to occur. According to the dispersion relation of the plasmon, as shown in Figure 2.1, the wavenumber of a light field propagating in a lower refractive index medium is not coincident with that of the plasmon except at the origin of the coordinates, which means that it is impossible to generate a plasmon by using a light field propagating in a lower refractive index medium. On the other hand, the dispersion curve of the plasmon intersects that of a light field propagating into the substrate (higher refractive index material). The wavenumber of the plasmon is

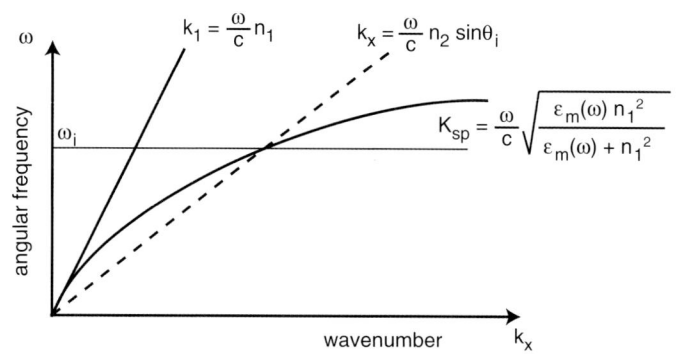

Figure 2.1 Dispersion relation of surface plasmon.

matched with that of the light field through the substrate at the crossing point, hence the plasmon can be generated by being coupled resonantly with the light field. Considering the wavenumber of the light field through the lower refractive index medium, the component perpendicular to the surface is imaginary while the component parallel to the surface is the same as that of the plasmon. The light field accompanied by the plasmon does not propagate into the lower refractive index medium but the field is localized on the boundary. Such a light field is called an evanescent field. By tuning the incident angle of the light field propagating into the substrate properly for coincidence with both the wavenumbers of the plasmon and the light field, plasmons can be generated in the three layer system (called the Kretchmann configuration) [4]. As the resonant condition, that is, the coincidence of the wavenumbers of both the plasmons and the light field, is very sensitive to the refractive index of the medium located on the metallic thin film, this configuration is widely used for refractive index measurement [5]. The sensor, which is called a surface plasmon resonance (SPR) sensor, is applied as an immunosensor and so on.

A plasmon can be generated in a metallic nanoparticle [1]. When the nanoparticle is irradiated with a light field, free electrons in the metallic nanoparticle are forced to oscillate synchronously with the electric polarity of the light field. Polarization is then induced in the nanoparticle. The polarization P is given by the product of the incident light field E and the polarizability of the metallic nanoparticle α ($P = \alpha \cdot E$). Assuming that the nanoparticle is a sphere and that its diameter is much smaller than the wavelength of the light field, the polarizability of the nanoparticle is given by the following equation,

$$\alpha = \frac{4\pi r_m^3 \{\varepsilon_m(\omega) - \varepsilon_1\}}{\{\varepsilon_m(\omega) + 2\varepsilon_1\}} \tag{2.1}$$

Here, r_m, $\varepsilon_m(\omega)$, ε_1, and ω are the radius, the dielectric constant of the nanoparticle, the dielectric constant of the medium surrounding the nanoparticle, and the angular frequency, respectively. Accordingly, the polarization is diverged if the denominator of Eq. (2.1) approaches zero ($\text{Re}\{\varepsilon_m(\omega)\} + 2\varepsilon_1 = 0$). The divergence is equivalent to the resonant phenomenon which corresponds to surface plasmon resonance generated in the metallic nanoparticle. Compared with the surface plasmons generated in a thin metallic film, the plasmons are localized around the nanoparticle because they do not propagate into the space. The light field or evanescent field is accompanied by the plasmons in the same manner as the surface plasmons on a metallic film. This type of plasmon is called a localized surface plasmon (or local plasmon). As can be deduced from Eq. (2.1), the resonant frequency of the local plasmons is determined by the dielectric constants of the nanoparticle material and the medium surrounding the nanoparticle provided that the diameter of the nanoparticle is negligible compared to the wavelength. If the diameter is not negligible then Mie scattering theory gives the precise resonant frequency for the local plasmons of a metallic sphere. In this case, the size of the metallic particle is an additional parameter in the determination of the resonant frequency.

2.3
Near-Field Optical Microscopy Using a Metallic Nano-Tip

Optical microscopy is used for observation, analysis, fabrication, and manipulation in a wide number of fields ranging from basic science to industrial applications. The light field that is focused with an objective lens on the sample plane interacts with the sample (e.g., absorption, scattering, fluorescence) at a micro or sub-micro scale. As the focusing or imaging phenomena of the light field are a result of the nature of the wave, the spatial resolution of optical microscopy is limited to about half the wavelength [6]. This is the so-called diffraction limit of a light wave. How do we overcome the limitation of spatial resolution? One answer is the use of a metallic nanoparticle or nanostructure which confines the light field around the nanostructure by generating local plasmons. The nanostructure works as a nano-light-source.

In 1985, Wessel proposed such a new type of microscope which used a submicrometer-sized metal particle attached to a glass substrate, as shown in Figure 2.2 [7]. As the metal particle confines and enhances the light field by coupling with local plasmons, the light field can be interacted with the sample in a much smaller region than the wavelength and the light field can be detected with high sensitivity. Optical mapping and imaging of a sample can be achieved with super-resolved power by scanning the metal particle on the sample surface. Wessel called this technique surface-enhanced optical microscopy. In 1989, Pohl and Fisher reported experimental results of such optical microscopy for the first time [8]. They formed a metallic nano-protrusion by evaporating gold film onto a polystyrene particle adsorbed on a glass substrate. When the protrusion approached the sample surface, surface plasmons generated on the gold film were coupled with local plasmons. Here the local plasmon is generated on the gold protrusion due to matching of the resonant condition with the relative refractive index of the environment configured with the metallic film, protrusion and sample. Since plasmon resonance is highly sensitive to changes in the refractive index of the sample, the optical properties of the sample surface were measured by detecting the light field

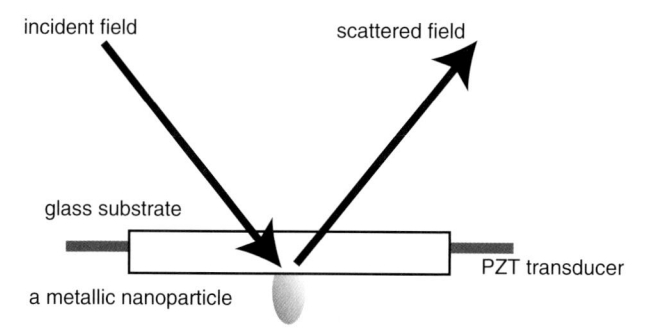

Figure 2.2 Optical probe of surface-enhanced microscopy proposed by J. Wessel. A metallic nanoparticle attached to a glass substrate confines and enhances the light field.

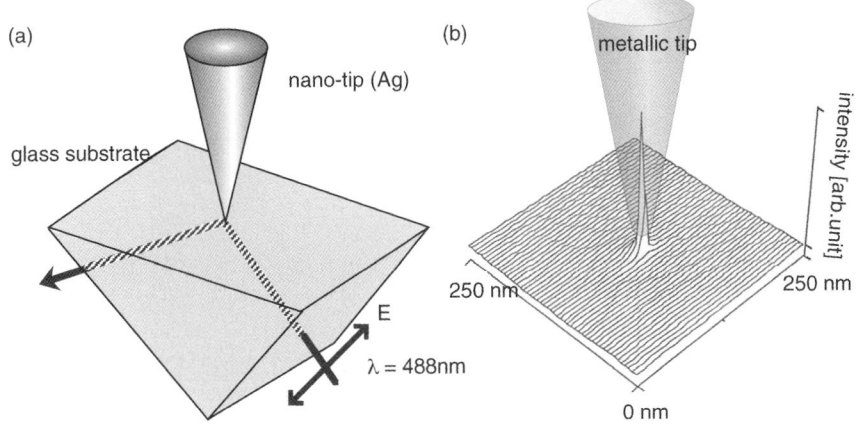

Figure 2.3 Numerical analysis of a nano-light-source generated by a metallic nano-tip. (a) Model for numerical analysis. (b) Intensity distribution of light scattered by the metallic nano-tip.

scattered from the nano-protrusion. In 1994, we proposed the use of a metallic needle with a nano-sized apex which enables us to scan the probe on the sample surface easily by combining a scanning tunneling microscope and an atomic force microscope to regulate the gap between the tip apex and the sample surface [2].

Next, we explain how to confine the light field at a metallic nano-tip by using electromagnetism. Figure 2.3a shows a model for the calculation of light field scattering at the nano-tip. The metallic nano-tip is located on the glass substrate while plane light propagates in the glass substrate towards the boundary at an incident angle of 45°, as shown in the figure. Here, we assume that the material of the tip is silver and the diameter of the tip apex is 20 nm. The finite-differential time-domain (FDTD) method was employed in the calculation. Polarization of the incident field is parallel to the incident plane, which corresponds to p-polarization. Figure 2.3b shows the intensity distribution of a scattered light field just under the metallic nano-tip. As shown in the figure, the light field is confined under the nano-tip due to generation of the local plasmon polaritons. As the dimension of field confinement coincides with the size of the nano-tip, the localized field works as the nano-light-source. Furthermore, the light field is enhanced by a factor of ~100 or more, depending on the structure and size of the tip apex, due to the resonant effect of the plasmon polaritons. Assuming that the incident field is perpendicular to the incident plane, that is s-polarization, confinement and enhancement of the light field do not occur because the light field oscillating in this direction does not couple with the collective oscillation of electrons at the nano-tip. Even if the incident light field possesses p-polarization, a dielectric or sharpened glass fiber nano-tip exhibits much less confinement and enhancement than a metallic nano-tip. To summarize, a metallic nano-tip and p-polarized incident light field are requisite to nano-imaging with high sensitivity, which indicates that local plasmons play an important role in the enhancement of the electric field.

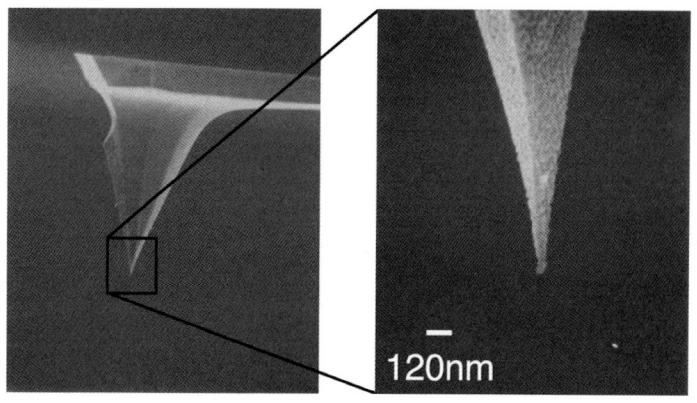

Figure 2.4 A scanning electron microscopy image of an AFM cantilever tip covered with a thin silver film.

As a near-field probe, many types of nano-tips are proposed and employed, for example, an STM needle tip [2], AFM tips covered with metallic thin film [9], tetrahedral tips [10], laser-trapped metallic nanoparticles [11], bow-tie antennas [12], and so on. In general, metal-coated AFM cantilevers are used for near-field observation and analysis due to their simple preparation and ease of handling. Figure 2.4 shows a scanning electron microscopy image of an AFM tip covered with a thin silver film. As the film was coated by thermal evaporation in vacuum with a thickness of 20 nm, a metallic nanostructure of this size is formed at the tip apex. The evaporation rate should be set carefully to avoid bending of the lever due to additional tension. Chemical reduction is another method for the creation of a metallic nanostructure or thin film on the apex. In order to obtain extreme enhancement of the light field, the design or optimization of nano-tips should be taken into account. For example, a triangular structure having a certain shape and size shows, from electromagnetic theory, an enhancement factor of the light field intensity up to $\sim 10^6$ under a well-tuned resonant condition of local plasmon polaritons [12].

2.4
Tip-Enhanced Near-Field Raman Spectroscopy and Imaging

A nano-light-source generated on the metallic nano-tip induces a variety of optical phenomena in a nano-volume. Hence, nano-analysis, nano-identification and nano-imaging are achieved by combining the near-field technique with many kinds of spectroscopy. The use of a metallic nano-tip applied to nanoscale spectroscopy, for example, Raman spectroscopy [9], two-photon fluorescence spectroscopy [13] and infrared absorption spectroscopy [14], was reported in 1999. We have incorporated Raman spectroscopy with tip-enhanced near-field microscopy for the direct observation of molecules. In this section, we will give a brief introduction to Raman spectroscopy and demonstrate our experimental nano-Raman spectroscopy and imaging results. Furthermore, we will describe the improvement of spatial resolution by introducing nonlinear Raman spectroscopy into the tip-enhanced near-field technique.

2.4.1
Raman Spectroscopy

Molecules vibrate due to displacement of the relative position of the atoms since covalent bonds among the atoms work as springs. Each vibration mode has a specific frequency. When laser light hits molecules, the energy of the laser light is partly transferred to the molecules due to light scattering and the molecules begin to vibrate by acquiring the energy. The frequency of the laser light decreases (accordingly, the wavelength becomes longer) because the laser light loses the energy corresponding to the molecular vibration energy due to inelastic scattering. Such a phenomenon is called Raman scattering (more precisely, Stokes Raman scattering). Since each vibrational mode has an intrinsic frequency shift of the Raman scattering, the scattering spectrum indicates which kind of vibration modes a molecule possesses or what kind of molecules are present in a sample.

Raman spectroscopy is a very powerful tool for the analysis of molecules and their dynamics, like infrared absorption spectroscopy and fluorescence spectroscopy. While transition among electronic states plays a part in fluorescence spectroscopy, vibrational transition occurs in Raman scattering and mid-infrared absorption. Accordingly, molecular vibration is directly observed for the latter. Not all molecules emit fluorescence, on the other hand Raman scattering is induced for all molecules due to the vibration of chemical bonding. Furthermore, the quenching and photobleaching phenomena that are sometimes seen in fluorescence spectroscopy do not occur in Raman and IR absorption spectroscopy. Visible lasers are available in Raman spectroscopy while a broadband light source is required for IR absorption spectroscopy.

However, as Raman scattering is a two-photon process, the probability of the Raman scattering process is lower than that of fluorescence and IR absorption processes. The cross section of Raman scattering is $\sim 10^{-30}$ cm^2, which is much smaller than that of fluorescence ($\sim 10^{-16}$ cm^2) and IR absorption ($\sim 10^{-20}$ cm^2). When we detect Raman scattering at the nanoscale, the number of photons obtained is less than with the usual micro-Raman spectroscopy due to reduction in the detection area or the number of molecules. To overcome this problem, we need to devise a method for amplification of Raman scattering.

2.4.2
Near-Field Nano-Raman Microscopy

With regard to the confinement and enhancement ability of a metallic nano-tip, we have proposed near-field Raman microscopy using a metallic nano-tip [9]. The metallic nano-tip is able to enhance not only the illuminating light but also the Raman scattered light [9, 15, 16]. Figure 2.5 illustrates our nano-Raman microscope that mainly comprises an inverted microscope for illumination and collection of Raman

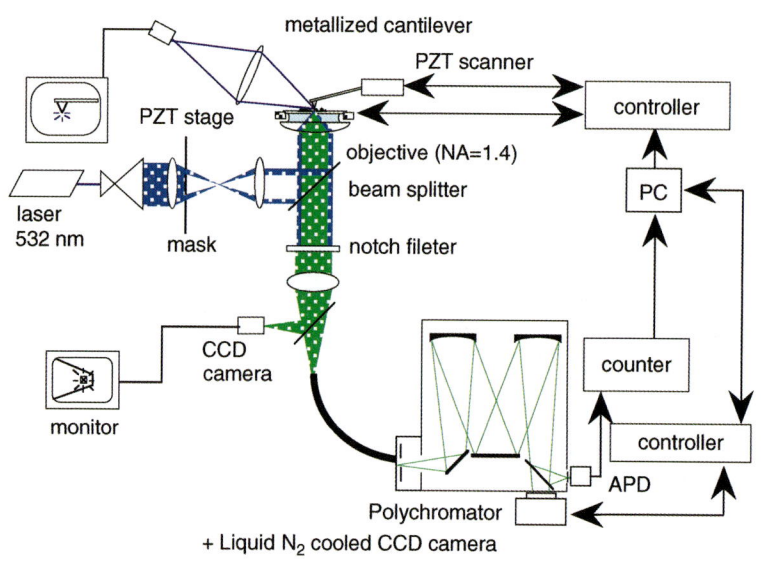

Figure 2.5 A tip-enhanced near-field Raman microscope which we have developed. The microscope is based on AFM for control of the metallic nano-tip, an inverted optical microscope for illumination/collection of the light field and a polychromator for measurement of the Raman signal.

scattering from the nano-tip, an atomic force microscope (AFM) for control of the nano-tip and a dispersive spectrophotometer for detection of the Raman spectrum [17]. Laser light (wavelength: 532 nm) is focused onto the sample plane through an objective lens (NA: 1.4, magnification: ×100). A focused spot is formed by using an annular mask in the pupil plane to reject any light component for which NA is less than 1.0 [18]. When put into the focused spot, the metallic nano-tip produces a nano-light-source at an apex of the tip and induces Raman scattering from molecules located under the tip. Raman scattered light is collected via the same objective and guided to the spectrophotometer after passing a notch filter to eliminate Rayleigh scattering. An AFM cantilever, the tip of which is coated with a thin silver film of thickness 40 nm, is operated in contact mode.

2.4.3
Tip-Enhanced Near-Field Raman Spectroscopy and Imaging

In this section, we will describe some experiments which we have performed using the above-mentioned nano-Raman microscope. Figure 2.6a shows the Raman spectrum of an adenine nanocrystal of height 7 nm and width 30 nm [19]. Several Raman bands are observed as the probe tip is near enough to the sample (AFM operation is made in contact mode). These bands, except the one appearing at 924 cm^{-1}, are assigned as the vibrational modes, inherent to the adenine molecule, according to the molecular orbital calculation. For examples, two major bands, one at

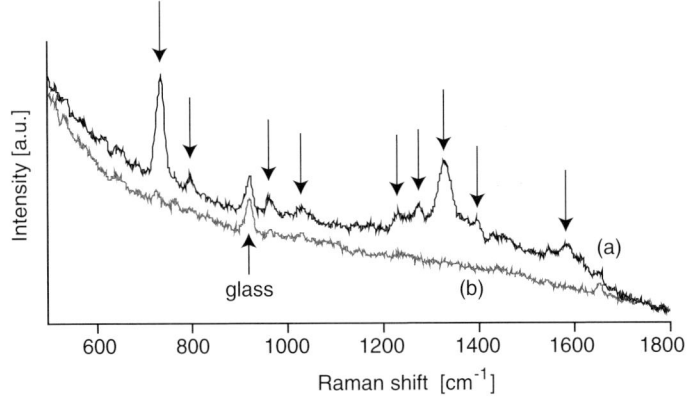

Figure 2.6 Raman spectrum of an adenine nanocrystal obtained (a) with and (b) without the metallic tip. Spectrum (a) corresponds to the tip-enhanced near-field Raman spectrum while spectrum (b) shows the conventional micro-Raman spectrum.

739 cm^{-1} and the other at 1328 cm^{-1} are the ring breathing mode, and a combination of the C–N stretching mode and C–C stretching mode, respectively. The spectral peak at 924 cm^{-1} is assigned as a Raman band of the glass substrate. This was proved from the experimental result shown in Figure 2.6b, which is the same as Figure 2.6a except that measurement was made when the tip was far from the sample. This indicates not only that the bands, except that at 924 cm^{-1}, are all due to Raman scattering by adenine molecules located near the probe tip, but also that the Raman spectrum is detected only when the probe is in the near-field of the molecules of interest, otherwise the photon field is not enhanced enough to scatter Raman-shifted photons. The laser power for illumination was 2.5 mW and the exposure time for obtaining the Raman spectra was 1 min. Comparing the intensity per unit area of the Raman band at 739 cm^{-1} in Figure 2.6a and b, the enhancement factor afforded by the metallic nano-tip is estimated at 2700. We carried out spectral mapping of the nanocrystals in order to evaluate the spatial response of the measurement. Figure 2.7a shows the spectral mapping attained by scanning the sample at intervals of 30 nm. Near-field Raman spectra were detected for 10 s at each position using a nitrogen-cooled CCD camera. The optical response of the two major Raman bands at 739 cm^{-1} and 1328 cm^{-1}, taken from Figure 2.7a, is shown in Figure 2.7b. The intensity of the two Raman bands exhibits a similar optical response and the minimum spatial response is 30 nm. This value corresponds to the size of the metallic nano-tip. This experimental result agrees well with the numerical analysis shown in Figure 2.3.

Nonlinear optical phenomena, as well as near-field optics, provide us with super resolving capability [20]. The probability of nonlinear optical phenomena is proportional to the number of photons which participate in the phenomenon. For example, the intensity distribution of two-photon excited fluorescence corresponds to the square of the excitation light. Thus, we proposed a combination of the field

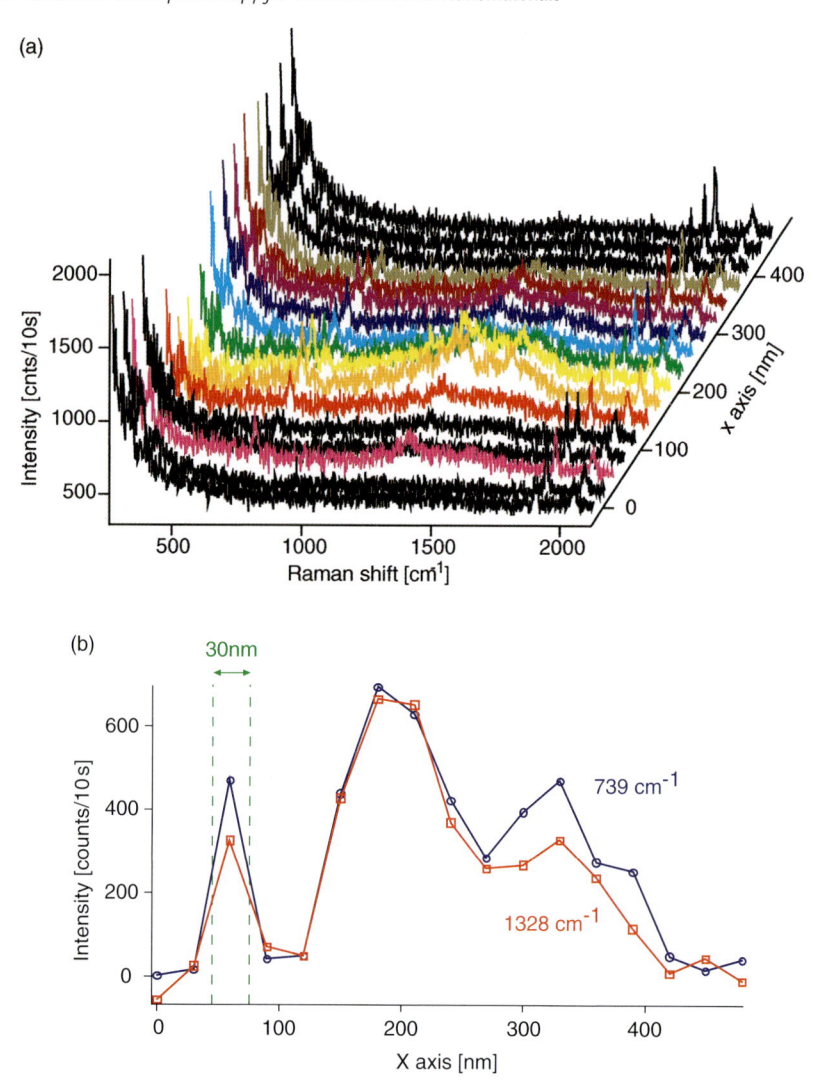

Figure 2.7 (a) Tip-enhanced near-field Raman spectral mapping of the adenine nanocrystal at 30 nm intervals. (b) Raman intensity distribution of two major bands at 739 cm^{-1} and 1328 cm^{-1}.

enhancement effect of a metallic nano-tip and coherent anti-Stokes Raman scattering (CARS) spectroscopy, a third-order nonlinear Raman spectroscopy [21]. With the tip enhancement of CARS, the excitation of CARS polarization can be further confined and highly enhanced at the very end of the probe tip owing to its third-order nonlinearity, providing higher spatial resolution capability. In addition, because of the nonlinear responses, even a small enhancement of the excitation field can lead to a huge enhancement of the emitted signal, allowing a reduction of the far-field background.

CARS spectroscopy utilizes three incident fields including a pump field (ω_1), a Stokes field (ω_2; $\omega_2 < \omega_1$), and a probe field ($\omega'_1 = \omega_1$) to induce a nonlinear polarization at $\omega_{CARS} = 2\omega_1 - \omega_2$. When $\omega_1 - \omega_2$ coincides with one of the molecular-vibration frequencies of a given sample, the anti-Stokes Raman signal is resonantly generated [22, 23]. We induce the CARS polarization by the tip-enhanced field at the metallic tip end of the nanometric scale.

In our tip-enhanced near-field CARS microscopy, two mode-locked pulsed lasers (pulse duration: 5 ps, spectral width: 4 cm^{-1}) were used for excitation of CARS polarization [21]. The sample was a DNA network nanostructure of poly(dA-dT)-poly(dA-dT) [24]. The frequency difference of the two excitation lasers ($\omega_1 - \omega_2$) was set at 1337 cm^{-1}, corresponding to the ring stretching mode of diazole. After the "on-resonant" imaging, ω_2 was changed such that the frequency difference corresponded to none of the Raman-active vibration of the sample ("off-resonant"). The CARS images at the on- and off- resonant frequencies are illustrated in Figure 2.8a and b, respectively. A spontaneous Raman spectra is shown in Figure 2.8d in which the on- and off-resonant frequencies are indicated. The DNA bundles are observed at the resonant frequency, as shown in Figure 2.8a, while they cannot be seen at the off-resonant frequency in Figure 2.8b. This indicates that the observed contrast is dominated by the vibrationally resonant CARS signals. Figure 2.8c shows a cross-section of Figure 2.8a denoted by two solid arrows, which were acquired with a ~5 nm step. The FWHM of

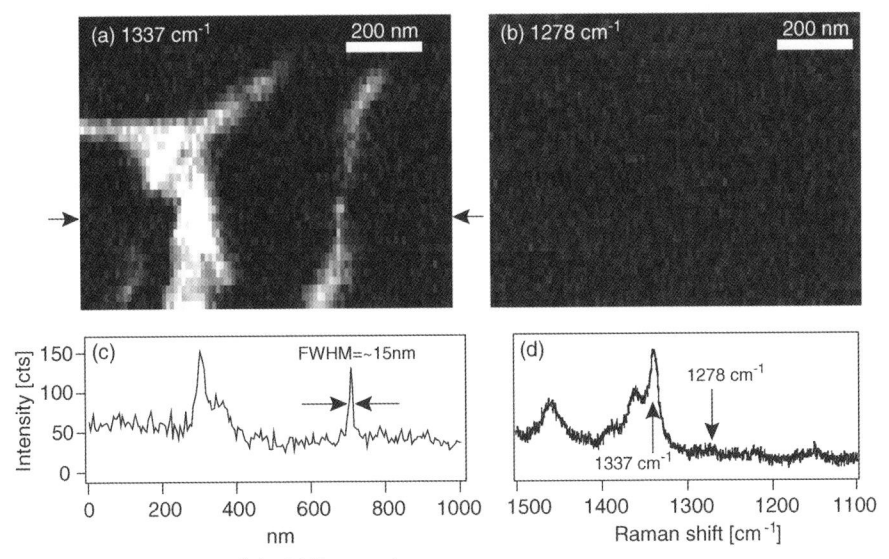

Figure 2.8 CARS images of the DNA network structure. (a) A tip-enhanced CARS image in the on-resonant condition ($\omega_1 - \omega_2 = 1337$ cm^{-1}). (b) A tip-enhanced CARS image in the off-resonant condition ($\omega_1 - \omega_2 = 1278$ cm^{-1}). (c) Line profile of the row indicated by the solid arrows in (a). (d) A spontaneous Raman spectrum of the DNA sample, in which the arrows indicate the frequencies adopted for the on- and off- resonant conditions.

the narrowest peak is found to be ~15 nm, as shown in Figure 2.8c. This spatial response is better than that of the spontaneous Raman scattering appearing in Figure 2.7b. The improvement of spatial resolution is attributed to the nonlinearity of the CARS process. The size of the effective excitation volume of the DNA structure and the enhancement factor of the CARS signal was estimated to be ~1 zeptoliter ~10^6, respectively. This huge enhancement factor is also attributed to the nonlinear effect.

2.5
Tip Effect on Near-Field Raman Scattering

Tip-enhanced near-field Raman scattering is the same phenomenon as surface-enhanced Raman scattering (SERS) except that only one metallic nanostructure participates in tip-enhanced Raman scattering while many metallic nanostructures are associated with SERS. Two enhancement mechanisms, that is, electromagnetic enhancement and chemical enhancement, exist in TERS as in SERS. The former is related to local plasmon polaritons, as described earlier, the latter is caused by formation of metal-molecule complexes by chemical adsorption and the change in the electronic state of the molecule [25]. Specific changes in Raman spectra are often seen in SERS spectroscopy due to the first layer effect of chemical adsorption; for example, peak shift and/or a huge enhancement of specific Raman bands and the appearance of new Raman bands. We observed such phenomena in tip-enhanced near-field Raman spectroscopy. Furthermore, we found a new phenomenon in which particular Raman bands in TERS spectra of adenine nanocrystals shift to higher wavenumber than those observed in SERS [19]. In this section, we will demonstrate such phenomena appearing in TERS.

Figure 2.9a shows a TERS spectrum of a single adenine nanocrystal. Several Raman bands, including two intense bands at 739 and 1328 cm^{-1} due to the ring breathing mode of the whole molecule and the ring-stretching mode of diazole, were detected. These peaks were not observed in the far-field Raman spectra of the same sample obtained by retracting the tip from the sample. The TERS spectrum is compared with an ordinary SERS spectrum (Figure 2.9b) and an ordinary near-infrared (NIR) Raman spectrum (Figure 2.9c) both of which we measured. The peak frequency of the RBM (ν_{RBM}) is 739, 733, and 723 cm^{-1} in the TERS, SERS and normal Raman spectra, respectively. The peak frequency of the TERS spectrum shifted to higher frequency than those of SERS and the normal Raman spectra. It is also seen that the peak frequency in the SERS spectrum shifted with respect to the normal Raman spectrum. These frequency shifts are strongly related to the chemical interaction between the silver and the molecules. However, the Raman band shift appearing in TERS is definitely different from that of SERS. Why were both spectra not coincident with each other? In TERS, a metallic nano-tip is controlled with an AFM which applies a constant force onto the sample to get the topography of the sample surface while observing the tip-enhanced near-field Raman scattering from the sample. As the nano-tip causes unidirectional pressure, the anisotropy in our techniques allows us to change the molecular bond lengths in one direction in a controlled manner, and the simultaneous

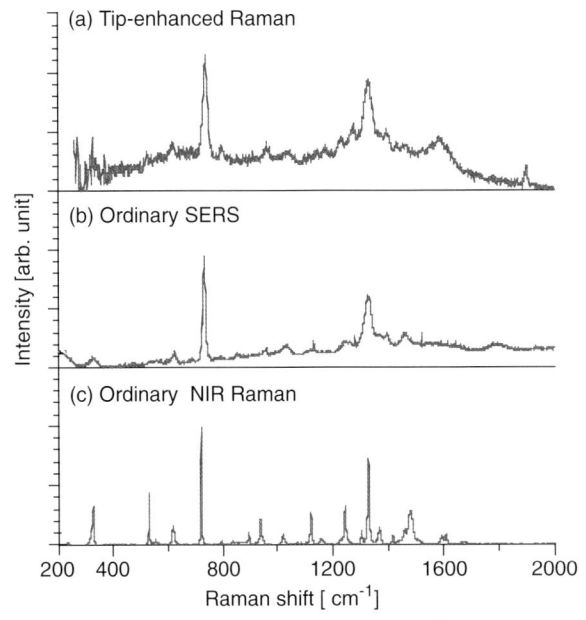

Figure 2.9 Spectra of a single adenine nanocrystal. (a) TERS spectrum, (b) ordinary SERS spectrum, and (c) ordinary near-infrared (NIR) Raman spectra. For the SERS measurement, a silver island film was used. For the NIR Raman measurement, a thick sample of adenine was used with a 1 h exposure.

near-field Raman scattering measurement allows us to perform an *in situ* nano-analysis. The uniaxial pressure effect shows up in interesting spectral changes, such as peak shifts, peak broadening and new peak appearance, as the AFM-controlled tip force changes. Accordingly, the difference between TERS and SERS is interpreted as a result of a "mechanical" pressure effect due to the dynamic contact of the metallic nano-tip and the molecules in addition to the chemical effect.

The chemical and mechanical effects were analyzed by using quantum chemical calculations of the simplest adenine-silver complex model where a single silver atom represents the silver nano-tip employed in our experiment. In order to improve the accuracy of the calculation for the vibrational frequencies, we use a complex model consisting of an adenine molecule and a silver cluster (quadrimer). In the model, a silver cluster is adjacent to the nitrogen atom at N3 (Ad-N3) (Figure 2.10a and b), which was found to be the best model to give good agreement with the experimental results. The vibrational properties were calculated using the UB3LYP/6-311 + G** (for adenine)/SDD (for Ag) [26]. For analysis of the mechanical effect, the bond distance between the N3 atom and the adjacent silver atom was changed, and the vibrational frequencies were calculated for the different bond distances. The calculated frequency shifts of ν_{RBM} as well as the calculated potential curves (binding energy) are plotted as a function of the bond distance between the silver atom and the N3 nitrogen of adenine in Figure 2.10c. The calculated frequency ν_{RBM} demonstrated a significant shift towards

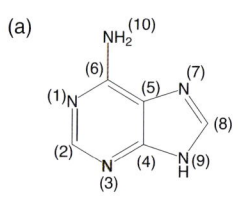

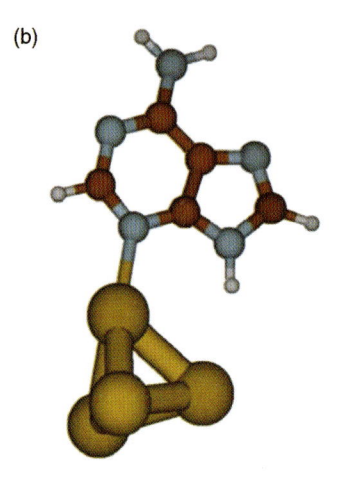

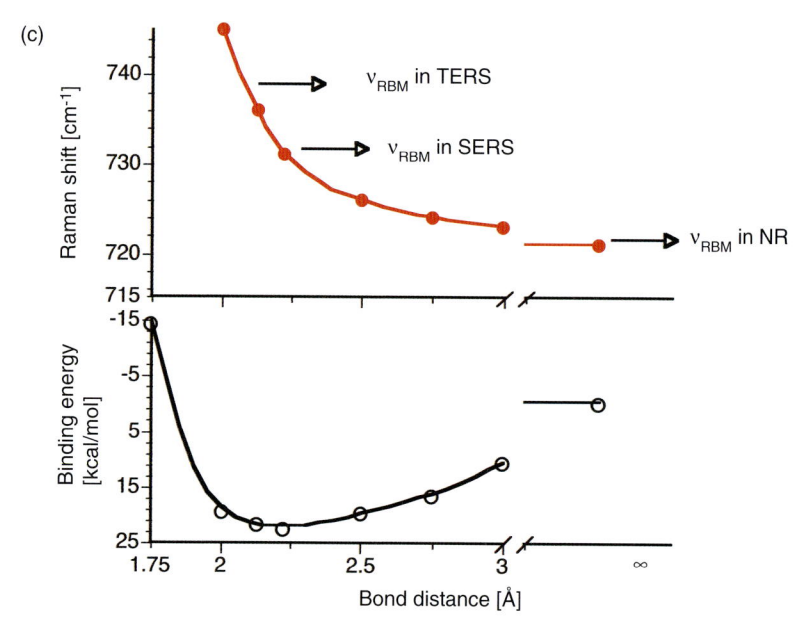

Figure 2.10 (a) Molecular structure and atomic numbering of adenine. (b) The calculated model of the adenine-silver quadrimer complex. (c) The calculated frequency shifts n_{RBM} of the Ad-N3 Ag quadrimer and the calculated binding energy as a function of the bond distance for the Ag–N linkage.

higher frequency with the contraction of the bond distance. The calculated frequency agrees with the corresponding band of the experimental TERS spectra of adenine when the bond distance of the Ag–N linkage is ~5% contracted from the equilibrium. In our TERS experiment, the adenine nanocrystals are pressurized by the nano-tip with a constant atomic force (ca. 1–5 pN molecule^{-1}). When the bond distance of the Ag–N linkage is reduced by ~5%, the repulsive force of 6 pN molecule^{-1} is derived from a harmonic oscillation of the binding energy difference. In this case, the repulsive force obtained from the calculated potential surface of the adenine silver quadrimer coincides quantitatively with the atomic force. These results support the idea that the frequency shifts occur due to the deformation of adenine molecules by the silver tip.

Moreover, we have found temporal fluctuation in TERS spectra of an adenine nanocrystal when we left the silver-coated cantilever (operated in contact mode) on the surface of the nanocrystal for 600 s [27]. Figure 2.11a shows a waterfall plot of a

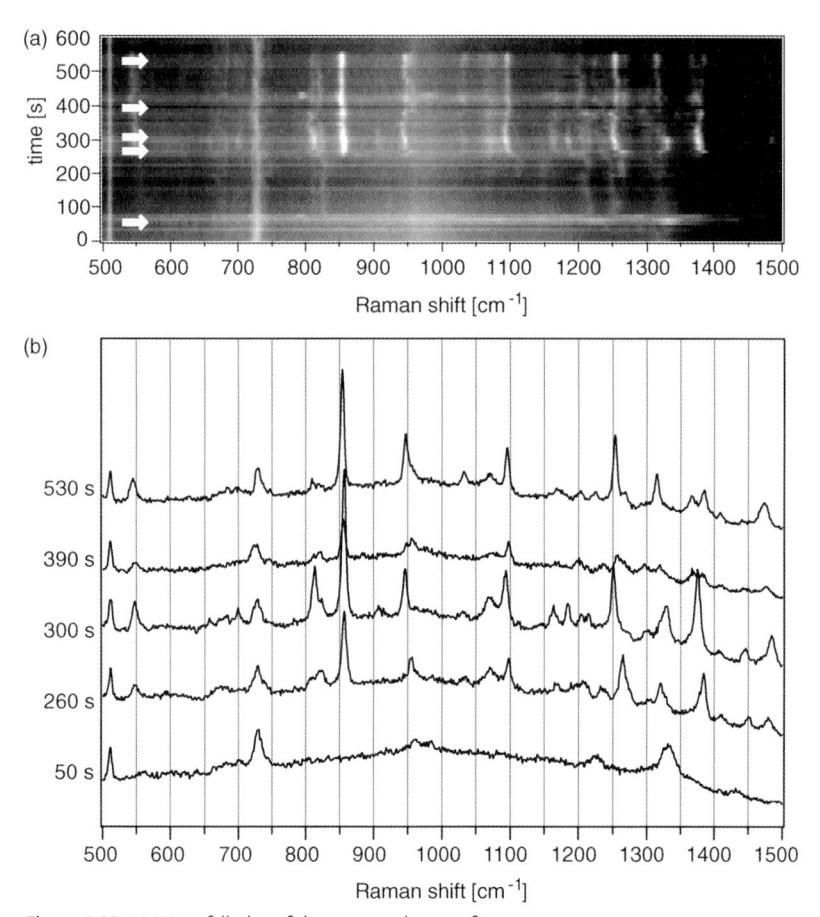

Figure 2.11 (a) Waterfall plot of the time evolution of Raman spectra of adenine polycrystal. (b) Five spectra taken from (a). The exposure time for each spectrum is 10 s.

time series of tip-enhanced near-field Raman spectra of the nanocrystal. Five characteristic spectra, indicated by arrows in Figure 2.11a, are shown in Figure 2.11b for facile comparison of their spectral shapes. Most of the Raman peaks observed in Figure 2.11a and b are assigned to adenine except for one peak at 511 cm^{-1} which is assigned to silicon, the material of the cantilever. It is obvious from Figure 2.11 that the intensities and frequencies of many peaks of adenine fluctuate temporally. This is quite a contrast to the stable peak of the silicon at 511 cm^{-1}. Moreover, several Raman peaks suddenly appeared at 250 s and disappeared at 550 s. This phenomenon can be referred to as "blinking". Furthermore, the relative intensities of several Raman peaks, for example, at 800 cm^{-1}, ~855 cm^{-1}, and ~945 cm^{-1} are exceptionally strong in the time range 250–550 s, although these peaks are relatively weak in the normal Raman spectrum of adenine. The observed phenomena in the tip-enhanced near-field Raman spectra, including fluctuation, blinking, and extraordinary enhancement of several peaks, are analogous to those observed in previous studies on surface-enhanced Raman spectroscopy of a single molecule. We suppose that this phenomenon was caused by temporal fluctuation of molecular adsorption on the silver tip. Then, we analyzed the vibrational modes of an adenine molecule for different polarization of the incident field by using density functional theory, as shown in Figure 2.12. The calculated result indicates that the Raman spectrum of the adenine molecule is strongly dependent on the polarization

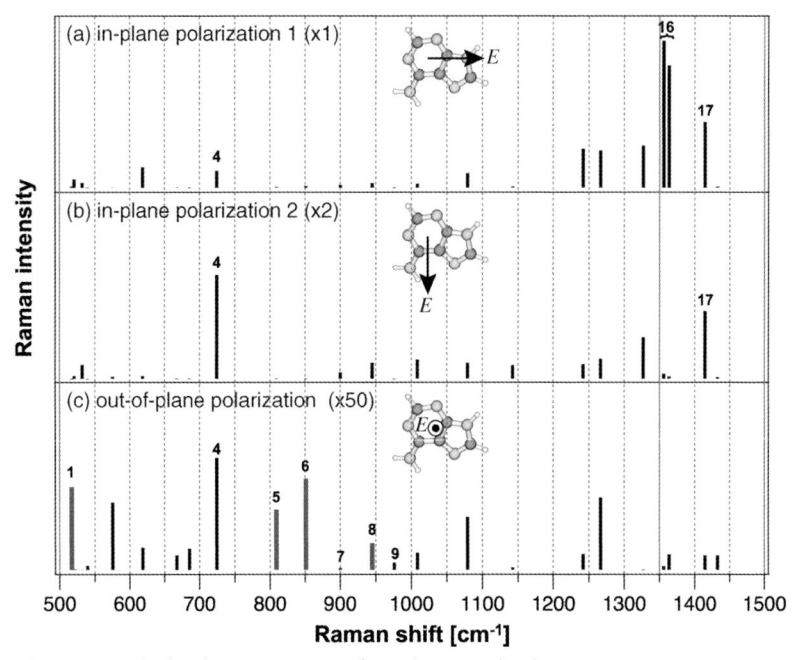

Figure 2.12 Calculated Raman spectra of an adenine molecule with three orthogonal polarization directions (a), (b): in-plane polarization, c: out-of-plane polarization).

of the incident field denoted by 'E' in the figure. As the tip-enhanced field has polarization parallel to the axis of the nano-tip, orientation of the molecules is supposed to change relative to the polarization. The experimental and calculated results support the fact that temporally fluctuating tip-enhanced near-field Raman spectra are affected by the molecular orientation. Hence, TERS spectroscopy enables us to determine the orientation of a molecule with high sensitivity as well as to analyze, identify and image molecules at the nanometric scale.

The force effect is applicable to investigation of the mechanical properties of nanomaterials [28, 29]. We measured TERS spectra of a single wall carbon nanotube (SWCNT) bundle with a metallic tip pressing a SWCNT bundle [28]. Figure 2.13a–e show the Raman spectra of the bundle measured *in situ* while gradually applying a force up to 2.4 nN by the silver-coated AFM tip. Raman peaks of the radial breathing

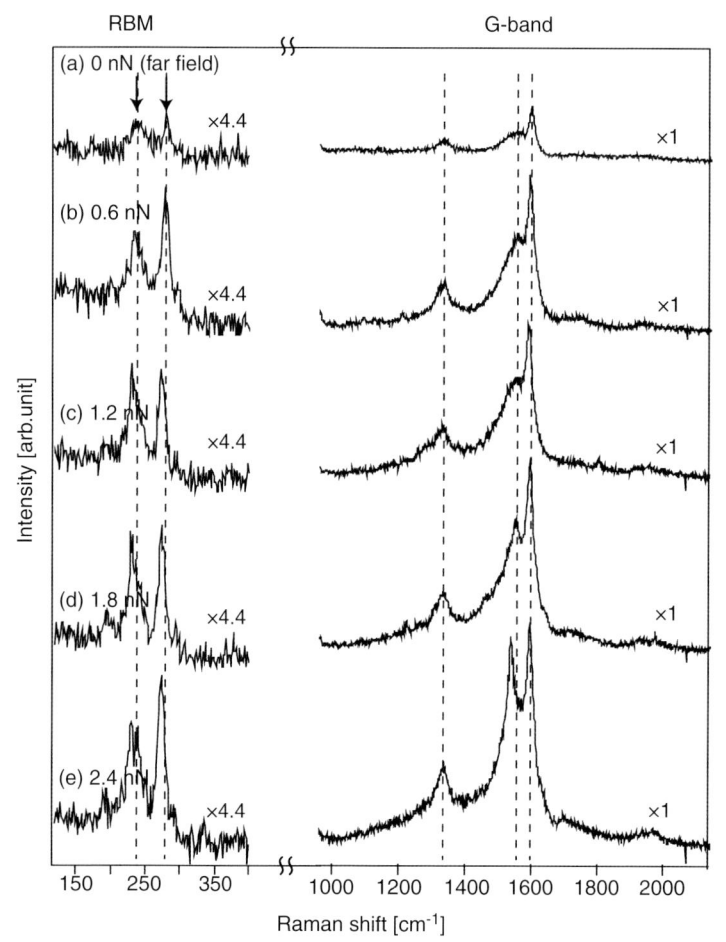

Figure 2.13 TERS spectra of an SWCNT bundle measured with an applied tip-force up to 2.4 nN.

mode bands and the lower-frequency Raman peak (stretching mode along the circumferential direction) of the G-band down-shifted by as much as $18\,\mathrm{cm}^{-1}$ as the tip force increased while no change was observed in the higher-frequency Raman peak of the G-band (the mode along the axial direction). The peak shift was caused by radial deformation of SWCNTs in the bundle. More interestingly, the Raman intensity of the two peaks in the G-band increased with increasing force. The intensity increase is attributed to the resonant Raman effect caused by modification of the electronic band gap energies of the SWCNTs.

2.6
Conclusion

Local plasmon polaritons that are generated in the vicinity of a metallic nano-tip work as a nano-light-source for revealing the nano-world with light. As the light field is strongly enhanced as well as confined at nano-volume due to the resonance effect of local plasmon, the metallic nano-tip is of benefit to Raman spectroscopy and imaging with spatial resolution at the nanoscale. Metallic atoms of the nano-tip are able to form complexes with molecules in the near-field measurement thanks to chemical adsorption. This phenomenon changes the Raman spectrum drastically, depending on the orientation of the molecules, thus the method is suitable for determination of the molecular orientation. The nano-tip also provides perturbative force on the sample at the nano/atomic level owing to the use of AFM for position control of the nano-tip. Hence, high-pressure Raman spectroscopy is feasible without any special equipment for high pressure. As the chemical and force phenomena are generated at the molecular or atomic scale, spatial resolution of the tip-enhanced Raman spectroscopy may reach the molecular or atomic levels. Nano-spectroscopy with the metallic nano-tip opens the way to real nano-imaging and nano-analysis of biosamples with a gentle and safe light.

References

1 Kawata, S. (2001) *Near-Field Optics and Surface Plasmon Polaritons*, Springer, Heidelberg.

2 Inouye, Y. and Kawata, S. (1994) Near-field scanning optical microscope using a metallic probe tip. *Opt. Lett.*, **19**, 159–161.

3 Raether, H. (1988) *Surface Plasmons on Smooth and Rough Surfaces and on Gratings*, Springer-Verlag, Heidelberg.

4 Kretschmann, E. (1971) Die Bestimmung Optischer Konstanten von Metallen durch Anregung von Oberflächenplasmaschwingungen. *Z. Phys.*, **241**, 313–324.

5 Matsubara, K., Kawata, S. and Minami, S. (1988) Optical chemical sensor based on surface plasmon measurement. *Appl. Opt.*, **27**, 1160–1163.

6 Born, M. and Wolf, E. (1980) *Principle of Optics*, 6th edn, Pergamon Press, Oxford.

7 Wessel, J. (1985) Surface-enhanced optical microscopy. *J. Opt. Soc. Am. B*, **2**, 1538–1540.

8 Fischer, U. Ch. and Pohl, D. W. (1989) Observation of single-particle plasmons by near-field optical microscopy. *Phys. Rev. Lett.*, **62**, 458–461.

9 Inouye, Y., Hayazawa, N., Hayashi, K., Sekkat, Z. and Kawata, S. (1999) Near-field scanning optical microscope using a metallized cantilever tip for nanospectroscopy. *Proc. SPIE*, **3791**, 40–48.

10 Koglin, J., Fischer, U. C. and Fuchs, H. (1997) Material contrast in scanning near-field optical microscopy at 1–10 nm resolution. *Phys. Rev. B*, **55**, 7977–7984.

11 Sugiura, T., Okada, T., Inouye, Y., Nakamura, O. and Kawata, S. (1997) Gold-bead scanning near-field optical microscope with laser-force position control. *Opt. Lett.*, **22**, 1663–1665.

12 Kottmann, J. P., Martin, O. J. F., Smith, D. R. and Schultz, S. (2001) Non-regularly shaped plasmon resonant nanoparticle as localized light source for near-field microscopy. *J. Microsc.*, **202**, 60–65.

13 Sanchez, E. J., Novotny, L. and Xie, X. S. (1999) Near-field fluorescence microscopy based on two-photon excitation with metal tips. *Phys. Rev. Lett.*, **82**, 4014–4017.

14 Knoll, B. and Keilmann, F. (1999) Near-field probing of vibrational absorption for chemical microscopy. *Nature*, **399**, 134–137.

15 Stockle, R. M., Suh, Y. D., Deckert, V. and Zenobi, R. (2000) Nanoscale chemical analysis by tip-enhanced Raman spectroscopy. *Chem. Phys. Lett.*, **318**, 131–136.

16 Anderson, M. S. (2000) Locally enhanced Raman spectroscopy with an atomic force microscope. *Appl. Phys. Lett.*, **76**, 3130–3132.

17 Hayazawa, N., Inouye, Y., Sekkat, Z. and Kawata, S. (2000) Metallized tip amplification of near-field Raman scattering. *Opt. Commun.*, **183**, 333–336.

18 Hayazawa, N., Inouye, Y. and Kawata, S. (1999) Evanescent field excitation and measurement of dye fluorescence using a high N.A. objective lens in a metallic probe near-field scanning optical microscopy. *J. Microsc.*, **194**, 472–476.

19 Watanabe, H., Ishida, Y., Hayazawa, N., Inouye, Y. and Kawata, S. (2004) Tip-enhanced near-field Raman analysis of tip-pressurized adenine molecule. *Phys. Rev. B*, **69**, 155418.

20 Denk, W., Strickler, J. H. and Webb, W. W. (1990) Two-photon laser scanning fluorescence microscopy. *Science*, **248**, 73–76.

21 Ichimura, T., Hayazawa, N., Hashimoto, M., Inouye, Y. and Kawata, S. (2004) Tip-enhanced coherent anti-Stokes Raman scattering for vibrational nano-imaging. *Phys. Rev. Lett.*, **92**, 220801.

22 Duncan, M. D., Reintjes, J. and Manuccia, T. J. (1982) Scanning coherent anti-Stokes Raman microscope. *Opt. Lett.*, **7**, 350–352.

23 Zumbusch, A., Holtom G. R. and Xie, X. S. (1999) Three-dimensional vibrational imaging by coherent anti-Stokes Raman scattering. *Phys. Rev. Lett.*, **82**, 4142–4145.

24 Tanaka, S., Cai, L. T., Tabata, H. and Kawai, T. (2001) Formation of two-dimensional network structure of DNA molecules on Si substrate. *Jpn. J. Appl. Phys.*, **40**, L407–L409.

25 Chang, R. K. and Furtak, T. E. (1982) *Surface Enhanced Raman Scattering*, Plenum Press, New York and London.

26 Becke, A. D. (1993) Density-functional thermochemistry. III. The role of exact exchange. *J. Chem. Phys.*, **98**, 5648–5652.

27 Ichimura, T., Watanabe, H., Morita, Y., Verma, P., Kawata, S. and Inouye, Y. (2007) Temporal fluctuation of tip-enhanced Raman spectra of adenine molecules. *J. Phys. Chem. C*, **111**, 9460–9464.

28 Verma, P., Yamada, K., Watanabe, H., Inouye, Y. and Kawata, S. (2006) Near-field Raman scattering investigation of tip effects on C60 molecules. *Phys. Rev. B*, **73**, 045416.

29 Yano, T., Inouye, Y. and Kawata, S. (2006) Nanoscale uniaxial pressure effect of a carbon nanotube bundle on tip-enhanced near-field Raman spectra. *Nano Lett.*, **6**, 1269–1273.

3
Near-Field Optical Imaging of Localized Plasmon Resonances in Metal Nanoparticles
Hiromi Okamoto and Kohei Imura

3.1
Introduction

Metal nanostructures show peculiar optical characteristics totally different from bulk metals. For example, metal nanoparticles show characteristic resonance peaks in extinction spectra, which are greatly affected by the nanoscale geometry differences [1, 2]. It is also known that strongly enhanced optical fields localized in a nanometric spatial scale are generated under certain conditions in the vicinity of metal nanostructures. Novel functions of metal nanostructures to enhance interaction between light and molecules can be developed based on this property. It is of fundamental importance to study the optical and spectroscopic characteristics of single metal nanostructures (nanoparticles in particular) and their origins, as a basis for nano-optics and physics and for chemical and biological applications. The optical characteristics of metal nanostructures originate primarily from the collective oscillation of conduction electrons coupled with electromagnetic fields, known as localized plasmon polariton resonance (hereafter simply called plasmon) [1, 2]. The peculiar optical characteristics of metal nanostructures arising from plasmon resonances are found in a spatial scale of the order of 100 nm or less. It is therefore essential to reveal the optical characteristics in a nanometric spatial scale in order to understand and control the physical and chemical natures of metal nanostructures.

In conventional optical microscopy in the visible wavelength region, the spatial resolution is restricted by the diffraction limit of light, which is a sub-micrometer regime in so far as the optical process concerned is linear. Consequently, it is difficult or impossible to apply such conventional microscopy to characterize the optical properties of materials in nanometric spatial resolution. For the purpose of topographic measurements, the electron microscope, scanning tunneling microscope, atomic force microscope, and so forth are very powerful and can achieve atomic level spatial resolution. However, these methods are not suitable for the study of spectroscopic characteristics. Scanning near-field optical microscopy (SNOM) has been developed together with the progress of the scanning probe microscopy technique, as a method to meet the demand for optical measurements with high spatial resolution [3–6]. The highest spatial resolution achieved by this method is around 10 nm. This is not as high as those of other probe microscopes, but is still much higher than that of the conventional optical microscope determined by the diffraction limit of light. No vacuum environment is needed for SNOM measurements and even sample surfaces covered by liquids can be measured under appropriate conditions. Since near-field microscopy is based on optical measurements, we can combine SNOM with various advanced techniques developed in the field of laser spectroscopy, such as nonlinear and time-resolved methods, to develop new microscopic methods. The near-field optical method is thus expected to provide various unique nanoscale imaging techniques.

In this chapter, we review the studies of fundamental optical and plasmon characteristics of metal nanoparticles by near-field optical imaging and spectroscopy. We will show, on gold nanorods, as typical non-isotropic metal nanoparticles, that wavefunctions of plasmon modes can be imaged by near-field microscopy. We will also show ultrafast time-resolved imaging of a gold nanorod to detect dynamic changes of plasmon waves induced by photoexcitation. For assembled spherical gold nanoparticles, we will demonstrate that strongly enhanced optical fields confined in the interstitial gaps between nanoparticles can be visualized using near-field imaging. The contribution of the enhanced fields to surface enhanced Raman scattering (SERS) will also be discussed.

3.2
Near-Field Spectroscopic Method

Two different types of near-field microscopic methods are currently used. In one, a tiny aperture in an opaque metal film, with diameter less than the wavelength of light, is used [3, 5]. When the aperture is irradiated with light from one side of the film, an optical near-field is generated in the vicinity of the aperture on the back side of the film, and this is used to locally excite the sample. For the other type, we use a localized optical field generated in the neighborhood of a metal tip (or a nanoparticle) by irradiation of light [4]. This method is called the "scattering type" or "apertureless type", whereas the former one is called the "aperture type". The spatial resolution is given approximately by the aperture diameter or the radius of curvature of the metallic tip, respectively, for the aperture type and the scattering type. Both methods have advantages and drawbacks which we do not discuss in detail in this chapter.

In the authors' laboratory, we use home-built aperture-type SNOM systems [7–12]. The apparatus is shown schematically in Figure 3.1. We use an apertured near-field probe made of a single-mode optical fiber. The core of the optical fiber is sharpened by chemical etching and coated with metal, and an aperture with a diameter of typically 50–100 nm is opened at the apex of the sharpened tip. High-efficiency probes with doubly tapered structure are available from JASCO Corp. The incident light on the sample is introduced from the other end of the optical fiber. We use a Xe discharge arc lamp as the light source for near-field transmission spectral measurements, as well as

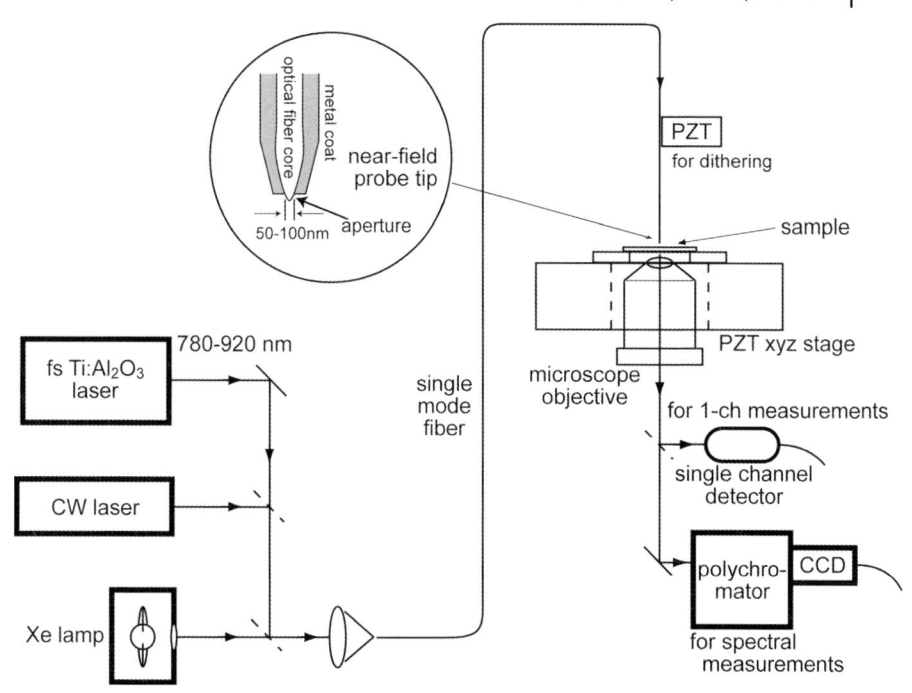

Figure 3.1 Schematic diagram of the near-field optical microscope system. The structure of the near-field probe tip is illustrated in the circle. (Reproduced with permission from Royal Society of Chemistry [10]).

various lasers for other purposes. The sample is set just beneath the probe tip, and the position is controlled by a piezo-driven high-resolution translation stage. The height of the probe tip from the sample surface is maintained at a few nanometers to about 10 nm by the shear-force feedback method, and scanned in the xy-plane. The samples for measurements are usually prepared on transparent substrates. The light from the probe excites the sample, and the radiation transmitted through, scattered by, or emitted from the sample is collected by a microscope objective lens beneath the substrate and conveyed to the detection system. For detection, a single-channel photodetector (such as a photodiode, avalanche photodiode, or photomultiplier tube) or a multi-channel spectral detector (such as a charge coupled device (CCD)) is used, depending on the purpose. To select the detected polarization, a polarizer is placed in front of the detector. If a CCD is adopted as a detector, it is possible to take two polarization components of spectra at the same time, by using a Wollaston prism to separate the two components into different directions. To select the polarization of the incident light on the sample, we adjust the ellipticity and polarization axis of the incident light by a quarter-wave plate and a half-wave plate prior to coupling to the optical fiber, so as to get the desired polarization at the tip of the fiber probe.

For near-field imaging based on nonlinear or ultrafast spectroscopy, light pulses from a femtosecond Ti:sapphire laser (pulse width ca. 100 fs, repetition rate ca.

80 MHz) is used as a light source. Since femtosecond laser pulses are spectrally broad, the dispersion effect arising from the fiber optics seriously broadens the pulse duration. This causes lower signal efficiency in nonlinear imaging and lower time resolution in ultrafast measurements. To avoid this effect, we place a grating pair device [13], to compensate for dispersion in the fiber, before coupling the beam into the fiber [7, 8, 10–12, 14]. The grating separation is adjusted to obtain the shortest pulse width at the tip of the probe.

3.3
Fundamental Spectroscopic Characteristics of Gold Nanoparticles

In this section, the fundamental spectroscopic characteristics of gold nanoparticles are described. The samples studied are crystalline gold nanoparticles synthesized in water solution. For spherical gold nanoparticles, commercially available colloidal solutions were used in most cases. Non-isotropic nanoparticles, such as rod-shaped particles (nanorods), were prepared by the seed-mediated growth method where a colloidal solution of small spherical nanoparticles was mixed with growth solutions containing a high concentration of surfactant molecules [15, 16]. As is well known, a colloidal water solution of spherical gold nanoparticle shows a strong extinction peak at about 530 nm [1, 2, 4]. This extinction band is attributed to the collective oscillation of conduction electrons in the nanoparticle, known as localized plasmon resonance. The extinction peak shifts toward the longer wavelength side and becomes broad when the diameter of the particle exceeds 100 nm.

In non-spherical or aggregated gold nanoparticles, strong extinction peaks are, in general, found in longer wavelength region. Gold nanorods, for example, show strong extinction peaks in the longer wavelength region [1, 2, 17], in addition to the peak at about 530 nm that is also found for the spherical nanoparticles. The peak shifts towards the longer wavelength side with increasing aspect ratio (ratio of the length of the rod to its diameter) (Figure 3.2). The peak at about 530 nm and that at a longer wavelength are attributed, respectively, to plasmon modes polarized along the short axis (hereafter called "transverse mode") and the long axis ("longitudinal mode") [14]. In a nanorod with large aspect ratio, the longitudinal plasmon shifts to the near-infrared region, and an additional extinction peak appears near the transverse-mode plasmon peak. This new peak is attributed to a higher longitudinal mode [18–21], where the direction of the electronic oscillation is dependent on the position on the rod, as will be explained later.

3.4
Wavefunction Images of Plasmon Modes of Gold Nanorod — Near-Field Transmission Method

We usually use white light from a Xe discharge arc lamp for the measurement of near-field transmission images and spectra [9]. The spectrum of transmitted light

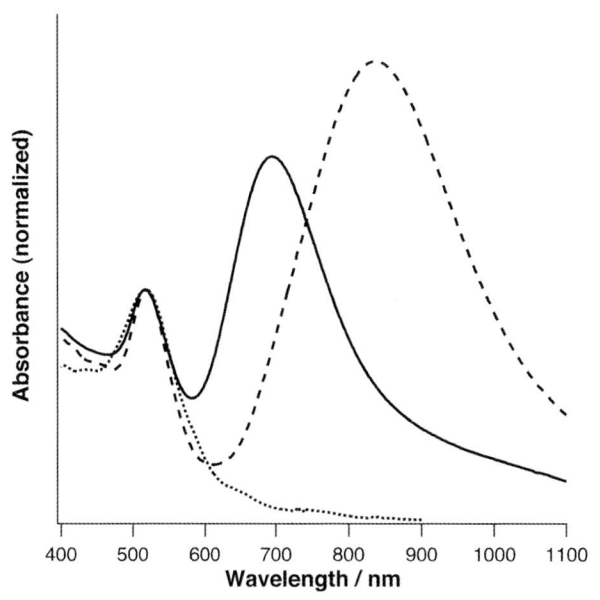

Figure 3.2 Extinction spectra of colloidal water solutions of gold nanospheres and nanorods. Dotted curve: nanospheres (diameter 15–25 nm). Solid curve: nanorods, low aspect ratio. Dashed curve: nanorods, high aspect ratio. Extinction is normalized at about 520 nm. (Reproduced with permission from Royal Society of Chemistry [10]).

from the sample is recorded as a function of lateral position by means of a polychromator equipped with a CCD. Figure 3.3 shows near-field transmission images of a gold nanorod (diameter 20 nm, length 510 nm), and transmission spectra taken at positions 1 and 2 on the rod [10, 22]. The near-field image consists of several dark spots along the long axis, instead of the uniform shadow of the rod. The number of spots decreases one by one on increasing the wavelength of observation. In the near-field transmission spectra, several resonant extinction peaks are found [20, 21]. The number of dark spots in the image is characteristic of each resonant peak. This result suggests that the images obtained correspond to visualization of wavefunctions of longitudinal plasmon modes. Near-field observation of wavefunctions was also reported for quantum well structures of semiconductors [23]

Plasmon modes of gold nanorods are shown schematically in Figure 3.4 [10, 12]. For both longitudinal and transverse modes, the direction of the collective electronic oscillation is uniform on the rod in the case of the fundamental dipolar mode. There are higher modes, in addition to the dipolar modes, where the direction of the electronic oscillation is dependent on the position on the rod. The oscillation amplitude as a function of the position gives a wavefunction of the plasmon mode [12]. The sign of the wavefunction does not alter for the dipolar mode, whereas the wavefunctions have nodes for higher plasmon modes. The oscillation frequency is approximately constant, regardless of the modes, for transverse modes. In contrast,

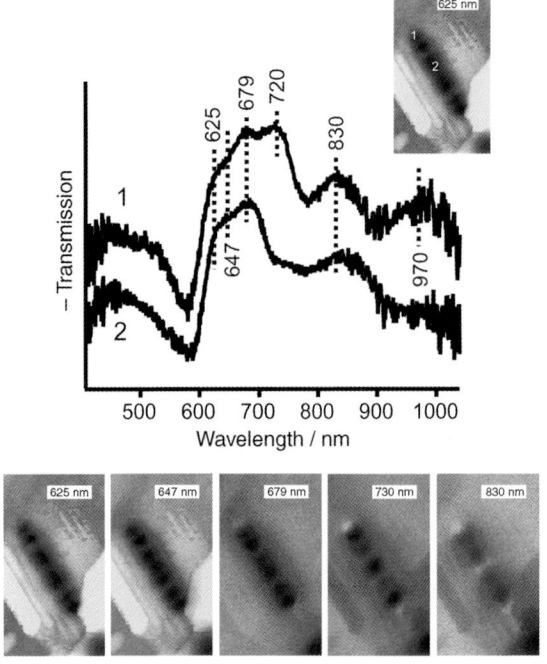

Figure 3.3 Near-field transmission spectra and images of a single gold nanorod (length 510 nm, diameter 20 nm). The two transmission spectra were obtained at positions 1 and 2 indicated in the inset. Each image was obtained at the resonance peak wavelength. (Reproduced with permission from Royal Society of Chemistry [10]).

the dipolar longitudinal mode gives the lowest frequency among the longitudinal modes, and the frequency becomes higher in higher modes, finally approaching the transverse-mode frequency as a limit.

The magnitude of optical extinction is given approximately by the square of the electronic oscillation amplitude, or the square modulus of the wavefunction. Consequently, the magnitude of the optical extinction is expected to oscillate along the long axis of the rod in the near-field images of higher longitudinal modes. The experimentally observed near-field transmission image in Figure 3.3 is in good accordance with this expectation. The number of dark spots increases in higher optical frequencies in Figure 3.3, which is again consistent with the scheme given in Figure 3.4. The longitudinal plasmon resonances interact with the radiation field polarized along the long axis of the nanorod, and interaction with radiation polarized along the short axis is negligible. In fact, the images shown in Figure 3.3 are not observable when the polarization is perpendicular to the long axis [14, 20]. These results strongly support that the transmission images observed can be interpreted as visualization of square moduli of wavefunctions of longitudinal plasmon modes. Theoretical simulation of plasmon-wavefunction amplitudes (or more precisely

Longitudinal modes

$m = 1$

$m = 2$

$m = 3$

higher frequency →

Transverse modes

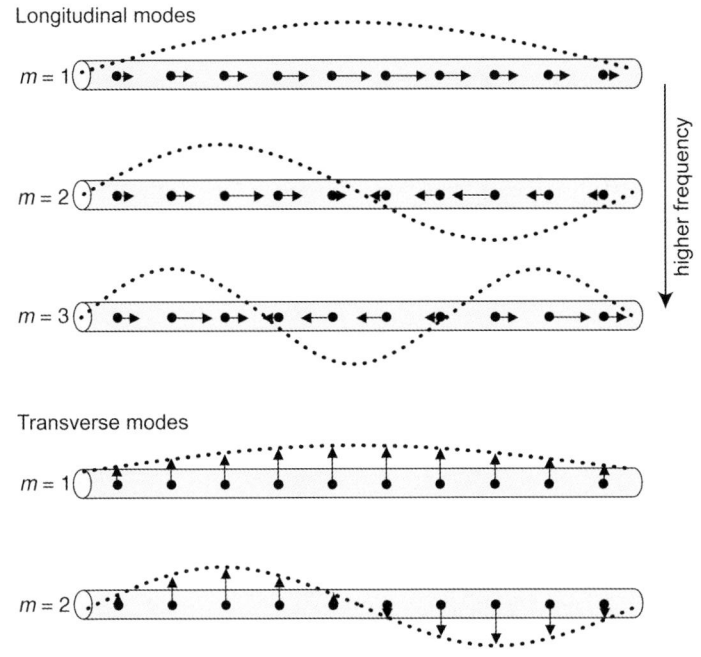

$m = 1$

$m = 2$

Figure 3.4 Schematic view of plasmon modes of a metal nanorod. (Reproduced with permission from The Japan Society of Applied Physics [12]).

photon local density of states) based on Green dyadic formalism [24–26] reproduces well the observed near-field images [14, 20].

In summary, it has been demonstrated that plasmon-mode wavefunctions of gold nanoparticles resonant with the incident light can be visualized by near-field transmission imaging.

3.5
Ultrafast Time-Resolved Near-Field Imaging of Gold Nanorods

Imaging of ultrafast processes occurring in the sample is realized by a pump–probe transient absorption scheme at the near-field probe tip, using femtosecond pulses as the incident light [7, 8]. In the present study, we adopted an equal-pulse transmission correlation method to get time-resolved images, with femtosecond pulses from a mode-locked Ti:sapphire laser at a wavelength of 780 nm [14]. The pulse from the laser was split into pump and probe pulses of approximately equal intensities by a beam splitter. The beams passed through optical delay lines to adjust the time separation between the pump and probe pulses, and then were combined again collinearly. The combined beam was incident on the optical fiber probe after a dispersion compensation device described in the previous section. The total intensity of the light transmitted through the sample was detected by a single-channel

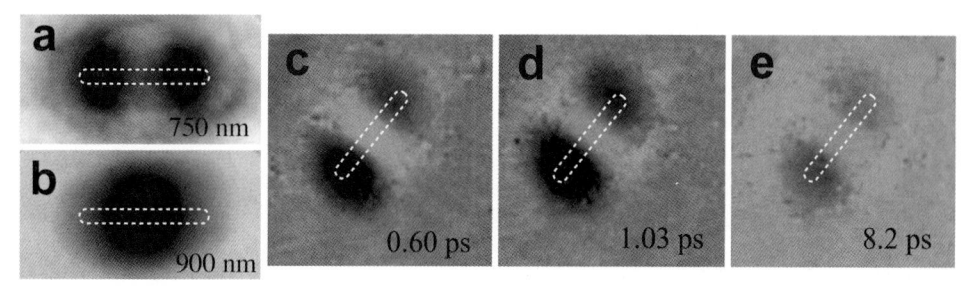

Figure 3.5 Near-field static ((a), (b)) and transient ((c)–(e)) transmission images of a single gold nanorod (length 300 nm, diameter 30 nm). Observed wavelengths are 750 nm (a), 900 nm (b), and 780 nm ((c)–(e)). The pump–probe delay times in ((c)–(e)) are 0.60, 1.03, and 8.2 ps, respectively. The dark and bright parts in (c)–(e) indicate the bleached and induced absorption regions, respectively. The approximate position of the nanorod is indicated by the broken line in each panel.

photodetector, while the sample position was scanned laterally. The transmission intensity difference between pump-on and -off was detected using a mechanical chopper and a lock-in amplifier, to get an image of transient transmission change induced by the pump pulse. By changing the delay time between the pump and probe pulses, variation of the image was recorded which gives information on the dynamics after photoexcitation.

Figure 3.5 shows time-resolved near-field transient transmission images of a gold nanorod (diameter 30 nm, length 300 nm) and a static near-field image of the same rod observed at 780 nm [14]. In the static image, the wavefunction of the plasmon resonant with the incident light is visualized (with a node in the center), as mentioned in the previous section. In the transient transmission immediately after the excitation (0.6 ps), a characteristic image with bleached and induced absorptions at the center and at both ends, respectively, is observed, for the single nanorod. The induced absorption at both ends rises up to about 1 ps, and then gradually decays to thermal equilibrium. Based on the previous ensemble measurements of ultrafast spectroscopy on gold nanoparticles, the rapid rise may be attributed to an electron–electron scattering process in the nanoparticle, whereas the slower decay may arise from electron–photon scattering.

To get insight into the origin of the characteristic image observed shortly after photoexcitation, we tried a numerical simulation of the transient near-field image based on electromagnetic theory [27]. We do not describe the details of the theoretical formulation here. We assumed that the energy of the pump pulse dissipates quickly in the particle after photoexcitation, and as a result the electronic temperature of the rod rises uniformly. The temperature rise causes temporal change of dielectric function of the material (gold), and induces changes in plasmon modes as a consequence. By taking this effect into account, a simulated transient image, as shown in Figure 3.6, was obtained. The simulated image corresponds well with the observed images of transient transmission changes. We sometimes observed inverted transient transmission images with respect to that of Figure 3.5 (i.e., induced absorption in the center and bleached absorptions at both ends), depending on the

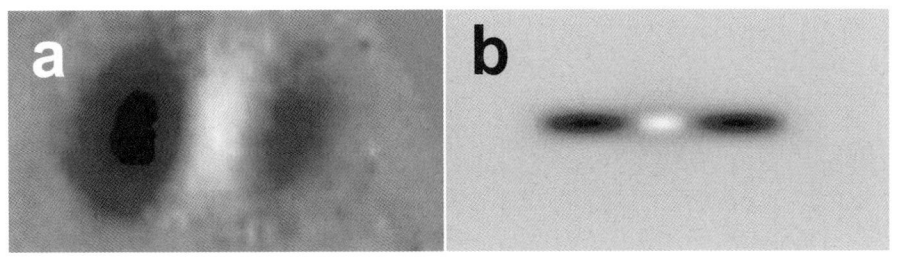

Figure 3.6 Near-field transient transmission image of a single gold nanorod observed at 0.6 ps (a) and corresponding simulated image (b). (Reproduced with permission from The American Physical Society [27]).

dimension of the rod [27]. In the simulation, this result can be reproduced as arising from the difference in the resonance condition that is determined by the dimension. This result supports the reasonableness of the simulation. From these facts, it has been clarified that the characteristic transient transmission image observed reflects changes in plasmon wavefunctions accompanied by electronic temperature rise induced by the photoexcitation.

3.6
Near-Field Two-Photon Excitation Images of Gold Nanorods

Some kinds of gold nanoparticles (or assemblies) show quite strong two-photon induced photoluminescence (TPI-PL) in the region between 500 and 700 nm, when excited by femtosecond pulses at around 800 nm (i.e., Ti:sapphire laser wavelength) [28–32]. By detecting this TPI-PL, we can obtain near-field two-photon excitation probability images of the samples [28–30, 33, 34]. Figure 3.7 shows typical examples of two-photon excitation images of gold nanorods [28, 29]. In Figure 3.7a, an image corresponding to the steady-state wavefunction of the plasmon resonant with the incident light (780 nm) is observed, in a similar way to the near-field transmission images in Figure 3.3. For another rod, shown in Figure 3.7b, in contrast, no wavefunction is visible, and two-photon excitation probability is found only in the edge region of the rod. The electric fields are expected to be localized in the edges of metal rods due to the so-called "lightning rod effect" [35, 36]. The observed image in Figure 3.7b corresponds well with the optical-field distribution expected for this effect. The results suggest that images corresponding to the spatial distribution of optical fields are obtainable by near-field two-photon excitation imaging.

As shown in Figure 3.7, some rods give images of steady-state plasmon wavefunctions, while others show localized excitation probability at the edges. Since the resonance frequency is strongly dependent on the rod dimension, the resonance condition with the incident light is different for each nanorod. The difference in the observed images illustrated in Figure 3.7 probably originates from such an effect. In the next section, we will apply this near-field two-photon excitation imaging method to investigate enhanced-field distribution in nanoparticle assemblies.

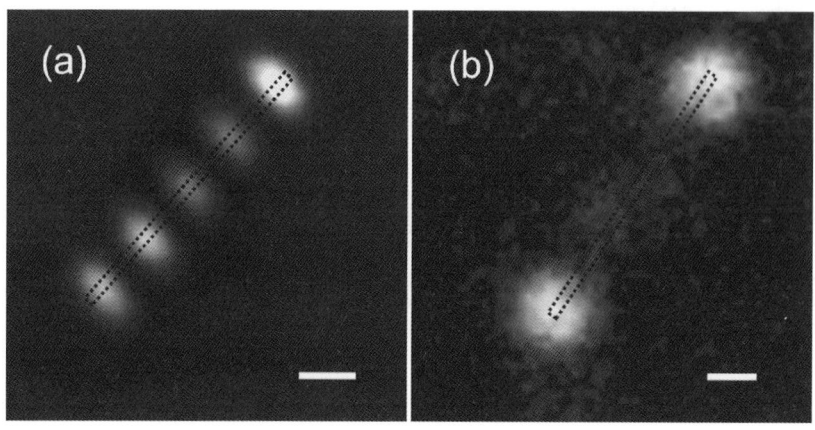

Figure 3.7 Near-field two-photon excitation images of single gold nanorods detected by two-photon induced photoluminescence. Nanorod dimensions (length, diameter) are 540 nm, 20 nm for (a) and 565 nm, 21 nm in (b). Scale bars: 100 nm. (Reproduced with permission from The Chemical Society of Japan [11]).

3.7
Enhanced Optical Fields in Spherical Nanoparticle Assemblies and Surface Enhanced Raman Scattering

In late 1990s, Raman scattering from adsorbed species on noble metal (principally silver) nanoparticle aggregates was reported to be so enormously enhanced that even single-molecular detection might be possible [37, 38]. This discovery is important not only for analytical applications, but the mechanism of the enhancement has also attracted fundamental interest from many researchers. Many reports have been published since then on this topic. A number of studies have also been devoted to simulation of model systems to explain the enhancement based on electromagnetic theory. The general conclusion of these simulation studies is as follows. When an assembly of noble metal nanoparticles is irradiated by light, a strongly enhanced electric field is induced in the gap between the nanoparticles [39–41]. This makes the optical field applied to the molecule adsorbed in the gap very strong, and very strong Raman scattering is induced. The scattered Raman radiation is further influenced by the field enhancement effect in the gap. As a result, the Raman-scattered field is enhanced with a factor given by approximately the fourth power of the local electric field enhancement.

Enhanced electric-field distribution is illustrated schematically in Figure 3.8, based on reported electromagnetic simulations, for a dimer of a noble metal spherical nanoparticle. The optical field enhancement at the gap site occurs only when the incident polarization is parallel to the interparticle axis of the dimer.

According to the electromagnetic simulation, the enhancement factor of the electric field can be as high as a few thousand under some conditions, and

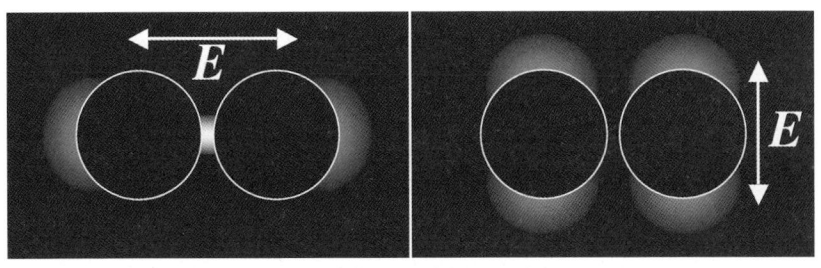

Figure 3.8 Schematic view of enhanced field distribution in the vicinity of a dimer of noble metal nanospheres. (Reproduced with permission from The Japan Society of Applied Physics [12]).

consequently Raman enhancement factor can reach as high as 10^{11} [41]. Such a field enhancement in the gap site of a noble metal nanoparticle aggregate is called a "hot spot", and is believed to be a major origin of the extremely high Raman enhancement. However, it is practically impossible to prove this mechanism by observation of optical field distribution with a conventional optical microscope, since the spatial scale of the whole aggregate is in the subwavelength region. To observe the optical field distribution in the aggregates, optical measurements on a subwavelength scale are essential.

We tried to visualize the hot spot using the near-field two-photon excitation imaging, on a dimeric assembly of spherical gold nanoparticles as a model of nanoparticle aggregates [33, 34]. The dimer sample was prepared on a glass substrate by self-assembly of spherical gold nanoparticles (diameter 100 nm). Near-field two-photon excitation images of the sample were obtained with a Ti:sapphire laser (at 785 nm) as an incident light source. Figure 3.9 shows the near-field two-photon excitation images of the sample obtained with two mutually perpendicular incident polarization directions, together with the topograph image.

In the topograph image, a few dimers and isolated nanoparticles are found. In the two-photon excitation images, high two-photon excitation probability is found in the vicinity of the gap sites of the dimers when the incident polarization is parallel to the interparticle axis. This spatial distribution of two-photon excitation probability is consistent with the theoretically predicted structure of the electric field for the hot spot of the spherical metal nanoparticle dimer (Figure 3.8). In contrast, enhanced two-photon excitation is scarcely observed in the neighborhood of isolated nanoparticles. From these results, it becomes evident that we can visualize nanoscale optical field distributions in nanoparticle assemblies by the near-field method, and that the hot spot predicted theoretically can be observed as a real image by this method.

We also tried measurements to demonstrate that hot spots make significant contributions to surface enhanced Raman scattering [34]. For this purpose, the sample of nanoparticle assembly was doped with Raman active molecules by a spin-coating method, and near-field excited Raman scattering from the sample was recorded. We adopted Rhodamine 6G dye as a Raman active material, which is

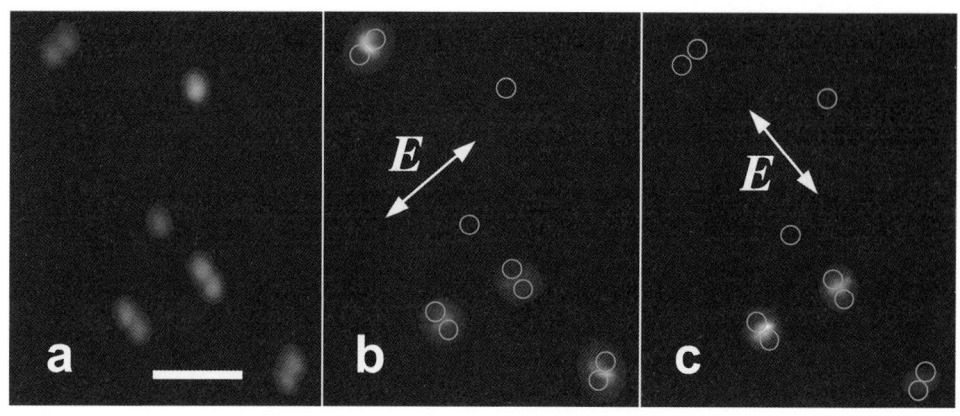

Figure 3.9 Near-field two-photon excitation images of gold nanosphere dimers. (a) Topography. Scale bar: 500 nm. (b) and (c) Two-photon excitation images. The excitation wavelength is 780 nm. Incident polarization directions are indicated by arrows. The approximate positions of the particles are indicated by circles. (Reproduced with permission from The Japan Society of Applied Physics [12]).

sometimes used for examination of surface enhanced Raman scattering. In the near-field Raman experiment, the Raman excitation light (785 nm, cw) was illuminated from the near-field probe, and the spectrum of the scattered photons in the far field was recorded. The result is shown in Figure 3.10. When the near-field probe for excitation is positioned at the dimer of the nanoparticle, strong Raman scattering was observed when the incident polarization was parallel to the interparticle axis. The observed Raman bands are attributed to Rhodamine molecules by comparing with previous reports. We could not observe any Raman scattering signal when the probe was positioned at an isolated nanoparticle or at the substrate, regardless of the incident polarization. This indicates that Raman scattering is enhanced only at the dimeric sites.

We can take a near-field Raman excitation probability image by scanning the sample laterally while monitoring the Raman-band intensity at around $1600\,\mathrm{cm}^{-1}$, as shown in Figure 3.10. These images clearly demonstrate that Raman enhancement is observed only when the incident polarization is parallel to the dimer axis, and is localized at the interparticle gap site. The result is again excellently consistent with the prediction of the hot spot model, and strongly supports the idea that electric field enhancement at hot spots makes a major contribution to the mechanism of surface enhanced Raman scattering.

To summarize, we have shown here that enhanced electric-field distribution in metal nanoparticle assemblies can be visualized on the nanoscale by a near-field two-photon excitation imaging method. By combining this method and near-field Raman imaging, we have clearly demonstrated that hot spots in noble metal nanoparticle assemblies make a major contribution to surface enhanced Raman scattering.

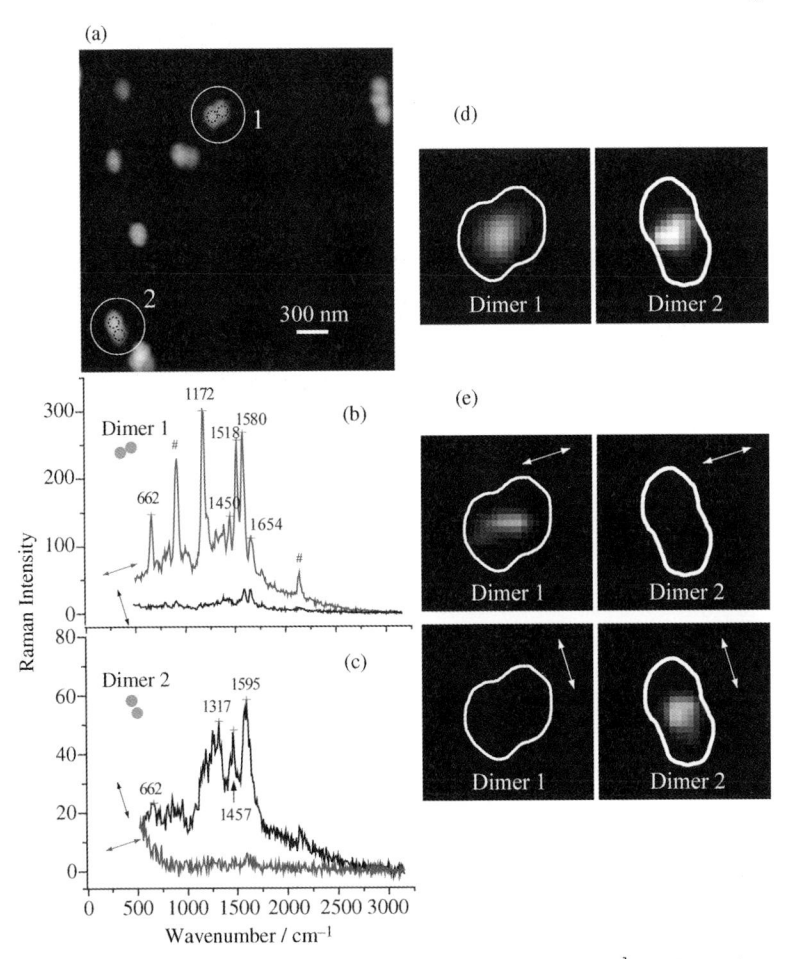

Figure 3.10 (a) Topography of the sample. (b), (c) Near-field excited Raman spectra at dimers 1 and 2, respectively, taken at two different incident polarizations. The peaks marked with # are unassigned. (d) Near-field two-photon excitation images of dimers 1 and 2. (e) Near-field Raman excitation images of dimers 1 and 2 obtained for bands near 1600 cm^{-1}. Incident polarizations are indicated by arrows. White lines in (d) and (e) indicate the approximate shapes of the dimers. The average diameter of the spheres is about 100 nm. (Reproduced with permission from The American Chemical Society [34]).

3.8
Concluding Remarks

In this chapter, we have provided an overview of near-field imaging and spectroscopy of noble metal nanoparticles and assemblies. We have shown that plasmon-mode wavefunctions and enhanced optical fields of nanoparticle systems can be visualized. The basic knowledge about localized electric fields induced by the plasmons may lead to new innovative research areas beyond the conventional scope of materials.

The contribution of near-field microscopy is not limited to studies on plasmon-based nanomaterials, but may also provide valuable fundamental information on novel functions of various nanomaterials.

The theoretical framework for near-field imaging, on the other hand, is not as straightforward as that for the far-field optical measurements. This is primarily because the effects of perturbation from the near-field probe on the optical characteristics of the samples are not well known. Further developments in theoretical treatments and practical and precise simulation methods for realistic near-field measurement systems are desired.

Acknowledgments

The authors are grateful to Drs T. Nagahara, J. K. Lim, N. Horimoto, M. K. Hossain, T. Shimada, and Professor M. Kitajima for their contributions to this work and fruitful discussion. The authors also thank the Equipment Development Center of IMS for their collaboration in the construction of the near-field apparatus. This work was supported by Grants-in-Aid for Scientific Research (Nos. 17655011, 18205004, 18685003) from the Japan Society for the Promotion of Science and for Scientific Research on Priority Areas (Area No. 432, No. 17034062) from the Ministry of Education, Culture, Sports, Science and Technology.

References

1 Bohren, C. F. and Huffman, D. R. (1983) *Absorption and Scattering of Light by Small Particles*, Wiley, New York.

2 Kreibig, U. and Vollmer, M. (1995) *Optical Properties of Metal Clusters*, Springer, Berlin.

3 Ohtsu, M. (ed.) (1998) *Near-Field Nano/Atom Optics and Technology*, Springer, Tokyo.

4 Kawata, S. (ed.) (2001) *Near-Field Optics and Surface Plasmon Polariton, Topics in Applied Physics*, vol. 81, Springer, Berlin.

5 Courjon, D. (2003) *Near-Field Microscopy and Near-Field Optics*, Imperial College Press, London.

6 Novotny, L. and Hecht, B. (2006) *Principle of Nano-Optics*, Cambridge University Press, Cambridge.

7 Nagahara, T., Imura, K. and Okamoto, H. (2003) Spectral inhomogeneities and spatially resolved dynamics in porphyrin J-aggregate studied in the near-field. *Chem. Phys. Lett.*, **381**, 368–375.

8 Nagahara, T., Imura, K. and Okamoto, H. (2004) Time-resolved scanning near-field optical microscopy with supercontinuum light pulses generated in microstructure fiber. *Rev. Sci. Instrum.*, **75**, 4528–4533.

9 Imura, K., Nagahara, T. and Okamoto, H. (2004) Characteristic near-field spectra of single gold nanoparticles. *Chem. Phys. Lett.*, **400**, 500–505.

10 Okamoto, H. and Imura, K. (2006) Near-field imaging of optical field and plasmon wavefunctions in metal nanoparticles. *J. Mater. Chem.*, **16**, 3920–3928.

11 Imura, K. and Okamoto, H. (2008) Development of novel near-field microspectroscopy and imaging of local excitations and wave functions of nanomaterials. *Bull. Chem. Soc. Jpn.*, **81**, 659–675.

References

12 Okamoto, H. and Imura, K. (2008) Near-field optical imaging of nanoscale optical fields and plasmon waves. *Jpn. J. Appl. Phys.*, **47**, 6055–6062.

13 Nechay, B. A., Siegner, U., Achermann, M., Bielefeldt, H. and Keller, U. (1999) Femtosecond pump-probe near-field optical microscopy. *Rev. Sci. Instrum.*, **70**, 2758–2764.

14 Imura, K., Nagahara, T. and Okamoto, H. (2004) Imaging of surface plasmon and ultrafast dynamics in gold nanorods by near-field microscopy. *J. Phys. Chem. B*, **108**, 16344–16347.

15 Busbee, B. D., Obare, S. O. and Murphy, C. J. (2003) An improved synthesis of high-aspect-ratio gold nanorods. *Adv. Mater.*, **15**, 414–416.

16 Murphy, C. J., Sau, T. K., Gole, A. M., Orendorff, C. J., Gao, J., Gou, L., Hunyadi, S. E. and Li, T. (2005) Anisotropic metal nanoparticles: Synthesis, assembly, and optical applications. *J. Phys. Chem. B*, **109**, 13857–13870.

17 Link, S. and El-Sayed, M. A. (1999) Spectral properties and relaxation dynamics of surface plasmon electronic oscillations in gold and silver nanodots and nanorods. *J. Phys. Chem. B*, **103**, 8410–8426.

18 Schider, G., Krenn, J. R., Hohenau, A., Ditlbacher, H., Leitner, A., Aussenegg, F. R., Schaich, W. L., Puscasu, I., Monacelli, B. and Boreman, G. (2003) Plasmon dispersion relation of Au and Ag nanowires. *Phys. Rev. B*, **68**, 155427 (4 pages).

19 Hohenau, A., Krenn, J. R., Schider, G., Ditlbacher, H., Leitner, A., Aussenegg, F. R. and Schaich, W. L. (2005) Optical near-field of multipolar plasmons of rod-shaped gold nanoparticles. *Europhys. Lett.*, **69**, 538–543.

20 Imura, K., Nagahara, T. and Okamoto, H. (2005) Near-field optical imaging of plasmon modes in gold nanorods. *J. Chem. Phys.*, **122**, 154701 (5 pages).

21 Lim, J. K., Imura, K., Nagahara, T., Kim, S. K. and Okamoto, H. (2005) Imaging and dispersion relations of surface plasmon modes in silver nanorods by near-field spectroscopy. *Chem. Phys. Lett.*, **412**, 41–45.

22 Imura, K. and Okamoto, H. (2006) Reciprocity in scanning near-field optical microscopy: illumination and collection modes of transmission measurements. *Opt. Lett.*, **31**, 1474–1476.

23 Matsuda, K., Saiki, T., Nomura, S., Mihara, M., Aoyagi, Y., Nair, S. and Takagahara, T. (2003) Near-field optical mapping of exciton wave functions in a GaAs quantum dot. *Phys. Rev. Lett.*, **91**, 177401 (4 pages).

24 Girard, C. and Dereux, A. (1996) Near-field optics theories. *Rep. Prog. Phys.*, **59**, 657–699.

25 Girard, C., Weeber, J.-C., Dereux, A., Martin, O. J. F. and Goudonnet, J.-P. (1997) Optical magnetic near-field intensities around nanometer-scale surface structures. *Phys. Rev. B*, **55**, 16487–16497.

26 Girard, C. (2005) Near fields in nanostructures. *Rep. Prog. Phys.*, **68**, 1883–1933.

27 Imura, K. and Okamoto, H. (2008) Ultrafast photoinduced changes of eigenfunctions of localized plasmon modes in gold nanorods. *Phys. Rev. B*, **77**, 041401(R) (4 pages).

28 Imura, K., Nagahara, T. and Okamoto, H. (2004) Plasmon mode imaging of single gold nanorods. *J. Am. Chem. Soc.*, **126**, 12730–12731.

29 Imura, K., Nagahara, T. and Okamoto, H. (2005) Near-field two-photon-induced photoluminescence from single gold nanorods and imaging of plasmon modes. *J. Phys. Chem. B*, **109**, 13214–13220.

30 Imura, K., Nagahara, T. and Okamoto, H. (2006) Photoluminescence from gold nanoplates induced by near-field two-photon absorption. *Appl. Phys. Lett.*, **88**, 023104 (3 pages).

31 Mühlschlegel, P., Eisler, H.-J., Martin, O. J. F., Hecht, B. and Pohl, D. W. (2005) Resonant optical antennas. *Science*, **308**, 1607–1609.

32 Ueno, K., Juodkazis, S., Mizeikis, V., Sasaki, K. and Misawa, H. (2008) Clusters

of closely spaced gold nanoparticles as a source of two-photon photoluminescence at visible wavelengths. *Adv. Mater.*, **20**, 26–30.

33 Imura, K., Okamoto, H., Hossain, M. K. and Kitajima, M. (2006) Near-field imaging of surface-enhanced Raman active sites in aggregated gold nanoparticles. *Chem. Lett.*, **35**, 78–79.

34 Imura, K., Okamoto, H., Hossain, M. K. and Kitajima, M. (2006) Visualization of localized intense optical fields in single gold-nanoparticle assemblies and ultrasensitive Raman active sites. *Nano Lett.*, **6**, 2173–2176.

35 Gersten, J. and Nitzan, A. (1980) Electromagnetic theory of enhanced Raman scattering by molecules adsorbed on rough surfaces. *J. Chem. Phys.*, **73**, 3023–3037.

36 Mohamed, M. B., Volkov, V., Link, S. and El-Sayed, M. A. (2000) The 'lightning' gold nanorods: fluorescence enhancement of over a million compared to the gold metal. *Chem. Phys. Lett.*, **317**, 517–523.

37 Nie, S. and Emory, S. R. (1997) Probing single molecules and single nanoparticles by surface-enhanced Raman scattering. *Science*, **275**, 1102–1106.

38 Kneipp, K., Wang, Y., Kneipp, H., Perelman, L. T., Itzkan, I., Dasari, R. R. and Feld, M. S. (1997) Single molecule detection using surface-enhanced Raman scattering (SERS). *Phys. Rev. Lett.*, **78**, 1667–1670.

39 Kelly, K. L., Coronado, E., Zhao, L. L. and Schatz, G. C. (2003) The optical properties of metal nanoparticles: The influence of size, shape, and dielectric environment. *J. Phys. Chem. B*, **107**, 668–677.

40 Futamata, M., Maruyama, Y. and Ishikawa, M. (2003) Local electric field and scattering cross section of Ag nanoparticles under surface plasmon resonance by finite difference time domain method. *J. Phys. Chem. B*, **107**, 7607–7617.

41 Xu, H., Aizpurua, J., Käll, M. and Apell, P. (2000) Electromagnetic contributions to single-molecule sensitivity in surface-enhanced Raman scattering. *Phys. Rev. E*, **62**, 4318–4324.

4
Structure and Dynamics of a Confined Polymer Chain Studied by Spatially and Temporally Resolved Fluorescence Techniques

Hiroyuki Aoki

4.1
Introduction

A polymer molecule is a chain-like molecule consisting of thousands of monomer units. The polymer chain has numerous degrees of freedom and can take various conformations in a three-dimensional space. The variety of the chain conformation is the origin of the characteristic properties of polymeric materials such as rubber elasticity and viscoelasticity. The fabrication of polymer materials with an ordered structure on a nanometric scale has been studied intensively. In a nanostructured polymeric material, the polymer chain is confined in a local space smaller than its unperturbed dimension and the degree of freedom of the polymer chain is greatly suppressed. Consequently, the physical properties of polymer nanomaterials will be different from the bulk systems, and it is important to understand the fundamental aspects of polymer systems confined in a nanometric space [1–5]. One of the simplest systems of polymer confined in a small space is a polymer thin film. In an ultra-thin film with a thickness less than the unperturbed size of a polymer chain, the chain dimension is regulated between the interfaces with air and a substrate, and loses the degree of freedom in the height direction. Another example of a restricted polymer is a graft polymer chain, one end of which is tethered on a solid substrate. When the polymer chain is grafted at a high density on the substrate, the graft chains interact with each other. The high-density graft chain is restricted not only by the fixation of a chain end on a substrate but also by the strong inter-chain repulsion.

For the investigation of polymer systems under spatial confinement, fluorescence microscopy is a powerful method providing valuable information with high sensitivity. A fluorescence microscopy technique with nanometric spatial resolution and nanosecond temporal resolution has been developed, and was used to study the structure and dynamics of polymer chains under spatial confinement: a polymer chain in an ultra-thin film and a chain grafted on a solid substrate. Studies on the conformation of the single polymer chain in a thin film and the local segmental motion of the graft polymer chain are described herein.

Molecular Nano Dynamics, Volume I: Spectroscopic Methods and Nanostructures
Edited by H. Fukumura, M. Irie, Y. Iwasawa, H. Masuhara, and K. Uosaki
Copyright © 2009 WILEY-VCH Verlag GmbH & Co. KGaA, Weinheim
ISBN: 978-3-527-32017-2

4.2
Conformation of a Confined Polymer Chain

4.2.1
Polymer Ultra-Thin Film

The conformation of polymer chains in an ultra-thin film has been an attractive subject in the field of polymer physics. The chain conformation has been extensively discussed theoretically and experimentally [6–11]; however, the experimental technique to study an ultra-thin film is limited because it is difficult to obtain a signal from a specimen due to the low sample volume. The conformation of polymer chains in an ultra-thin film has been examined by small angle neutron scattering (SANS), and contradictory results have been reported. With decreasing film thickness, the radius of gyration, R_g, parallel to the film plane increases when the thickness is less than the unperturbed chain dimension in the bulk state [12–14]. On the other hand, Jones et al. reported that a polystyrene chain in an ultra-thin film takes a Gaussian conformation with a similar in-plane R_g to that in the bulk state [15, 16].

The real-space imaging of a single chain provides direct information on the chain conformation in an ultra-thin film. Fluorescence microscopy would be a powerful tool to examine the conformation of a single polymer chain because it can distinguish the contour of a single chain embedded in a bulk medium. However, the spatial resolution of the conventional fluorescence microscopy is limited to a half of the wavelength due to the diffraction limit of light. Fluorescence microscopy has been used to observe a huge bio-macromolecule such as DNA, while conventional optical microscopy is not applicable to the imaging of a synthetic polymer chain with chain dimensions of the order of 10–100 nm. In 1990s, scanning near-field optical microscopy (SNOM) was developed as an optical microscopy to overcome the diffraction limit [17, 18]. SNOM allows fluorescence detection with nanometric spatial resolution; therefore, it is a powerful tool for the real-space imaging of a single polymer chain. In the next section, the conformation of a single polymer chain in a thin film less than the unperturbed dimension, is studied by fluorescence imaging with near-field optical microscopy [19, 20].

4.2.2
Near-Field Optical Microscopy

Scanning near-field optical microscopy is a novel imaging technique to achieve high spatial resolution free from the diffraction limit. When the light is incident on an object smaller than its wavelength, there arises not only scattering but also a non-propagating electric field. The non-propagating "light" is called the optical near-field, and it is restricted to the vicinity of the object. SNOM illuminates the sample with the optical near-field localized in a nanometric area. Figure 4.1a illustrates an experimental set-up of SNOM, which uses a probe tip with an aperture much smaller than the wavelength of light. The SEM image of the sub-wavelength aperture is shown on the in Figure 4.1c. The typical size of the aperture is less than 100 nm. When light is

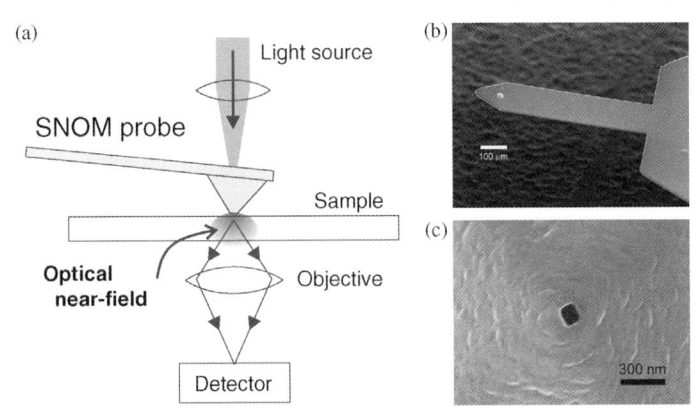

Figure 4.1 Block diagram of a SNOM apparatus (a), SEM images of a cantilever-type SNOM probe (b) and the aperture at the tip end (c).

delivered to the backside of the aperture at the probe end, the optical near-field generates around the aperture. The near-field is confined to an area as large as the aperture size; therefore, SNOM allows us to illuminate an area much smaller than the wavelength by approaching the specimen with the tip end. The scattering or fluorescence from the sample is collected by an objective lens and detected by a photodetector. The optical signal is recorded as a function of the position of the scanning probe to construct a microscope image. SNOM provides not only the signal intensity, but also spectroscopic and time-resolved information from a nanometric area.

Figure 4.2 shows the fluorescence SNOM image of a single Rhodamine 6G (Rh6G) molecule. The molecular size of Rh6G is much smaller than the SNOM aperture; therefore, the fluorescence image of a single Rh6G molecule corresponds to a point spread function of the apparatus. The single Rh6G molecule, which had a diameter of 300 nm when observed by conventional confocal microscopy, was observed as a circular bright spot with a diameter of 75 nm. This indicates the high spatial resolution of SNOM, beyond the diffraction limit of light. The unperturbed dimension of a single polymer chain with a molecular weight of 10^6 is of the order

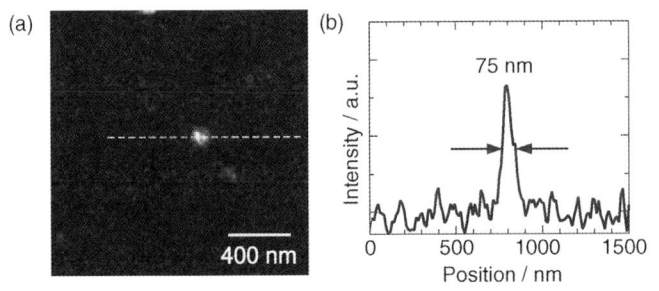

Figure 4.2 Fluorescence SNOM image of a single Rhodamine 6G molecule. Panel (b) indicates the cross-section profile for the dashed line in panel (a).

of 100 nm; therefore, SNOM is applicable to the observation of the conformation of individual polymer chains.

4.2.3
Structure of a Single Polymer Chain

The conformation of poly(methyl methacrylate) (PMMA) was investigated. For the SNOM measurement, PMMA was labeled with a fluorescent moiety of perylene. The perylene-labeled PMMA (PMMA-Pe) was prepared by the copolymerization of methyl methacrylate and 3-perylenylmethyl methacrylate [21]. The number-average molecular weight was 4.16×10^6 with a relatively narrow molecular weight distribution, and the fraction of perylene was less than 0.8% to avoid the influence of the introduction of the dye moiety on the properties of PMMA. Ultra-thin films with a thickness of 1–100 nm were prepared on glass plates by spin-coating and Langmuir–Blodgett methods from a mixed benzene solution of PMMA and PMMA-Pe, where the amount of PMMA-Pe was less than 0.01% of the unlabeled PMMA.

Figure 4.3 shows the fluorescence SNOM images of single PMMA-Pe chains embedded in the unlabeled PMMA matrix. In the SNOM image, each PMMA-Pe chain was observed as an isolated fluorescent spot. The molecular weight of each chain can be estimated from the integrated fluorescence intensity [19], and Figure 4.3

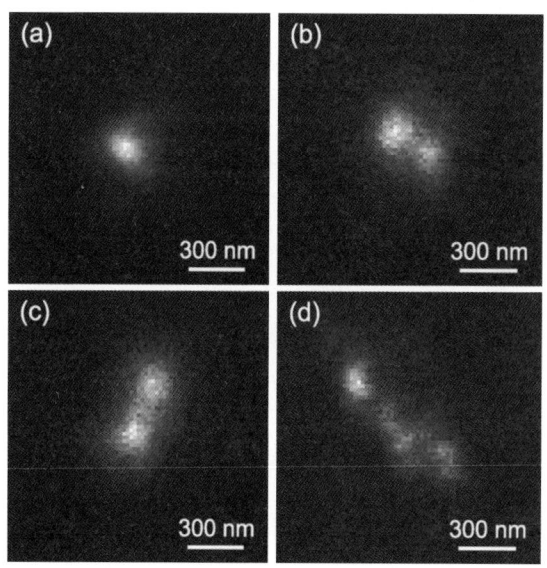

Figure 4.3 SNOM images of the single PMMA-Pe chains in an ultra-thin film. The scanned area is $1.4 \times 1.4\ \mu m^2$ for each image. The molecular weight (M) and the value of R_{xy} for each chain were evaluated as follows: (a) $M = 4.2 \times 10^6$ and $R_{xy} = 122$ nm; (b) $M = 4.2 \times 10^6$ and $R_{xy} = 160$ nm; (c) $M = 4.0 \times 10^6$ and $R_{xy} = 180$ nm; (d) $M = 3.6 \times 10^6$ and $R_{xy} = 256$ nm. Reproduced with permission from The Society of Polymer Science, Japan.

shows four PMMA chains with a molecular weight of about 4×10^6. The four chains have a similar chain length, but different chain conformations. The PMMA chain shown in Figure 4.3a has a shrunken shape and that in Figure 4.3d adopts a stretched conformation. The broad distribution of the chain conformation indicates the flexibility of the PMMA chain. For the statistical analysis, the chain dimension, R_{xy}, was evaluated for each chain according to the following equations.

$$R_{xy}^2 = \frac{\sum_i I_i(\mathbf{r}_i - \mathbf{r}_{cm})^2}{\sum_i I_i} \tag{4.1}$$

$$\mathbf{r}_{cm} = \frac{\sum_i I_i \mathbf{r}_i}{\sum_i I_i} \tag{4.2}$$

where $\mathbf{r}_i$ and I_i are the position vector of the ith pixel in the SNOM image and the fluorescence intensity therein, respectively. R_{xy} corresponds to the radius of gyration for the projection of the polymer chain in the film plane.

Figure 4.4 shows the histogram of R_{xy}, which corresponds to the probability distribution function of the chain dimension. Information on the distribution was not available from the previous experiments in inverse space. The average radius of gyration, $\langle R_{xy} \rangle$, was 138, 145, and 143 nm for the PMMA chains in thin films with thickness 15, 50, and 80 nm, respectively. The thickness of 15–80 nm is relatively

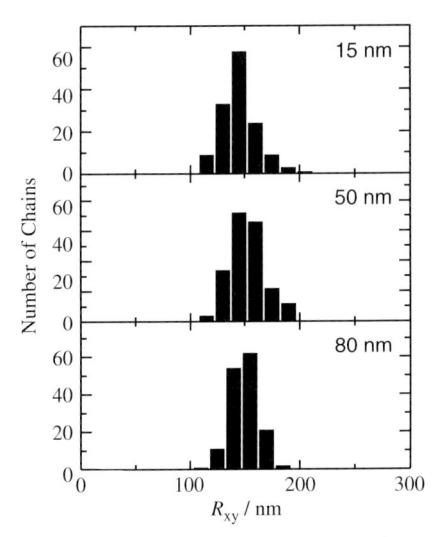

Figure 4.4 Histogram of the lateral chain dimension for the PMMA-Pe chains in ultra-thin films with thickness 15, 50, and 80 nm. The PMMA chains with a molecular weight of 4×10^6 were selected in the SNOM images and analyzed to construct the histogram. Reproduced with permission from The Society of Polymer Science, Japan.

small compared to the unperturbed chain size of ~100 nm in the bulk state [22]. This thickness range corresponds to an intermediate region between the two- and three-dimensional systems. Then we evaluated also the conformation in the model systems of two- and three-dimensional limits, which are a Langmuir–Blodgett monolayer and a 1-μm-thick film, respectively. The values of $\langle R_{xy} \rangle$ were 139 nm for the monolayer and 138 nm for the thick film. This indicates that the chain dimension in the lateral direction is not dependent on the film thickness in the range from a monolayer to the bulk. Although a polymer chain is spatially restricted in the height direction, the radius of gyration normal to the restriction remains a similar dimension to that in the unperturbed state. This result revealed by the direct observation with SNOM is consistent with the SANS result reported by Jones et al. [15, 16].

In a three-dimensional bulk state, the polymer chain adopts a random coil conformation. Considering the molecular size of the single chain (larger than 100 nm for a chain with a molecular weight of 4×10^6), there is much free space to allow the intrusion of the neighboring chains, as shown schematically by the solid curve in Figure 4.5a. In the bulk system, such free space is filled with the surrounding chains (the gray curves in Figure 4.5a), resulting in entanglement of the polymer chains. In an ultra-thin film, the SNOM revealed that the radius of gyration in the film plane is similar to that in the bulk state, whereas the chain dimension in the height direction is regulated by the film thickness. This indicates the reduction of the pervaded volume of a single chain, as shown in Figure 4.5b. Consequently, the surrounding chains are squeezed out and the entanglement among the chains is reduced in the thin film. The characteristic behavior of a two-dimensional polymer chain was predicted by the scaling theory [6]. The characteristic properties of the polymer ultra-thin film, different from those of the bulk state, have been attributed to lowered entanglement in a thin film state [23, 24]. SNOM provides direct information on the relationship between the structure at the single chain level and the macroscopic properties of polymer materials.

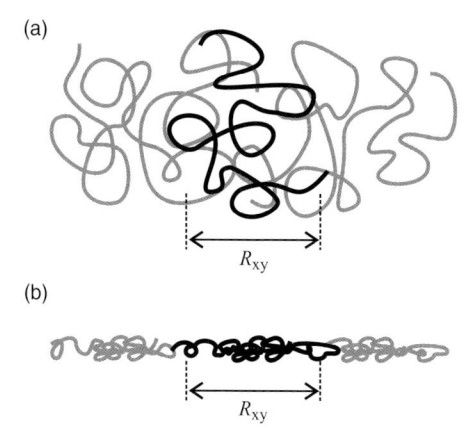

Figure 4.5 Schematic drawing of the chain conformations in the bulk (a) and ultra-thin film (b). To clarify the contour of the single chain, one chain is indicated as the solid curve. Reproduced with permission from The Society of Polymer Science, Japan.

4.3
Dynamics of a Confined Polymer Chain

4.3.1
Polymer Brush

Grafting of polymer chains onto a solid substrate is a common method to control the surface properties of materials [25–27]. At a low density of the graft chain on the substrate, the polymer chain adopts a "mushroom" conformation resembling the coiled conformation of a linear chain in solution. When the graft density increases and the average distance between the fixed ends of the graft chains is smaller than the unperturbed chain dimension, the polymer chains overlap. Due to the repulsive interaction among the graft polymer chains, the chain conformation is altered from a mushroom conformation. The graft chain with a high density is stretched away from the substrate due to the repulsion among the chains, which is called the "polymer brush". Recent development of the living polymerization technique allows us to increase the graft density up to ~0.7 chains nm^{-2} [28, 29]. In such a high-density polymer brush, the structure and dynamics of the polymer chain are greatly constrained due to strong interaction among the brush chains. The high-density polymer brush shows unique properties, which are different from the equivalent polymer film; for example, a high glass transition temperature and a low plate compressibility [30, 31].

The dynamics of the polymer brush has been extensively studied theoretically and numerically to elucidate the characteristic properties of polymer brushes [32–37]. The collective motion of polymer brushes in the time range of micro- to milliseconds has been extensively studied by the dynamic light scattering technique [38–40]; however, there are few studies especially on the chain dynamics of the polymer brush in the time range of nanoseconds because of the experimental difficulty. The time-resolved fluorescence depolarization method provides us with direct information on the dynamics of molecules with high sensitivity and high sub-nanosecond temporal resolution. In the following sections, the dynamics of a high-density polymer brush, studied by the fluorescence depolarization technique, is discussed [41].

4.3.2
Fluorescence Depolarization Method

The fluorescence depolarization technique excites a fluorescent dye by linearly polarized light and measures the polarization anisotropy of the fluorescence emission. The fluorescence anisotropy, r, is defined as

$$r = \frac{I_{\parallel} - I_{\perp}}{I_{\parallel} - 2I_{\perp}} \tag{4.3}$$

where $I_{\parallel}$ and $I_{\perp}$ are the polarization components of the fluorescence in the directions parallel and perpendicular to the excitation polarization, respectively. The fluorescence anisotropy is dependent on the difference of the angle between the electronic transition moments for excitation and emission. When the transition moments for excitation and emission are the same, the fluorescence anisotropy corresponds to the angular variation within the fluorescence lifetime, that is, the rotational motion of the molecule in the time range of nanoseconds. The following equation expresses the relationship between the fluorescence anisotropy, $r(t)$, and the transition moment, $\mathbf{M}(t)$, at time t after the instantaneous excitation of a fluorescent dye.

$$r(t) = r_0 \frac{3\langle\{\mathbf{M}(0)\cdot\mathbf{M}(t)\}^2\rangle - 1}{2} \tag{4.4}$$

where r_0 is the anisotropy at $t=0$. This indicates that the fluorescence anisotropy is equivalent to the second-order orientational auto-correlation function of the transition dipole; therefore, the time-resolved fluorescence depolarization method is able to directly probe the rotational relaxation of the molecule.

Figure 4.6 shows an apparatus for the fluorescence depolarization measurement. The linearly polarized excitation pulse from a mode-locked Ti-Sapphire laser illuminated a polymer brush sample through a microscope objective. The fluorescence from a specimen was collected by the same objective and input to a polarizing beam splitter to detect $I_\parallel$ and $I_\perp$ by photomultipliers (PMTs). The photon signal from the PMT was fed to a time-correlated single photon counting electronics to obtain the time profiles of $I_\parallel$ and $I_\perp$ simultaneously. The experimental data of the fluorescence anisotropy was fitted to a double exponential function,

$$r_{\text{model}}(t) = r_0 \left\{ x\exp\left(-\frac{t}{\tau_1}\right) + (1-x)\exp\left(-\frac{t}{\tau_2}\right) \right\} \tag{4.5}$$

where $0 \leq x < 1$. The mobility of the chain segment was discussed in terms of the correlation time for the anisotropy relaxation, τ_c, which is defined as

$$\begin{aligned}\tau_c &= r_0^{-1}\int_0^\infty r(t)dt \\ &= x\tau_1 + (1-x)\tau_2\end{aligned} \tag{4.6}$$

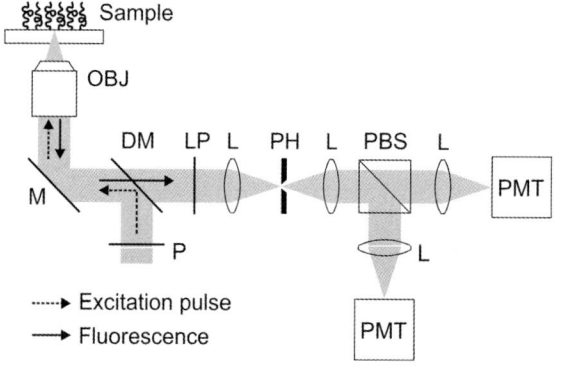

Figure 4.6 Block diagram of the apparatus for the fluorescence depolarization measurement. The dashed and solid arrows indicate the light paths of the excitation pulse and the fluorescence from the sample. OBJ: microscope objective, M: mirror, L: lens, DM: dichroic mirror, LP: long-pass filter, PH: pin-hole, PBS: polarizing beam splitter, P: polarizer, PMT: photomultiplier.

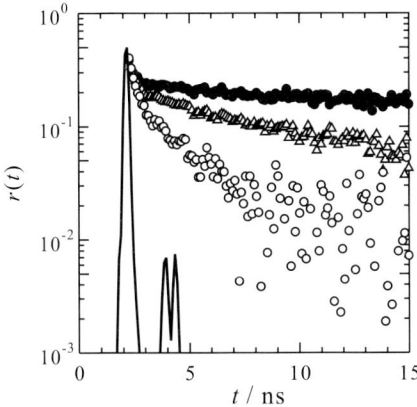

Figure 4.7 Fluorescence anisotropy decay curves for the PMMA brush swollen in benzene (filled circles) and the free PMMA chain in benzene solution at concentrations of 0.33 (triangles) and 2.9×10^{-3} g L^{-1} (open circles). The graft density of the brush is 0.46 chains nm^{-2}. The solid curve indicates the instrument response function. Reproduced with permission from the American Chemical Society.

4.3.3
Dynamics of a Polymer Brush

The dynamics of a perylene-labeled PMMA brush which was prepared by surface-initiated atom transfer radical polymerization is discussed here [42]. The molecular weight of the sample polymer was 1.0×10^5. Figure 4.7 shows the fluorescence anisotropy decay for the linear PMMA chain in benzene solution and the PMMA brush with a graft density of 0.46 chains nm^{-2} swollen in benzene. The linear chain solution showed the concentration dependence of the chain dynamics. The solution at a concentration of 0.33 g L^{-1} showed the decay of fluorescence anisotropy with a correlation time of 9.9 ns slower than that in the dilute solution ($\tau_c = 2.3$ ns). The motion of the polymer chains is suppressed by the inter-chain interaction at the concentration of 0.33 g L^{-1}. This concentration corresponds to the average polymer concentration in the PMMA brush layer with a graft density of 0.46 chains nm^{-2}. Even at the same polymer concentration, the brush chain showed a large correlation time of 31 ns compared to the corresponding solution of the linear PMMA. Such low mobility of the brush chain can be attributed to the effect of fixation of the chain end onto the solid substrate. For the linear polymer chain in solution, both chain ends show high mobility compared with the center segment in the main chain. One end of the polymer brush is chemically bound to the substrate. Therefore, the segmental motion is greatly suppressed at the fixed end, resulting in a large correlation time compared to the free chain in a solution.

Figure 4.8 shows the fluorescence anisotropy decay curves for PMMA brushes with various graft densities swollen in benzene and acetonitrile. Benzene and acetonitrile are good and Θ solvents for PMMA. As clearly shown in this figure,

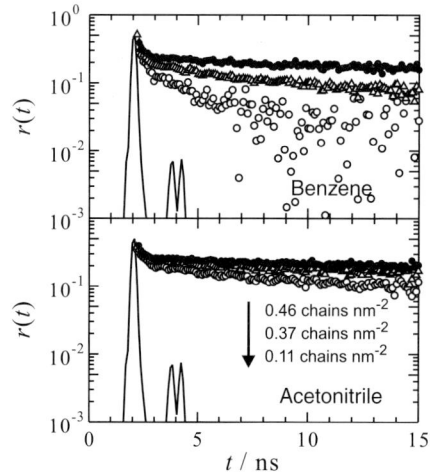

Figure 4.8 Fluorescence anisotropy decay curves for the PMMA brush swollen in benzene (a) and acetonitrile (b). The filled circles, triangles, and open circles indicate the data for the brush sample with graft densities of 0.46, 0.37, and 0.11 chains nm^{-2}, respectively. The solid curves indicate the instrument response function. Reproduced with permission from the American Chemical Society.

the chain mobility of the PMMA brush chain decreases with increasing graft density. A linear PMMA chain with a molecular weight of 10^5 shows an unperturbed dimension of 7 nm in a Θ solvent acetonitrile [43]. Therefore, at a density higher than $\sim$0.02 chains nm^{-2}, the distance between the brush chain is smaller than the chain dimension in solution. The steric interaction among the brush chains suppresses the chain motion. The higher the graft density, the stronger the inter-chain interaction, resulting in low segmental mobility of the high-density brush.

Next, the solvent dependence of the dynamics of the PMMA brush is discussed. Figure 4.9 summarizes the correlation time of the fluorescence anisotropy decay for the PMMA brushes of 0.11–0.46 chains nm^{-2} swollen in poor and good solvents. The solvent dependence of the PMMA brush at the graft density of 0.11 chains nm^{-2} was different from that at 0.46 chains nm^{-2}. Although the mobility of the low-density brush increases increasing solvent quality, the dynamics of the high-density brush is not dependent on the solvent. The solvent dependence of the low-density PMMA brush is similar to that of a linear chain in dilute solutions. The local chain motion of a linear PMMA in solutions depends on the solvent quality [44]. A polymer chain adoptss a relatively shrunken conformation in a poor solvent; therefore, the chain motion of the polymer chain is suppressed by an intra-chain interaction among the chain segments. In a good solvent, the polymer chain adopts an expanded conformation, resulting in high chain mobility due to little intra-chain interaction. Thus the chain dynamics of the linear chain in solution depends on the solvent quality. The similar solvent dependence of the linear chain solution and the low-density brush indicates that the nanosecond dynamics of the PMMA brush is affected by the chain

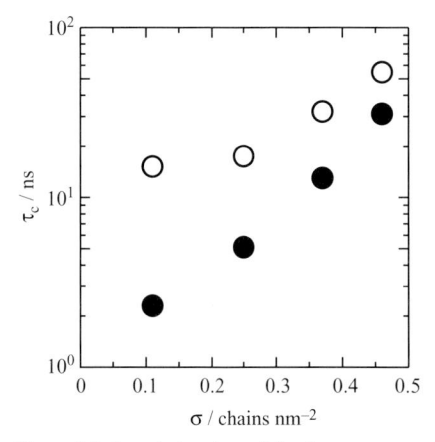

Figure 4.9 Correlation time of the fluorescence anisotropy decay for the PMMA brush. The open and closed circles indicate the correlation times for the brush in acetonitrile and benzene, respectively.

conformation dependent on the solvent quality. On the other hand, the high-density PMMA brush did not show significant solvent dependence. This implies that the conformation of the brush chain with a high graft density is not dependent on the solvent quality.

The chain dimension in the height direction was evaluated as the thickness of the brush layer, L, relative to the chain contour length, L_0, by atomic force microscopy (AFM). Figure 4.10 shows the solvent dependence of the conformation of the PMMA brush. Whereas the brush chain changes its conformation in response to the solvent quality at the low graft density, the high-density PMMA brush does not show

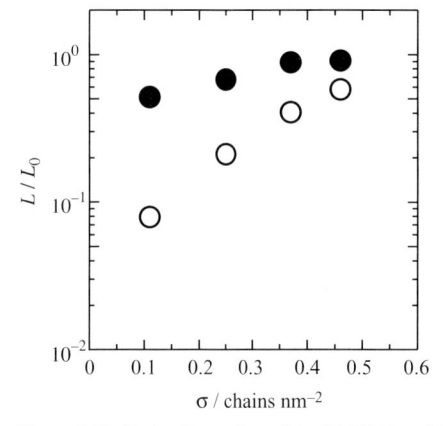

Figure 4.10 Chain dimension of the PMMA brush in the normal direction to the substrate. L and L_0 are the thickness of the brush layer and the contour length of the brush chain. The open and filled circles indicate the chain dimension in acetonitrile and benzene.

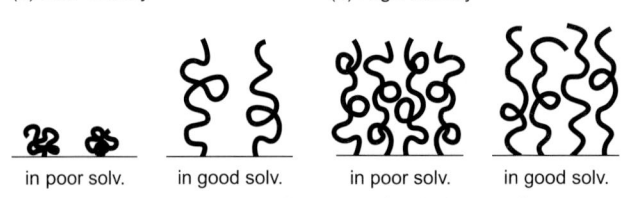

Figure 4.11 Schematic drawing of the PMMA brush chain swollen in solvents. (a) brush with low graft density, (b) brush with high graft density. The left- and right-hand sides of each image illustrate the brush chains in poor and good solvents, respectively. Reproduced with permission from the American Chemical Society.

significant dependence of the chain dimension on the solvent. For the PMMA brush with a low graft density of 0.11 chain nm^{-2}, the chain conformation in the dimension normal to the substrate is greatly altered by the swelling solvent quality: the chain dimension in benzene is more than five times larger than that in acetonitrile. Figure 4.11a shows a schematic drawing of the chain conformation of the PMMA brush with a low graft density. In the case of the low-density brush, the interaction among the brush chains is relatively weak and the brush chain can easily change its conformation, resulting in a sensitive response of the chain conformation to the solvent quality. On the other hand, it should be noted that the difference between the chain dimensions in both solvents is only 36% for the high-density PMMA brush. The chain dimension normal to the substrate is 161 nm in acetonitrile whereas the unperturbed dimension is estimated to be about 7 nm. Because of the strong repulsive inter-chain interaction, the brush chain is extremely stretched, even in the poor solvent. The conformation of the high-density brush chain is affected mainly by the strong inter-chain interaction rather than by the solvent quality. Thus, the chain conformation of the high-density PMMA brush does not show significant solvent dependence. The brush chain conformation in poor and good solvents is illustrated schematically in Figure 4.11b. As mentioned above, the segmental mobility of the polymer chain is dependent on the chain expansion in a solvent. Due to the similar chain conformations in the poor and good solvents, the micro-Brownian motion of the high-density PMMA brush chain is not affected by the solvent quality, whereas the low-density brush chain changes its mobility depending on the solvent.

Next we discuss the effect of the structure of the brush layer on the chain dynamics. The spatially resolved fluorescence depolarization measurement was performed using a high NA objective. Figure 4.12a shows the fluorescence image of the perylene-labeled PMMA brush. The bright area on the left-hand side indicates the PMMA brush layer, and the dark region on the right-hand side corresponds to the quartz substrate exposed by scratching the brush layer. The fluorescence anisotropy decay was observed in the polymer brush layer and at the scratched edge indicated by the filled and open circles in Figure 4.12a, respectively. The correlation time for the anisotropy decay inside the brush layer was 68 ns. On the other hand, the fluorescence anisotropy at the scratched edge showed the decay with a correlation time of 23 ns, indicating that the chain mobility of the PMMA brush increases at the

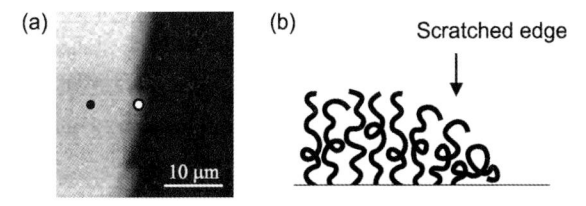

Figure 4.12 Fluorescence image of PMMA brush layer (a) and schematic drawing of the brush chain (b). The dark region (a) corresponds to the substrate surface exposed by scratching off the brush layer. The filled and open circles indicate the points where the fluorescence anisotropy decay was acquired.

scratched edge. This spatial heterogeneity of the chain dynamics results from the conformation of the brush chain dependent on the position in the brush layer. Figure 4.12b illustrates the conformation of brush chains near a scratched region. As mentioned above, a brush chain with a high graft density is surrounded by the neighboring chains and shows slow dynamics due to the strong steric inter-chain interaction. As shown in Figure 4.12b, a brush chain at the scratched edge is under less constraint from the neighboring chains because there is no brush chain on the right-hand side of the figure. Consequently, the degree of freedom of the brush chain at the edge increases, resulting in a relatively expanded conformation. The high segmental mobility at the scratched edge region is attributed to the relaxation of the constraint of the brush chain by the structural inhomogeneity of the brush layer.

4.4 Summary

The structure and dynamics of polymer systems confined in a small space was studied by the spatially and temporally resolved fluorescence methods, scanning near-field optical microscopy and the fluorescence depolarization technique. The conformation of PMMA was studied by optical imaging with nanometric resolution by near-field microscopy. The direct observation of the single chain of PMMA revealed the chain conformation characteristics of a thin film with a thickness less than the unperturbed size of the polymer chain. In an ultra-thin film, the chain dimension in the normal direction to the film plane is similar to that in a three-dimensional bulk state. The dynamics of the polymer brush was investigated by scanning optical microscopy combined with the time-resolved fluorescence depolarization technique. The PMMA brush with a high graft density adopts an extremely stretched conformation in the normal direction to the substrate and shows low segmental mobility due to the strong interaction among the brush chains. The fluorescence technique with nanosecond temporal resolution and nanometric spatial resolution would provide indispensable information from a specimen and the further development of novel fluorescence techniques would reveal more insight into polymer systems on the molecular scale.

Acknowledgments

The author appreciates support by a Grant-in-Aid for Scientific Research in Priority Area "Molecular Nano Dynamics" from the Ministry of Education, Culture, Sports, Science and Technology (MEXT) and a Grant-in-Aid from the Japan Society for the Promotion of Science (JSPS), Japan. The Innovative Techno-Hub for Integrated Medical Bio-imaging Project of the Special Coordination Funds for Promoting Science and Technology from MEXT is also acknowledged.

References

1 Grohens, Y., Hamon, L., Reiter, G., Soldera, A. and Holl, Y. (2002) Some relevant parameters affecting the glass transition of supported ultra-thin polymer films. *Eur. Phys. J. E*, **8**, 217–224.

2 Kajiyama, T., Tanaka, K. and Takahara, A. (1997) Surface molecular motion of the monodisperse polystyrene films. *Macromolecules*, **30**, 280–285.

3 Keddie, J. L., Jones, R. A. L. and Cory, R. A. (1994) Interface and Surface Effects on the Glass-Transition Temperature in Thin Polymer-Films. *Faraday Discuss.*, **98**, 219–230.

4 Keddie, J. L., Jones, R. A. L. and Cory, R. A. (1994) Size-Dependent Depression of the Glass-Transition Temperature in Polymer-Films. *Europhys. Lett.*, **27**, 59–64.

5 Tsui, O. K. C. and Zhang, H. F. (2001) Effects of chain ends and chain entanglement on the glass transition temperature of polymer thin films. *Macromolecules*, **34**, 9139–9142.

6 de Gennes, P. G. (1979) *Scaling Concepts in Polymer Physics*, Cornell University Press, Ithaca, New York.

7 Cavallo, A., Muller, M., Wittmer, J. P., Johner, A. and Binder, K. (2005) Single chain structure in thin polymer films: corrections to Flory's and Silberberg's hypotheses. *J. Phys. Condens. Matter*, **17**, S1697–S1709.

8 Kumar, S. K., Vacatello, M. and Yoon, D. Y. (1988) Off-Lattice Monte-Carlo Simulations of Polymer Melts Confined between 2 Plates. *J. Chem. Phys.*, **89**, 5206–5215.

9 Muller, M. (2002) Chain conformations and correlations in thin polymer films: a Monte Carlo study. *J. Chem. Phys.*, **116**, 9930–9938.

10 Reiter, J., Zifferer, G. and Olaj, O. F. (1989) Monte-Carlo studies of polymer-chain dimensions in the melt in 2 dimensions. *Macromolecules*, **22**, 3120–3124.

11 Yethiraj, A. (2003) Computer simulation study of two-dimensional polymer solutions. *Macromolecules*, **36**, 5854–5862.

12 Kraus, J., Muller-Buschbaum, P., Kuhlmann, T., Schubert, D. W. and Stamm, M. (2000) Confinement effects on the chain conformation in thin polymer films. *Europhys. Lett.*, **49**, 210–216.

13 Brulet, A., Boue, F., Menelle, A. and Cotton, J. P. (2000) Conformation of polystyrene chain in ultrathin films obtained by spin coating. *Macromolecules*, **33**, 997–1001.

14 Shuto, K., Oishi, Y., Kajiyama, T. and Han, C. C. (1993) Aggregation structure of a two-dimensional ultrathin polystyrene film prepared by the water casting method. *Macromolecules*, **26**, 6589–6594.

15 Jones, R. L., Kumar, S. K., Ho, D. L., Briber, R. M. and Russell, T. P. (1999) Chain conformation in ultrathin polymer films. *Nature*, **400**, 146–149.

16 Jones, R. L., Kumar, S. K., Ho, D. L., Briber, R. M. and Russell, T. P. (2001) Chain conformation in ultrathin polymer films using small-angle neutron scattering. *Macromolecules*, **34**, 559–567.

17 Ohtsu, M. (ed.) (1998) *Near-Field Nano/Atom Optics and Technology*, Springer, Tokyo.

18 Paesler, M. A. and Moyer, P. J. (1996) *Near-Field Optics: Theory, Instrumentation, and Applications*, John Wiley & Sons, New York.

19 Aoki, H., Anryu, M. and Ito, S. (2005) Two-dimensional polymers investigated by scanning near-field optical microscopy: conformation of single polymer chain in monolayer. *Polymer*, **46**, 5896–5902.

20 Aoki, H., Morita, S., Sekine, R. and Ito, S. (2008) Conformation of single poly(methyl methacrylate) chains in ultra-thin film studied by scanning near-field optical microscopy. *Polym. J.*, **40**, 274–280.

21 Aoki, H., Tanaka, S., Ito, S. and Yamamoto, M. (2000) Nanometric inhomogeneity of polymer network investigated by scanning near-field optical microscopy. *Macromolecules*, **33**, 9650–9656.

22 O'Reilly, J. M., Teegarden, D. M. and Wignall, G. D. (1985) Small-angle and intermediate-angle neutron-scattering from stereoregular poly(methyl methacrylate). *Macromolecules*, **18**, 2747–2752.

23 Sato, N., Ito, S. and Yamamoto, M. (1998) Molecular weight dependence of shear viscosity of a polymer monolayer: evidence for the lack of chain entanglement in the two-dimensional plane. *Macromolecules*, **31**, 2673–2675.

24 Si, L., Massa, M. V., Dalnoki-Veress, K., Brown, H. R. and Jones, R. A. L. (2005) Chain entanglement in thin freestanding polymer films. *Phys. Rev. Lett.*, **94**, 127801.

25 Kato, K., Uchida, E., Kang, E. T., Uyama, Y. and Ikada, Y. (2003) Polymer surface with graft chains. *Prog. Polym. Sci.*, **28**, 209–259.

26 Senaratne, W., Andruzzi, L. and Ober, C. K. (2005) Self-assembled monolayers and polymer brushes in biotechnology: Current applications and future perspectives. *Biomacromolecules*, **6**, 2427–2448.

27 Uyama, Y., Kato, K. and Ikada, Y. (1998) Surface modification of polymers by grafting. *Adv. Polym. Sci.*, **137**, 1–39.

28 Pyun, J., Kowalewski, T. and Matyjaszewski, K. (2003) Synthesis of polymer brushes using atom transfer radical polymerization. *Macromol. Rapid Commun.*, **24**, 1043–1059.

29 Tsujii, Y., Ohno, K., Yamamoto, S., Goto, A. and Fukuda, T. (2006) Structure and properties of high-density polymer brushes prepared by surface-initiated living radical polymerization. *Adv. Polym. Sci.*, **197**, 1–45.

30 Yamamoto, S., Tsujii, Y. and Fukuda, T. (2002) Glass transition temperatures of high-density poly(methyl methacrylate) brushes. *Macromolecules*, **35**, 6077–6079.

31 Urayama, K., Yamamoto, S., Tsujii, Y., Fukuda, T. and Neher, D. (2002) Elastic properties of well-defined, high-density poly(methyl methacrylate) brushes studied by electromechanical interferometry. *Macromolecules*, **35**, 9459–9465.

32 Lai, P. Y. and Binder, K. (1992) Structure and dynamics of polymer brushes near the theta point – a Monte-Carlo simulation. *J. Chem. Phys.*, **97**, 586–595.

33 Wittmer, J., Johner, A., Joanny, J. F. and Binder, K. (1994) Chain desorption from a semidilute polymer brush - a Monte-Carlo simulation. *J. Chem. Phys.*, **101**, 4379–4390.

34 Semenov, A. N. (1995) Rheology of polymer brushes - Rouse model. *Langmuir*, **11**, 3560–3564.

35 Semenov, A. N. and Anastasiadis, S. H. (2000) Collective dynamics of polymer brushes. *Macromolecules*, **33**, 613–623.

36 He, G. L., Merlitz, H., Sommer, J. U. and Wu, C. X. (2007) Static and dynamic properties of polymer brushes at moderate and high grafting densities: A molecular dynamics study. *Macromolecules*, **40**, 6721–6730.

37 Dimitrov, D. I., Milchev, A. and Binder, K. (2007) Polymer brushes in solvents of variable quality: Molecular dynamics simulations using explicit solvent. *J. Chem. Phys.*, **127**, 084905.

38 Fytas, G., Anastasiadis, S. H., Seghrouchni, R., Vlassopoulos, D., Li, J. B.,

Factor, B. J., Theobald, W. and Toprakcioglu, C. (1996) Probing collective motions of terminally anchored polymers. *Science*, **274**, 2041–2044.

39 Michailidou, V. N., Loppinet, B., Vo, D. C., Prucker, O., Ruhe, J. and Fytas, G. (2006) Dynamics of end-grafted polystyrene brushes in theta solvents. *J. Polym. Sci. B Polym. Phys.*, **44**, 3590–3597.

40 Yakubov, G. E., Loppinet, B., Zhang, H., Ruhe, J., Sigel, R. and Fytas, G. (2004) Collective dynamics of an end-grafted polymer brush in solvents of varying quality. *Phys. Rev. Lett.*, **92**, 115501.

41 Aoki, H., Kitamura, M. and Ito, S. (2008) Nanosecond dynamics of poly(methyl methacrylate) brushes in solvents studied by fluorescence depolarization method. *Macromolecules*, **41**, 285–287.

42 Ejaz, M., Yamamoto, S., Ohno, K., Tsujii, Y. and Fukuda, T. (1998) Controlled graft polymerization of methyl methacrylate on silicon substrate by the combined use of the Langmuir-Blodgett and atom transfer radical polymerization techniques. *Macromolecules*, **31**, 5934–5936.

43 Arai, T., Sawatari, N., Yoshizaki, T., Einaga, Y. and Yamakawa, H. (1996) Excluded-volume effects on the hydrodynamic radius of atactic and isotactic oligo- and poly(methylmethacrylate)s in dilute solution. *Macromolecules*, **29**, 2309–2314.

44 Horinaka, J., Ono, K. and Yamamoto, M. (1995) Local chain dynamics of syndiotactic poly(methyl methacrylate) studied by the fluorescence depolarization method. *Polym. J.*, **27**, 429–435.

5
Real Time Monitoring of Molecular Structure at Solid/Liquid Interfaces by Non-Linear Spectroscopy

Hidenori Noguchi, Katsuyoshi Ikeda, and Kohei Uosaki

5.1
Introduction

Many important processes such as electrochemical reactions, biological processes and corrosion take place at solid/liquid interfaces. To understand precisely the mechanism of these processes at solid/liquid interfaces, information on the structures of molecules at the electrode/electrolyte interface, including short-lived intermediates and solvent, is essential. Determination of the interfacial structures of the intermediate and solvent is, however, difficult by conventional surface vibrational techniques because the number of molecules at the interfaces is far less than the number of bulk molecules.

Recently, sum frequency generation (SFG) spectroscopy has been shown to be a very powerful technique to probe molecular structure at an interface [1–6]. SFG is a second-order nonlinear optical (NLO) process, in which two photons of frequencies ω_1 and ω_2 generate one photon of the sum frequency ($\omega_3 = \omega_1 + \omega_2$) [2]. Second-order NLO processes, including SFG, are prohibited in media with inversion symmetry under the electric dipole approximation and are allowed only at the interface between these media where the inversion symmetry is necessarily broken. By using visible light of fixed wavelength and tunable IR light as the two input light sources, SFG spectroscopy can be a surface sensitive vibrational spectroscopy as the SFG signal is resonantly enhanced when the energy of the IR beam becomes equal to that of a vibrational state of the surface species [2]. SFG spectroscopy is particularly useful for studying the structure of water molecules, the most common solvent, at various interfaces where the presence of a much larger amount of bulk water than interfacial water makes the measurement of interfacial water by other vibrational techniques very difficult. Furthermore, SFG is free from the ambiguity associated with the choice of reference spectrum as required for linear spectroscopy applied to interfaces such as surface-enhanced IR spectroscopy (SEIRAS), [7, 8] and surface-enhanced Raman scattering (SERS) [9]. Furthermore, time-resolved vibrational spectroscopy based on SFG is possible because a short pulse laser is used. Thus,

Molecular Nano Dynamics, Volume I: Spectroscopic Methods and Nanostructures
Edited by H. Fukumura, M. Irie, Y. Iwasawa, H. Masuhara, and K. Uosaki
Copyright © 2009 WILEY-VCH Verlag GmbH & Co. KGaA, Weinheim
ISBN: 978-3-527-32017-2

SFG spectroscopy is an ideal technique to investigate the mechanism of interfacial processes at solid/liquid interfaces [5, 6, 10–16].

Hyper-Raman scattering (HRS), which is inelastic scattering in the second-order nonlinear optical process, is also expected to be a useful spectroscopic technique to obtain vibrational information [17–24]. According to the molecular symmetry selection rules, IR-active vibrational modes, which are usually not evident in normal Raman spectra, are all active in HRS. Moreover, so-called "silent mode" vibration, which is not observed in either Raman or IR absorption spectra, can be revealed in HRS [17, 18]. However, since the HRS effect is extremely weak, enhancement of the HRS signal is essential for HRS to be a useful spectroscopic technique. Similar to the enhancement of Raman scattering, SERS, surface-enhanced hyper-Raman scattering, SEHRS, has been reported, utilizing metal colloids or a roughened electrode surface [21, 23, 24].

Here, we describe the basic principles and detailed experimental arrangement of SFG spectroscopy and present several examples of SFG study at solid/liquid interfaces. HRS is also described briefly.

5.2
Sum Frequency Generation Spectroscopy

5.2.1
Brief Description of SFG

Non-linear optical phenomena that can be observed with static electric and magnetic fields such as the Pöckels and Faraday effects have been known since the nineteenth century. Frequency conversion such as second harmonic generation (SHG) and SFG requires intense optical fields and its realization had to wait until the birth of the pulsed laser at the end of the 1950s. SHG was first observed in 1961 by Franken et al. in a bulk quartz crystal [25]. The foundation of the theory of NLO was laid by Bloembergen in the early 1960s [26]. Surface SHG was detected for the first time in 1974 from the surface of silver [27]. The first observation of surface SFG was in 1986 by Shen et al. for the coumarin dye on glass [28]. SFG spectroscopy has become extremely attractive for analysis of interface science, thanks to the recent development of a tunable laser source.

SFG is one of the second-order NLO processes, in which two photons of frequencies ω_1 and ω_2 generate one photon of a sum frequency ($\omega_3 = \omega_1 + \omega_2$) as shown in Figure 5.1.

Second-order NLO processes, including SFG, are strictly forbidden in media with inversion symmetry under the electric dipole approximation and are allowed only at the interface between these media where the inversion symmetry is necessarily broken. In the IR–Visible SFG measurement, a visible laser beam (ω_{Vis}) and a tunable infrared laser beam (ω_{IR}) are overlapped at an interface and the SFG signal is measured by scanning ω_{IR} while keeping ω_{Vis} constant. The SFG intensity (I_{SFG}) is enhanced when ω_{IR} becomes equal to the vibration levels of the molecules at the interface. Thus, one can obtain surface-specific vibrational spectra at an interface

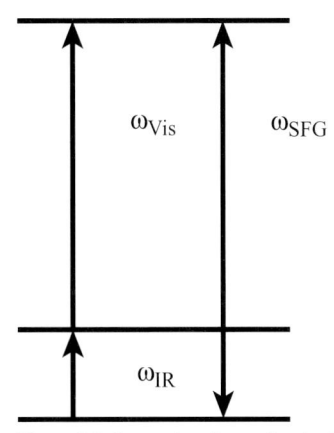

Figure 5.1 Energetic scheme for the SFG process.

between two phases with inversion symmetry, which cannot be obtained using traditional vibrational spectroscopy such as IR and Raman scattering.

In addition to the surface/interface selectivity, IR–Visible SFG spectroscopy provides a number of attractive features since it is a coherent process: (i) Detection efficiency is very high because the angle of emission of SFG light is strictly determined by the momentum conservation of the two incident beams, together with the fact that SFG can be detected by a photomultiplier (PMT) or CCD, which are the most efficient light detectors, because the SFG beam is in the visible region. (ii) The polarization feature that NLO intrinsically provides enables us to obtain information about a conformational and lateral order of adsorbed molecules on a flat surface, which cannot be obtained by traditional vibrational spectroscopy [29–32]. (iii) A pump and SFG probe measurement can be used for an ultra-fast dynamics study with a time-resolution determined by the incident laser pulses [33–37]. (iv) As a photon-in/photon-out method, SFG is applicable to essentially any system as long as one side of the interface is optically transparent.

5.2.2
Origin of SFG Process

The induced dipole ($\vec{\mu}$) of a molecule placed in an electric field ($\vec{E}$) is given by

$$\vec{\mu} = \vec{\mu}^0 + \overleftrightarrow{\alpha}\vec{E} + \overleftrightarrow{\beta}\vec{E}\vec{E} + \overleftrightarrow{\gamma}\vec{E}\vec{E}\vec{E} + \cdots \quad (5.1)$$

Where $\overleftrightarrow{\alpha}$, $\overleftrightarrow{\beta}$ and $\overleftrightarrow{\gamma}$ are the linear polarizability, the first- and second-hyper polarizabilities, respectively, and are represented by second, third and fourth rank tensors, respectively, and $\vec{\mu}^0$ is a static polarizability.

In condensed phases, it is more convenient to consider the dipole moment per unit volume or polarization ($\vec{P}$)

$$\vec{P} = \vec{P}^{(0)} + \vec{P}^{(1)} + \vec{P}^{(2)} + \vec{P}^{(3)} + \cdots$$
$$= \vec{P}^0 + \varepsilon_0(\overleftrightarrow{\chi}^{(1)}\vec{E} + \overleftrightarrow{\chi}^{(2)}\vec{E}\vec{E} + \overleftrightarrow{\chi}^{(3)}\vec{E}\vec{E}\vec{E} + \cdots) \quad (5.2)$$

where, $\overleftrightarrow{\chi}^{(1)}$, $\overleftrightarrow{\chi}^{(2)}$ and $\overleftrightarrow{\chi}^{(3)}$ are the first-, second-, and third-order susceptibilities, respectively, and ε_0 is the dielectric constant in vacuum. For a simple molecular material, the susceptibility depends on the number of molecules per unit volume, N, multiplied by the molecular polarizability averaged over all orientations of molecules in the material.

$$\overleftrightarrow{\chi}^{(1)} = N \langle \overleftrightarrow{\alpha} \rangle / \varepsilon_0, \quad \overleftrightarrow{\chi}^{(2)} = N \langle \overleftrightarrow{\beta} \rangle / \varepsilon_0, \quad \overleftrightarrow{\chi}^{(3)} = N \langle \overleftrightarrow{\gamma} \rangle / \varepsilon_0 \tag{5.3}$$

where the brackets $\langle \, \rangle$ denote the ensemble average of the molecular orientation. Since few materials have a static polarization, the first term in Eq. (5.2), $\vec{P}^{(0)}$ can be dropped.

$$\vec{P} = \varepsilon_0 \left(\overleftrightarrow{\chi}^{(1)} \vec{E} + \overleftrightarrow{\chi}^{(2)} \vec{E}\vec{E} + \overleftrightarrow{\chi}^{(3)} \vec{E}\vec{E}\vec{E} + \cdots \right) \tag{5.4}$$

When the electric field ($\vec{E}$) is weak, as in a non-laser light source, the second and third terms can be neglected and polarization ($\vec{P}$) can be expressed as below, corresponding to linear optics.

$$\vec{P} = \varepsilon_0 \overleftrightarrow{\chi}^{(1)} \vec{E} \tag{5.5}$$

When the electric field ($\vec{E}$) is strong, as in a laser light source, the second and third terms cannot be neglected and NLO processes take place. A remarkable consequence of the higher-order terms in Eq. (5.4) is that the frequency of the light can change. If we consider two electric fields with frequencies ω_1 and ω_2,

$$\vec{E}_1(\vec{r}, t) = \vec{E}_1(\vec{r})\cos(\omega_1 t), \quad \vec{E}_2(\vec{r}, t) = \vec{E}_2(\vec{r})\cos(\omega_2 t) \tag{5.6}$$

where $\vec{r}$ and t represent a position vector and time, respectively. The second-order polarization, $\vec{P}^{(2)}$, in a material in which $\overleftrightarrow{\chi}^{(2)}$ is non-zero is given by:

$$\begin{aligned}
\vec{P}^{(2)} &= \varepsilon_0 \overleftrightarrow{\chi}^{(2)} \vec{E}_1 \vec{E}_2 \\
&= \varepsilon_0 \overleftrightarrow{\chi}^{(2)} \vec{E}_1(\vec{r})\cos(\omega_1 t) \vec{E}_2(\vec{r})\cos(\omega_2 t) \\
&= 1/2\varepsilon_0 \overleftrightarrow{\chi}^{(2)} \vec{E}_1(\vec{r}) \vec{E}_2(\vec{r})\{\cos((\omega_1 + \omega_2)t) + \cos((\omega_1 - \omega_2)t)\}
\end{aligned} \tag{5.7}$$

This shows that there are now oscillating dipoles at frequencies ($\omega_1 + \omega_2$) and ($\omega_1 - \omega_2$), which give rise to SFG and difference frequency generation (DFG), respectively. When $\omega_1 = \omega_2$, Eq. (5.7) becomes:

$$\vec{P}^{(2)} = 1/2\varepsilon_0 \overleftrightarrow{\chi}^{(2)} \vec{E}_1(\vec{r}) \vec{E}_2(\vec{r})\{1 + \cos(2\omega_1 t)\} \tag{5.8}$$

The first term in the brackets represents a static electric field in the material and the second term represents a dipole oscillating at 2ω, twice the frequency of the incident light. This is a process known as SHG.

5.2.3
SFG Spectroscopy

In SFG vibrational spectroscopy, ω_1 is usually fixed in the visible region and ω_2 is scanned in the infrared region. In the most widely used geometry, the two laser

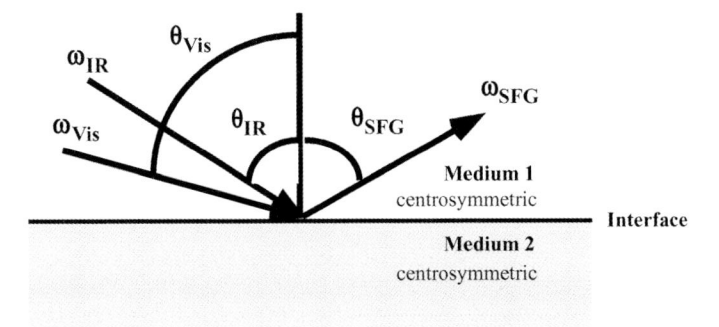

Figure 5.2 Schematic drawing of the optical arrangement of an SFG measurement.

beams are incident in either counter- or collinear-propagating geometry and the reflected SF light (ω_{sum}) is detected. Let us consider the case where visible and infrared beams are incident from medium 1 to the interface with medium 2, as depicted in Figure 5.2. The direction of emission is determined by conservation of momentum parallel to the surface.

$$\omega_{SFG} = \omega_{Vis} + \omega_{IR} \tag{5.9}$$

$$\omega_{SFG}\sin\theta_{SFG} = \omega_{Vis}\sin\theta_{Vis} + \omega_{IR}\sin\theta_{IR} \tag{5.10}$$

where θ_{SFG} is the emission angle of SFG light and θ_{Vis} and θ_{IR} are the incident angles of the visible and infrared light. Since the angles θ_{Vis}, θ_{IR} and the frequency ω_{Vis} are constant in a given SFG measurement, θ_{SFG} can be expressed as a function of ω_{IR}.

$$\theta_{SFG}(\omega_{IR}) = \sin^{-1}\left\{\frac{\omega_{Vis}\sin\theta_{Vis} + \omega_{IR}\sin\theta_{IR}}{\omega_{Vis} + \omega_{IR}}\right\} \tag{5.11}$$

As shown in this equation, the angle of emission changes as ω_{IR} is scanned.

The intensity of emitted SF light (I_{SFG}) is expressed by the following equation [38]

$$I_{SFG} = \frac{8\pi^3 \omega_{SFG}^2 (\sec\theta_{SFG})^2}{\hbar c^3 \cdot n_1(\omega_{SFG}) \cdot n_1(\omega_{Vis}) \cdot n_1(\omega_{IR})} \left|\overleftrightarrow{F} : \overleftrightarrow{\chi}^{(2)}\right|^2 I_{Vis} I_{IR} A T \tag{5.12}$$

where I_{vis} and I_{IR} are the intensities of incident visible and infrared light, respectively, c is the speed of light in vacuum, A is an overlapping beam cross section at the sample, T is a pulse duration, and $n_1(\omega_i)$ is the refractive index of medium 1 at frequency ω_i. and $\overleftrightarrow{F}$ is a Fresnel factor.

For a polar monolayer of molecules at an interface, the resonant $\overleftrightarrow{\chi}^{(2)}$ is typically $10^{-14} \sim 10^{-16}$ esu. If we take $\overleftrightarrow{\chi}^{(2)} \sim 10^{-15}$ esu, $I_{Vis} \sim 1$ mJ pulse^{-1} and $I_{IR} \sim 100$ μJ pulse^{-1}, $T \sim 25$ ps, $\omega_{Vis} = 5.64 \times 10^{14}$ s^{-1} (532 nm), $\omega_{IR} = 9.0 \times 10^{13}$ s^{-1} (3333 nm), $\omega_{SFG} = 6.55 \times 10^{14}$ s^{-1} (458.77 nm), $A \sim 10^{-3}$ cm^2, $\theta_{SFG} \sim 67°$, Eq. (5.12) predicts a signal of 1.5×10^4 photon pulse^{-1}. Such a signal can be readily detected by a PMT or CCD detector.

If neither ω_{Vis} nor ω_{SFG} is in resonance with an electric dipole transition in the material and only electric dipole transitions are considered, the hyperpolarizability,

β_{lmn}, near a vibrational transition in a molecule is given by [38]

$$\beta_{lmn} = \frac{1}{2\hbar^2} \cdot \sum_s \left(\frac{\langle g|\mu_l|s\rangle\langle s|\mu_m|v\rangle}{\omega_{SFG}-\omega_{sg}} - \frac{\langle g|\mu_m|s\rangle\langle s|\mu_l|v\rangle}{\omega_{Vis}+\omega_{sg}} \right)$$
$$\times \left(\frac{\langle v|\mu_n|g\rangle}{\omega_{IR}-\omega_{vg}+i\Gamma} \right) + \text{const.} \quad (5.13)$$

where $|g\rangle$ is the ground state, $|v\rangle$ is the excited vibrational state, $|s\rangle$ is any other state, Γ is the relaxation time of the excited vibrational state, const. is an off-resonant term, which can be considered as a constant and $\mu = er$ is the electric dipole operator. The first bracket is identified as a Raman transition dipole moment M_{lm}, and the term $\langle v|\mu_n|g\rangle$ as an IR transition dipole moment μ_n^{vg}. If we substitute ω_{vg} by ω_0, Eq. (5.13) becomes:

$$\beta_{ijk} = \frac{1}{2\hbar^2} \cdot \frac{a_{ijk}}{\omega_{IR}-\omega_0-i\Gamma} + \text{const.} \quad (5.14)$$

where

$$a_{ijk} = M_{ij} \cdot \mu_k^{vg}$$

From the above equation, it is clear that β_{ijk} takes a maximum value when the IR frequency is in resonance with the molecular vibration, that is, $\omega_{IR} = \omega_0$, and only the molecular vibration, which is both IR- and Raman-active is SFG-active.

If we introduce $\overleftrightarrow{A} = N/\varepsilon_n \langle \overleftrightarrow{a} \rangle$, the description of macroscopic resonant second-order susceptibility, $\overleftrightarrow{\chi}_R^{(2)}$, is given by

$$\overleftrightarrow{\chi}_R^{(2)} = \frac{1}{2\hbar^2} \frac{\overleftrightarrow{A}}{\omega_{IR}-\omega_0-i\Gamma} \quad (5.15)$$

The emitted SFG light is a sum of resonant contributions with nth vibrational states, $\overleftrightarrow{\chi}_{R,n}^{(2)}$ and nonresonant contributions, $\overleftrightarrow{\chi}_{NR}^{(2)}$.

$$\overleftrightarrow{\chi}^{(2)} = \overleftrightarrow{\chi}_{NR}^{(2)} + \sum_n \overleftrightarrow{\chi}_{R,n}^{(2)} = \overleftrightarrow{\chi}_{NR}^{(2)} + \sum_n \frac{1}{2\hbar^2} \frac{\overleftrightarrow{A}_n}{\omega_{IR}-\omega_n-i\Gamma_n} \quad (5.16)$$

Here, $\overleftrightarrow{\chi}_{NR}^{(2)}$ is the sum of nonresonant contributions from molecules ($\overleftrightarrow{\chi}_{NR,mol}$), substrate ($\overleftrightarrow{\chi}_{NR,sub}$) and the interaction between molecules and substrate ($\overleftrightarrow{\chi}_{NR,m-s}$).

$$\overleftrightarrow{\chi}_{NR} = \overleftrightarrow{\chi}_{NR,mol} + \chi_{NR,sub} + \chi_{NR,m-s} \quad (5.17)$$

Combining Eqs. (5.12) and (5.16), the SFG intensity can be written as:

$$I_{SFG} \propto \left| \overleftrightarrow{\chi}_{eff,NR}^{(2)} + \sum_n \frac{\overleftrightarrow{A}_{eff,n}}{\omega_{IR}-\omega_n-i\Gamma_n} \right|^2 \quad (5.18)$$

where $\overleftrightarrow{\chi}_{eff,NR}^{(2)} = \overleftrightarrow{F} : \overleftrightarrow{\chi}_{NR}^{(2)}$ and $\overleftrightarrow{A}_{eff,n} = \overleftrightarrow{F} : \overleftrightarrow{A}_n$. This is the equation generally used to analyze an observed SFG spectrum.

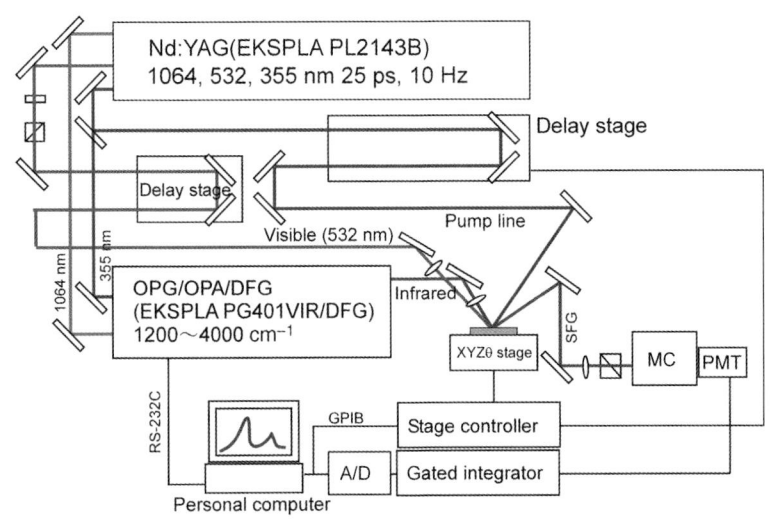

Figure 5.3 Schematic drawing of a picosecond SFG system.

5.2.4
Experimental Arrangement for SFG Measurements

5.2.4.1 Laser and Detection Systems

Figure 5.3 shows the schematic diagram of a picosecond SFG spectrometer used in this work. A passive mode-locked Nd:YAG laser system (EKSPLA, PL2143B) generates ~25 ps pulses of 1064 nm (fundamental), 532 nm (second harmonic: SHG) and 355 nm (third harmonic: THG) at a repetition rate of 10 Hz. The total output power was about 33~35 mJ pulse^{-1} and typical pulse energies of 1064 nm, 532 nm and 355 nm were 8, 8 and 6 mJ pulse^{-1}, respectively. The ratio of the harmonic generations can be adjusted to some extent by tuning the phase match angles of the mixing crystals. The 50% of SHG and 15% of THG were split and transferred to a pump line. The residual SHG was directed to a sample through a time-delay and used as the incident visible light for an SFG measurement. The power of the incident visible light can be adjusted by a combination of a half wave plate and a Gran–Taylor prism and the polarization can be changed by an additional half wave plate. The 85% of THG output (about 5 mJ pulse^{-1}) was used to pump an optical parametric generation and amplification (EKSPLA, PG401VIR/DFG) system, containing a LBO crystal and double gratings. The output from the OPG/OPA was mixed with the 1064 nm laser output in a nonlinear crystal, Ag_2GaS_2, to generate a tunable infrared output from 2.3 to 8.5 μm by the difference frequency generation (DFG) process. The power of the incident infrared light can be roughly adjusted by tuning a total power of the laser and the ratio of the harmonic generator. The polarization of the IR beam can be changed by exchanging the mirror sets, which contains three and four mirrors. The line-width of the tunable infrared output was ~6 cm^{-1}, determined by the double grating set-up in the OPA path.

The "visible" output from the laser and tunable infrared output from the DFG were incident and overlapped at the sample surface with angles of 70° and 50°, respectively. Either a 532 nm or a 1064 nm laser output was used as the "visible" light in the SFG measurement. A BK7 lens ($f=600$) and a ZnSe lens ($f=50$) were used to focus the visible and IR beams, respectively. The position of the ZnSe lens was controlled by an XY stage to perform the spatial overlapping of the two incident beams. The beam spot of the "visible" and infrared lights had an elliptical shape with dimensions of about $2.5 \times 0.9\,mm^2$ and $0.3 \times 0.2\,mm^2$, respectively. The SFG signal was separated from the reflected visible and IR lights by passing through irises and a monochromator (Oriel Instruments, MS257) and was detected by a photomultiplier tube (PMT: Hamamatsu, R630-10 or R3896) with a gated electronic system (Stanford Research System). When the 532 nm visible light was used, two holographic Super Notch filters (Kaiser Optical System, INC, HSPF-532-1.0) were placed in front of the monochromator to further reduce scattered visible light.

All SFG spectra were obtained by averaging 50~400 pulses per data point and were normalized against the intensity of the visible and infrared inputs, which were simultaneously monitored by a power meter (Oriel instruments, Model 70833 and 70811) and Si photodiode, respectively.

5.2.4.2 Spectroscopic Cells

Spectroelectrochemical Cell Figure 5.4 shows spectroelectrochemical cells used in electrochemical SFG measurements. An Ag/AgCl (saturated NaCl) and a Pt wire were used as a reference electrode and a counter electrode, respectively. The electrolyte solution was deaerated by bubbling high-purity Ar gas (99.999%) for at least 30 min prior to the electrochemical measurements. The electrode potential was controlled with a potentiostat. The electrode potential, current, and SFG signal were recorded by using a personal computer through an AD converter.

After introduction of the working electrode to the spectroelectrochemical cell, continuous potential cycling was performed to obtain a clean surface before each

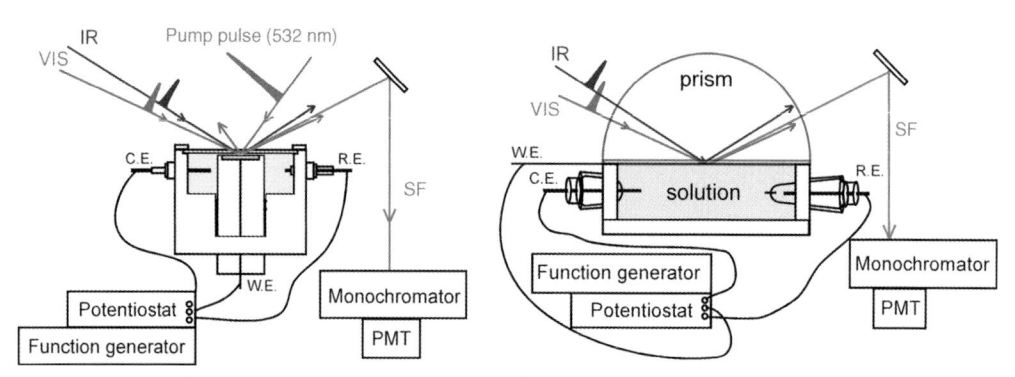

Figure 5.4 Spectroelectrochemical cells for electrochemical SFG measurements.

electrochemical SFG experiment until the cyclic voltammogram (CV) became that of a clean electrode.

When single- or poly-crystalline disk electrodes are used as a working electrode, the electrode must be gently pushed against the CaF_2 window of the spectroelectrochemical cell to achieve a thin layer (about 5 μm) configuration so that strong absorption of the IR beam by the electrolyte solution was avoided. A thin metal electrode deposited on a hemispherical prism can also be used as a working electrode. Strong absorption of the IR beam by the electrolyte solution can be more easily avoided in this configuration.

Flow Cell *In situ* SFG measurements were carried out in the internal reflection geometry to avoid the strong absorption of IR incident beam by water. Since the SFG signal from the quartz disk/water interface is very weak if a flat quartz disk is used as a window, due to the very large reflection loss of the incident beams at the air/quartz interface and the small Fresnel factors at the quartz/water interface, a hemi-cylindrical prism, where higher enhancement of the sensitivity is expected, was used for the *in situ* SFG measurements. The flat face of the quartz prism was in contact with electrolyte solution and the infrared and visible beams were incident from the quartz prism side and were overlapped at the quartz/electrolyte solution interface. The incident angle of the visible light was about 70°, which is near the critical angle of total reflection (θ_c) of quartz/water (72°), so that a strong surface field can be obtained. The incident angle of the infrared light was set at 50°, which is far from θ_c, so that a large change in the Fresnel factor can be avoided when the IR frequency is scanned over the OH-stretching vibration region.

A flow cell made of polychlorofluoroethylene, shown schematically in Figure 5.5, was used to change between gas or liquid during SFG measurements. Electrolyte solution was introduced into the flow cell by the pressure of the pure argon gas so that the solution could be exchanged without being exposed to the atmosphere and the optical alignment and the sample position were not affected when the solution was exchanged.

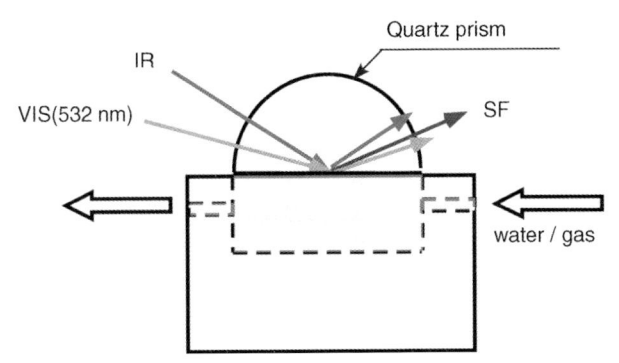

Figure 5.5 A flow cell for SFG measurements.

5.3
SFG Study of the Potential-Dependent Structure of Water at a Pt Electrode/Electrolyte Solution Interface

5.3.1
Introduction

Interfacial water molecules play important roles in many physical, chemical and biological processes. A molecular-level understanding of the structural arrangement of water molecules at electrode/electrolyte solution interfaces is one of the most important issues in electrochemistry. The presence of oriented water molecules, induced by interactions between water dipoles and electrode and by the strong electric field within the double layer has been proposed [39–41]. It has also been proposed that water molecules are present at electrode surfaces in the form of clusters [42, 43]. Despite the numerous studies on the structure of water at metal electrode surfaces using various techniques such as surface enhanced Raman spectroscopy [44, 45], surface infrared spectroscopy [46, 47], surface enhanced infrared spectroscopy [7, 8] and X-ray diffraction [48, 49], the exact nature of the structure of water at an electrode/solution interface is still not fully understood.

Here, we demonstrate the usefulness of SFG spectroscopy in the study of water structure at electrode/electrolyte solution interfaces by showing the potential dependent SFG spectra in the OH-stretching vibration region at a Pt/thin film electrode/0.1 M $HClO_4$ solution interface in internal reflection mode.

5.3.2
Results and Discussion

Figure 5.6 shows a typical CV of a thin Pt film electrode in a 0.1 M $HClO_4$ solution. The hydrogen waves in the potential range between −200 and about 50 mV were observed and surface oxidation and reduction peaks were observed in the positive potential region (>600 mV). This result confirmed that the conductivity of the Pt thin film was good enough to be used as an electrode.

Figure 5.7 shows SFG spectra in the OH-stretching region (2800–3800 cm^{-1}) obtained at the Pt electrode in a 0.1 M $HClO_4$ solution at various potentials. Two broad peaks were observed at about 3200 cm^{-1} and 3400 cm^{-1}. These peaks have been assigned to the vibration of OH oscillators of three-coordinated hydrogen-bonded water, that is, less ordered "liquid-like" water, molecules and that of the four-coordinated hydrogen-bonded water, that is, highly ordered "ice-like" water molecules, respectively, based on an IR study of water clusters [50]. Thus, the intensity ratio between these two peaks can be considered as an index of the order of the interfacial water [51]. Our previous work showed that the SFG spectra of Au thin film/0.05 M H_2SO_4 solution interface were dominated by the peak corresponding to the "liquid-like" water at all potentials we investigated [13]. Thus, water seems to be more highly oriented at the Pt electrode than at the Au electrode.

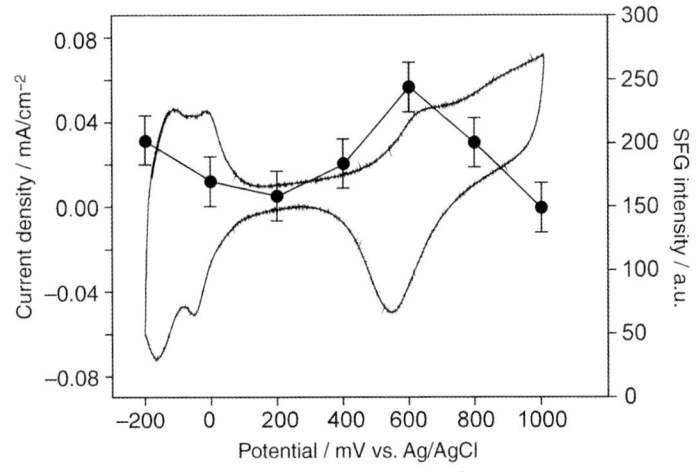

Figure 5.6 CV obtained with a sweep rate of 50 mV s^{-1} (solid line) and potential dependence of integrated SFG intensity in the OH-stretching region (•) of a Pt thin film electrode in 0.1 M HClO$_4$ solution.

While the shape of the SFG spectra did not change significantly with potential, the intensity depended on potential. To clarify the potential dependences of the SFG intensity, the integral intensities of SFG spectra of the Pt electrode between 2800 cm^{-1} and 3800 cm^{-1}, taken from Figure 5.7 were plotted against electrode potential,l as shown also in Figure 5.6. Parabolic behavior of the SFG intensity was observed between −200 and 600 mV with a minimum around 200 mV, which is close to the potential of zero charge, pzc, of Pt electrode in a HClO$_4$ solution [52, 53], although the value of the pzc of a Pt electrode is still debated [54].

Previously, we have proposed that SFG intensity due to interfacial water at quartz/water interfaces reflects the number of oriented water molecules within the electric double layer and, in turn, the double layer thickness based on the pH dependence of the SFG intensity [10] and a linear relation between the SFG intensity and (ionic strength)$^{-1/2}$ [12]. In the case of the Pt/electrolyte solution interface the drop in the potential profile in the vicinity of electrode become precipitous as the electrode becomes more highly charged. Thus, the ordered water layer in the vicinity of the electrode surface becomes thinner as the electrode is more highly charged. Since the number of ordered water molecules becomes smaller, the SFG intensity should become weaker at potentials away from the pzc. This is contrary to the experimental result.

When the electrolyte concentration is relatively high, the potential dependence of the double layer thickness is low and the potential dependence of the fraction of oriented water predominantly determines that of the SFG intensity. Since the polarization of IR in the present experiment is "p", water oriented normal to the surface is effectively detected by SFG. Water molecules are expected to lie parallel to the surface around the pzc and reorient from "oxygen up" to "oxygen down" as the surface charge of electrode surface changes from negative to positive so long as no specific adsorption of ions take place. IR study [7, 8] as well as computer simulation [55, 56] also suggest that

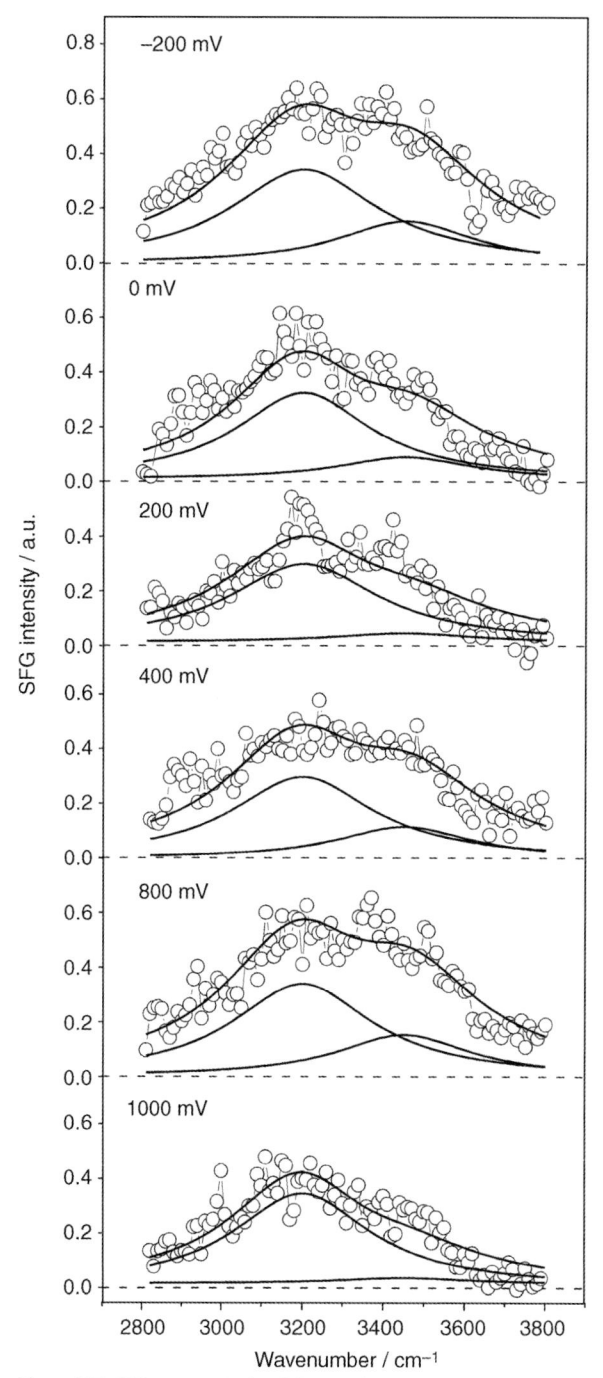

Figure 5.7 SFG spectra in the OH-stretching region at a Pt electrode at each potential in 0.1 M HClO$_4$ solution.

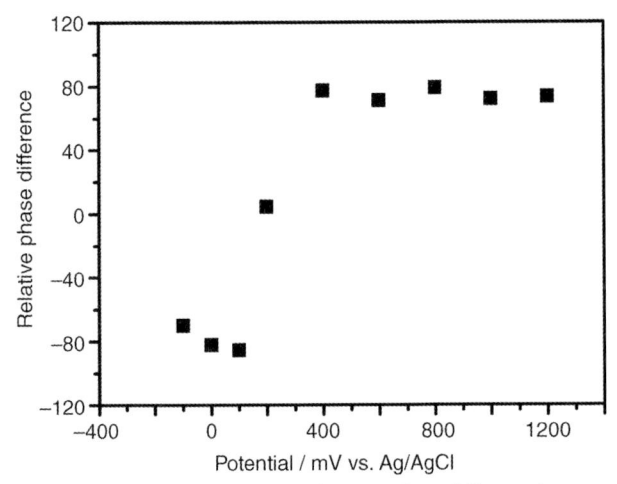

Figure 5.8 Potential dependence of relative phase difference between χ_{NR} and χ_R.

water molecules at the metal electrode surface have an oxygen up and oxygen down orientation on negatively and positively charged surfaces, respectively, on average.

For fixed non-resonant $\chi_{NR}^{(2)}$ sign, when the dipole moment of a molecule rotates by 180°, the relative phase between the resonant and non-resonant part of the signal also changes by 180° [57]. Thus, by analyzing the relative phase difference between $\chi_{NR}^{(2)}$ and $\chi_R^{(2)}$, the potential dependent orientation of water molecules on a Pt electrode surface can be determined.

Figure 5.8 shows the potential dependence of the relative phase difference between $\chi_{NR}^{(2)}$ and $\chi_R^{(2)}$. The relative phase was changed by about 180° at 200 mV, which is close to the pzc for a Pt electrode in $HClO_4$ electrolyte solution [52, 53]. This orientation change is most probably associated with a change in sign of the charge at the Pt surface. This clearly demonstrates that the orientation of water dipoles flips by 180° at the pzc.

SFG intensity in the OH-stretching region decreased as the potential became more positive where Pt oxide was formed, as shown in Figure 5.6. There are several possibilities for this decrease. One is the disruption of the well-ordered hydrogen bonded network structure of water molecules at a roughened Pt electrode surface compared to that at an atomically flat surface, since it is well known that the atomically flat surfaces of Pt were roughened by surface oxide formation [58]. The other possibility is the electric effect. Since Pt oxide is an insulating thin film [59], an additional potential drop takes place within the Pt oxide layer, resulting in a smaller electric field within the double layer. Furthermore, surface charge should be also affected by the oxide formation.

5.3.3
Conclusions

In conclusion, electrochemical SFG measurements showed that the SFG spectra in the OH-stretching region (2800–3800 cm^{-1}) at the Pt electrode in a 0.1 M $HClO_4$

solution showed two broad peaks at about $3200\,cm^{-1}$ and $3400\,cm^{-1}$, which are assigned to the vibration of OH oscillators of three-coordinated hydrogen-bonded water, that is, less ordered "liquid-like" water, molecules and that of the four-coordinated hydrogen-bonded water, that is, highly ordered "ice-like" water molecules, respectively, in contrast to the Au thin film/0.05 M H_2SO_4 solution interface where the spectra were dominated by the peak corresponding to the "liquid-like" water at all potentials, showing that water molecules are more highly oriented at the Pt electrode than at the Au electrode. The SFG intensity of the OH-stretching region of interfacial water at a Pt/electrolyte solution interface exhibits a strong dependence on electrode potential. Parabolic behavior of SFG intensity was observed between -200 and 600 mV with a minimum around 200 mV, which is close to the pzc for a Pt electrode in $HClO_4$ electrolyte solution. In the Pt oxide formation region (more positive than 600 mV), SFG intensity due to interfacial water decreased.

5.4
Photoinduced Surface Dynamics of CO Adsorbed on a Platinum Electrode

5.4.1
Introduction

The dynamics of interactions between molecules and a surface such as vibrational excitations, energy exchange, and relaxation are of fundamental importance in surface science [60, 61]. The time scale of these processes is in the ps to fs regime. Recent development of short pulse laser techniques has enabled direct observation of ultrafast surface dynamics, not only to identify the surface species but also to probe the transient species generated by the pump pulse in real time [62–64]. Although most of the SFG studies so far have been concerned with the static structure of molecules at interfaces, a more important contribution of SFG spectroscopy should be its high time resolution and time-resolved SFG (TR-SFG) is expected to be one of the most powerful methods for observing and identifying transient states of surface adsorbates [65–68].

Adsorbed CO on a metal surface is one of the simplest adsorbates and has attracted significant interest within the areas of fundamental surface science, catalysis, and electrochemistry. An understanding of the oxidation mechanism of adsorbed CO is important to design and develop electrocatalysts for fuel cells [69–73] and the surface dynamics of adsorbed CO on electrode surfaces in electrolyte solutions is, therefore, very important.

TR-SFG seems to be an ideal tool to study the surface dynamics of adsorbed CO at solid/liquid interfaces. Although there are several reports of TR-SFG studies on an electrode, they are only of investigations of vibrational relaxation lifetime by IR excitation [34, 65, 67].

In the present study, we investigated the SFG response of CO adsorbed on a Pt electrode surface in an electrolyte solution upon irradiation of intense visible pulses with time resolution of about 20 ps.

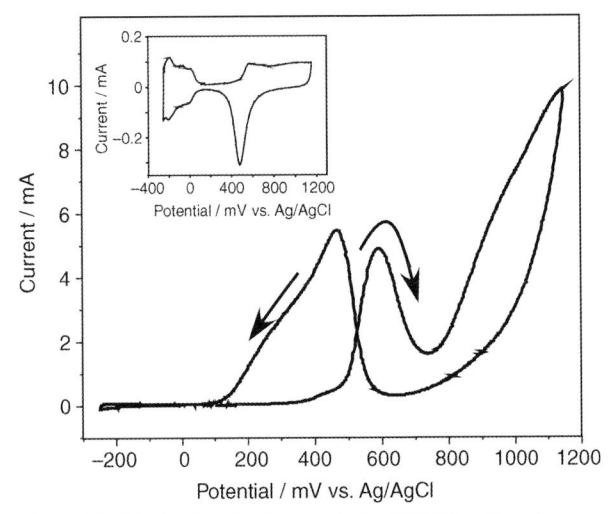

Figure 5.9 CVs of a Pt-poly electrode in 0.1 M HClO₄ with and without (inset) 0.1 M HCHO with sweep rate of 20 mV s⁻¹.

5.4.2
Results and Discussion

Figure 5.9 shows CVs of the Pt electrode in 0.1 M HClO$_4$ solution with and without (inset) 0.1 M HCHO. They are in good agreement with the results reported previously [74]. The hydrogen waves in the potential range between −250 and about 50 mV (inset) were suppressed in the solution containing HCHO, indicating the existence of a CO adlayer on the Pt electrode surface [72, 73].

SFG measurements were carried out in the potential range 0–300 mV, where the presence of adsorbed CO was expected, after a CO adlayer was formed at 0 V. Figure 5.10 shows SFG spectra of the Pt electrode in 0.1 M HClO$_4$ solution containing 0.1 M HCHO at various potentials. A peak centered at about 2055 cm^{-1} was observed, in agreement with results of previous studies by IR spectroscopy [75], and was assigned to the stretching vibration of CO (v_{CO}) adsorbed on a one-fold coordinated (atop) site of the Pt surface. As the potential became more positive, the peak position of this band shifted to a higher wavenumber by approximately 33 cm^{-1} V^{-1} up to 200 mV (Figure 5.10 inset), which also agrees with the previous IR results [75]. This shift has often been referred to as electrochemical Stark tuning. A decrease in intensity and a red shift of the SFG peak were observed at 300 mV, indicating the loss of adsorbed CO as a result of CO oxidation. When the potential was made more positive, the SFG peak was no longer observed (data not shown), indicating complete loss of the adsorbed CO from the Pt surface. The CO oxidation threshold potential observed by SFG in the present experiments was about 200 mV more negative than that expected from the CV (Figure 5.9). A similar discrepancy between the CO oxidation potential in CV and the potential of the disappearance of the SFG signal was reported previously [76, 77]. The difference in the data acquisition

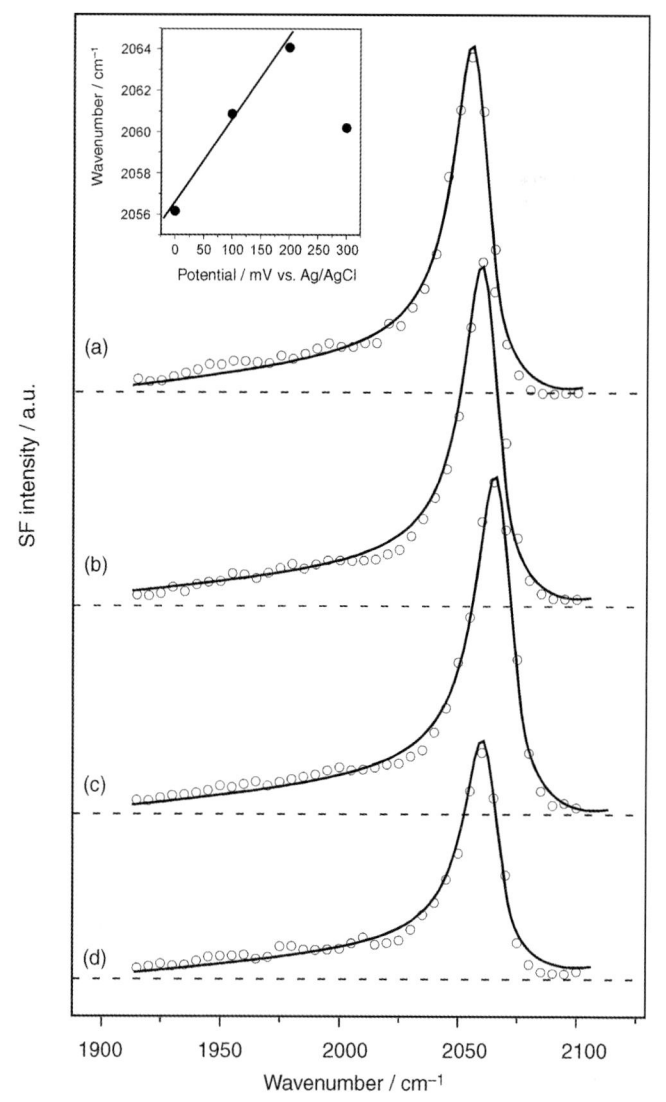

Figure 5.10 SFG spectra of the Pt surface in a solution of 0.1 M HCHO in 0.1 M HClO$_4$ at (a) 0, (b) 100, (c) 200, and (d) 300 mV.

time (SFG: 20–60 min for a spectrum at a given potential, CV: typically less than 1 min) is considered to be the origin of this discrepancy.

Figure 5.11 shows the temporal profile of the intensity change in the SFG signal at the peak of the ν_{CO} mode (2055 cm^{-1}) at 0 mV induced by visible pump pulse irradiation. The solid line is the least-squares fit using a convolution of a Gaussian function for the laser profile (FWHM = 20 ps) and a single exponential function for the recovery profile. The SFG signal fell to a minimum within about 100 ps and recovered

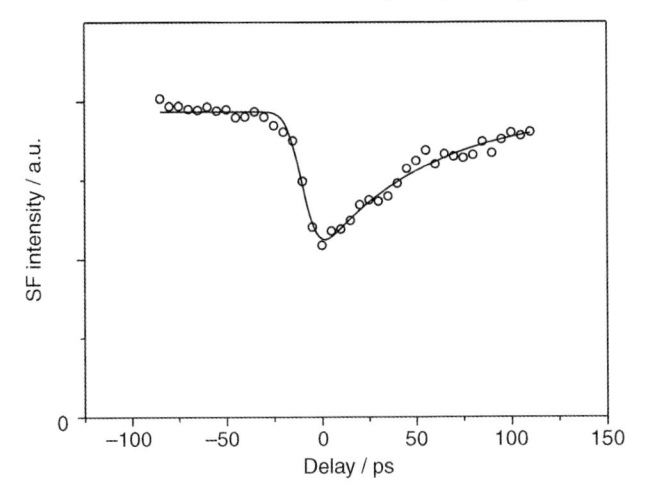

Figure 5.11 Temporal profile of SFG signal intensity at 2055 cm^{-1} at a potential of 0 mV. The solid line is the result of a least-squares fitting.

to the initial value. There was a linear correlation between the pump fluence and the intensity decrease induced by the pump pulse and no intensity change was observed when the pump fluence was less than that of the SFG probe (532 nm) pulse.

Figure 5.12 shows TR-SFG spectra at delay times of −80, 0, and 70 ps. The spectrum observed at −80 ps, that is, 80 ps before pumping, is the same as those observed without pumping (Figure 5.10), indicating that the irradiation-induced changes in the spectra were restored during the 0.1 s interval of pump pulse repetition (10 Hz). When the delay time was 0 ps, a maximum change in spectral features was observed. The height of the SFG peak at 2055 cm^{-1} was decreased, the peak position was slightly shifted to a lower wavenumber, and the peak was broadened. The values of FWHM derived from the fits are about 20 cm^{-1} and about 25 cm^{-1} at −80 ps and 0 ps delay, respectively.

At 0 ps, in addition to the changes in the peak at 2055 cm^{-1}, a new peak appeared at around 1980 cm^{-1}. One possible origin of the new peak is $v = 1 \rightarrow 2$ transition (hot band) of the stretching vibration of CO adsorbed on the Pt surface. Due to a vibrational anharmonicity, a hot band will give rise to a new peak in a lower wavenumber region than the fundamental peak ($v = 0 \rightarrow 1$). The frequency shift from the initial CO stretching band to the new broad peak observed in the present study was, however, about 88 cm^{-1}, which is much larger than the previously reported value for CO on a Pt surface (about 30 cm^{-1}) [78] and in the gas phase (about 27 cm^{-1}) [79]. Thus, it is unlikely that the new peak is due to excitation of the hot band generated by pump pulses. Another possible reason for the appearance of the new peak is the transient site migration of CO on the Pt surface. It has been reported that CO adsorbed on a multi-bonded [80] or asymmetric bridge site [81] gives a peak at about 1980 cm^{-1}, which is in agreement with the position of the transiently observed peak in the present study. Thus, it is reasonable to assume that the decrease

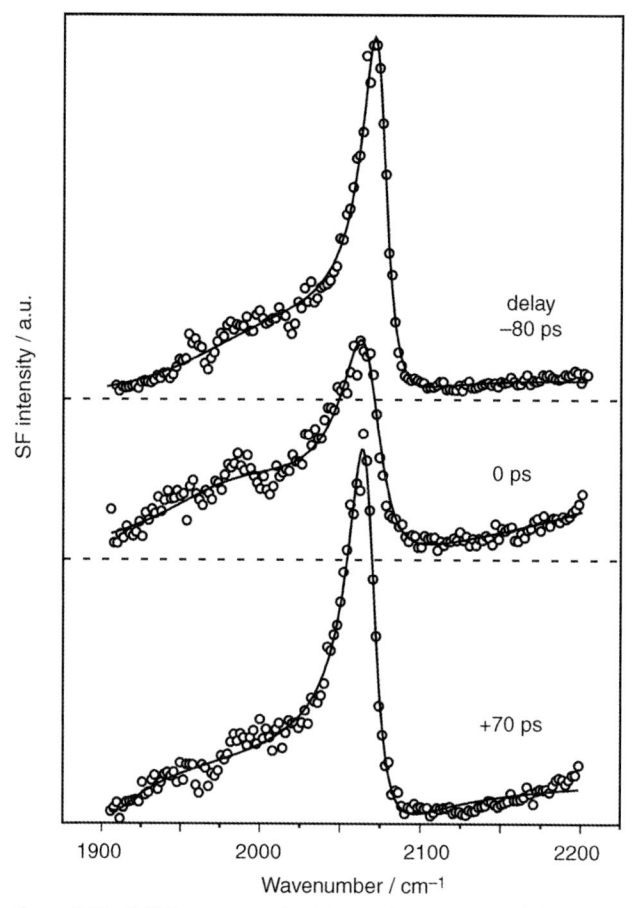

Figure 5.12 TR-SFG spectra in the CO-stretching region at delay times of −80, 0, and 70 ps.

in the peak at 2055 cm^{-1} and the appearance of the new peak at 1980 cm^{-1} were caused by reversible site migration of CO on the Pt surface from the on-top site to a multi-bonded or asymmetric bridge site induced by intense pump pulse irradiation.

The low-frequency shift and the broadening of the CO spectra at 0 ps suggest that the low-frequency modes of adsorbed CO, that is, stretching, frustrated rotation, and frustrated translation modes of Pt–CO, were thermally excited by pump pulses, as reported by Bonn et al. [82] Thus, it is concluded that the transient site migration of adsorbed CO on the Pt electrode surface was caused by a transient rise in the surface temperature of Pt induced by pump pulses.

5.4.3
Conclusions

TR-SFG measurements at a Pt electrode/electrolyte interface covered with a CO monolayer excited by the irradiation of picosecond visible pulses showed that the

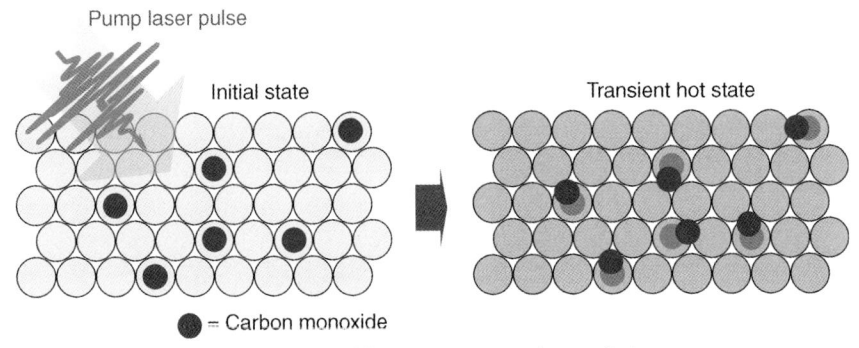

Figure 5.13 Transient CO migration induced by intense pump pulse irradiation.

population of on-top CO instantly decreased, accompanied by an increase in multi-bonded CO due to the transient temperature jump at the surface and the initial state was recovered within 100 ps, showing the transient reversible migration of CO molecules on the Pt surface under electrochemical conditions, as shown schematically in Figure 5.13.

5.5
Interfacial Water Structure at Polyvinyl Alcohol (PVA) Gel/Quartz Interfaces Investigated by SFG Spectroscopy

5.5.1
Introduction

Hydropolymer gel has been considered as a possible candidate for an artificial articular cartilage in artificial joints because it exhibits very low friction when it is in contact with a solid. The origin of such low friction is considered to be associated with the water absorbed in the gel [83–86], some of which is squeezed out from the gel under the load and serves as a lubricant layer between the gel and solid surface, resulting in hydrodynamic lubrication [87, 88]. Although the structural information about the interfacial water is important to understand the role of water for the low frictional properties of hydrogel in contact with a solid and the molecular structure of lubricants other than water at solid/solid interfaces have been investigated theoretically [89–91] and experimentally [92–98], no experimental investigations on water structure at gel/solid interfaces have been carried out due to the lack of an effective experimental technique.

Attractive or repulsive interaction between two solid surfaces should play an important role in the interfacial frictional behavior [87, 92–95]. From previous theoretical [89] and experimental investigations [87, 95], it was known that the attractive interaction result in a high friction and repulsive interaction results in low friction force. To characterize the interfacial molecular structure between two solids under electrostatic interaction is also important to elucidate the frictional properties of two solids.

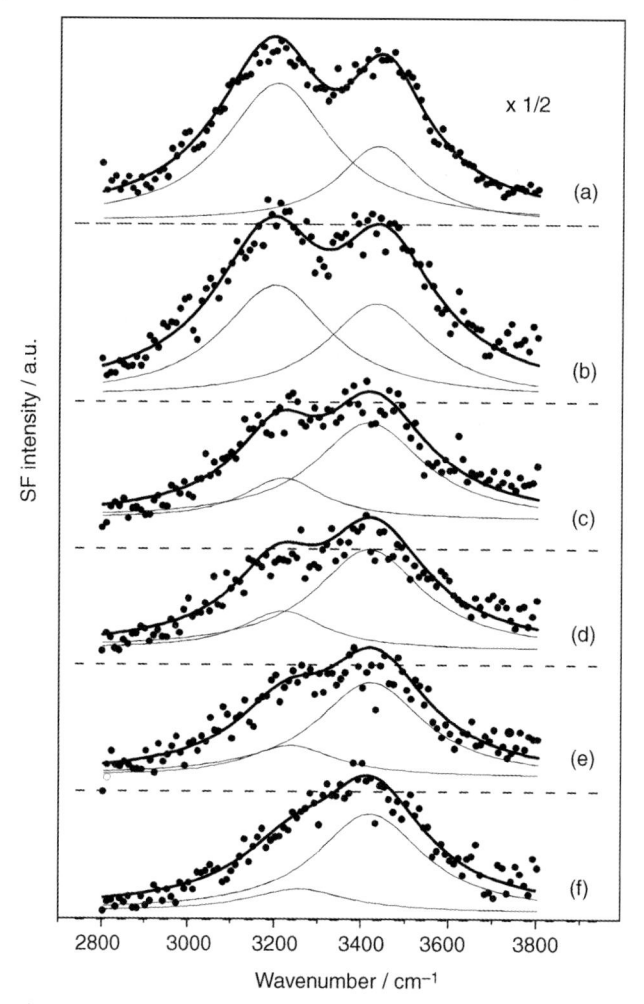

Figure 5.14 SFG spectra in the OH-stretching region (2800–3800 cm^{-1}) at a quartz surface in water before (a) and after contact of the PVA gel with applied pressure of (b) 0 (just in contact), (c) 0.2, (d) 0.4, (e) 0.6, and (f) 0.8 MPa.

Here, the structures of interfacial water at a fused quartz surface with and without contact of polyvinyl alcohol (PVA) were investigated by *in situ* SFG spectroscopy and their role in low friction between PVA and a fused quartz surface is discussed.

5.5.2
Results and Discussions

Figure 5.14 shows SFG spectra in the OH-stretching region (2800–3800 cm^{-1}) obtained at a quartz surface in water before (a) and after (b–f) the contact of PVA

5.5 Interfacial Water Structure at Polyvinyl Alcohol (PVA) Gel/Quartz Interfaces

with various applied pressures. Solid lines indicate least-square fits with "ice-like" and "liquid-like" water components using Eq. (5.1). At the quartz surface before the PVA gel was contacted (Figure 5.14a), two broad peaks at about $3200\,\text{cm}^{-1}$ and $3400\,\text{cm}^{-1}$ were observed. It is known that the higher the wavenumber, the lower the degree of hydrogen bonding. Based on IR studies on water clusters [50], the band at $3200\,\text{cm}^{-1}$ is attributed to the vibration of OH oscillators of three-coordinated hydrogen-bonded water molecules at the surface and that at $3400\,\text{cm}^{-1}$ to the four-coordinated hydrogen-bonded water molecules [6]. The intensity ratio between the two peaks can be considered as an index of the order of the interfacial water [51]. The SFG intensity decreased when PVA gel was in contact with the quartz surface. Weakening of SFG intensity is mostly due to the change in the Fresnel coefficients upon contact with the PVA gel [99].

The shape of the spectra also changed as the PVA gel was pressed toward the quartz surface and the pressure on the PVA gel was increased (Figure 5.14b–f). The SFG signal from the "liquid-like" water component became dominant as the pressure on the PVA gel surface was increased.

This trend is shown more clearly in Figure 5.15 where the intensity ratio between the two peaks, which can be considered as an indicator of the order of the interfacial water, is plotted against the pressure applied to the PVA gel.

The friction behavior of PVA gel and solid is strongly dependent on the nature of the solid surface. While the friction is low when the PVA gel is in contact with a quartz surface, which is hydrophilic, it is very high when the PVA gel is in contact with a

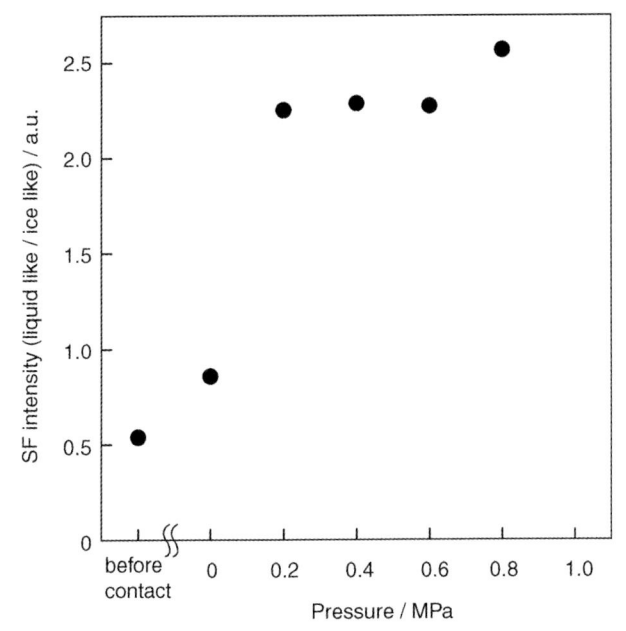

Figure 5.15 Effect of applied pressure on the intensity ratio between the SFG signals due to "ice-like" and "liquid-like" water components.

hydrophobic surface [100]. To further clarify the importance of water structure in interfacial friction behavior, SFG measurement was also carried out at an interface between PVA gel and a hydrophobic surface, which was prepared by modifying the quartz surface with a octadecyltrichlorosilane (OTS) monolayer [101].

Figure 5.16 shows SFG spectra in the region of 2800–3800 cm^{-1} obtained at the OTS modified quartz in water before (a, b) and after (c–f) the contact of PVA with various applied pressures. Solid lines indicate least-square fits with "ice-like" and "liquid-like" water components using Eq. (5.1), except for Figure 5.16a where fitting for CH-stretching peaks is also included. Before the PVA gel was contacted to the quartz surface (Figure 5.16a and b), a sharp peak was observed at 3680 cm^{-1} corresponding to non-hydrogen bonded OH, that is, free OH, in addition to a broad peak at about 3200 cm^{-1} with a shoulder at around 3400 cm^{-1}. A peak at 3400 cm^{-1} corresponding to the "liquid-like" water was significantly suppressed compared to the SFG spectrum of Figure 5.2a and the spectrum was similar to that of an ice/air interface, showing that the water molecules at the OTS/quartz interface are extremely well ordered. In addition to the peaks corresponding to OH-stretching, two large peaks were observed at 2879 and 2940 cm^{-1} (Figure 5.16a), which are attributed to the CH-stretching of the CH_3 group, indicating that the OTS monolayer maintained its highly ordered structure in water. These results are in agreement with those reported previously by our group [10]. No intensity change in the CH-stretching region was observed when pressure was applied to the PVA gel, although detailed discussion about these peaks is not possible because the SFG spectra were obtained in 10 cm^{-1} steps.

Although the total intensity of the SFG spectra decreased as the pressure on the PVA gel was increased, the intensity ratio between the peaks corresponding to "ice-like" water and "liquid-like" water was almost constant. Since the OTS-modified quartz surface was hydrophobic, the water squeezed from the bulk gel was "ice-like" at the PVA gel/OTS-modified quartz interface.

These results at the interfaces between the PVA gel and quartz surfaces, with and without modification by OTS, suggest that the weakly hydrogen bonded, that is, "liquid-like", water plays an important role for the low friction at the PVA gel/quartz interface.

5.5.3
Conclusions

The structure of water at the PVA/quartz interface was investigated by SFG spectroscopy. Two broad peaks were observed in the OH-stretching region at 3200 and 3400 cm^{-1}, due to "ice-like" and "liquid-like" water, respectively, in both cases. The relative intensity of the SFG signal due to "liquid-like" water increased when the PVA gel was pressed against the quartz surface. No such increase of the "liquid-like" water was observed when the PVA gel was contacted to the hydrophobic OTS-modified quartz surface where friction was high. These results suggest the important role of water structure for low friction at the polymer gel/solid interfaces.

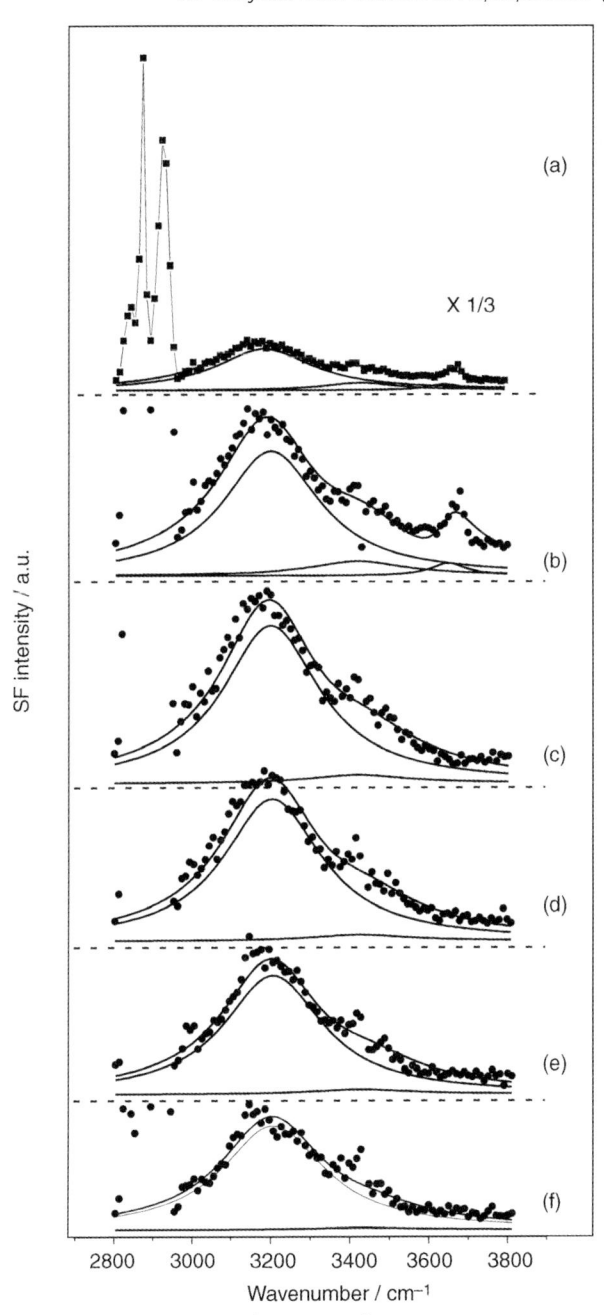

Figure 5.16 SFG spectra in the OH-stretching region (2800–3800 cm^{-1}) at the OTS-modified quartz surface in water before (a), (b) and after contact of the PVA gel with applied pressure of (c) 0 (just in contact), (d) 0.2, (e) 0.6, and (f) 0.8 MPa.

5.6
Hyper-Raman Spectroscopy

5.6.1
Selection Rules for Hyper-Raman Scattering

The dipole moment $\vec{\mu}$ induced in a molecule placed in an electromagnetic field is given by Eq. (5.1). As mentioned previously in Section 5.2, the third term $\overleftrightarrow{\beta}\vec{E}\vec{E}$ acts as the source of SFG. This SF radiation process may be accompanied by *inelastic* nonlinear scattering, which can be understood by analogy with Raman scattering. Such an inelastic scattering effect in the second-order NLO effects is called hyper-Raman scattering [102]. (Since practical hyper-Raman spectroscopy is usually carried out under the condition of input frequency-degeneracy, that is, SHG.) This is described by a first-order expansion of β in the normal coordinate of vibration Q:

$$\beta = \beta^0 + \sum_l (\partial\beta/\partial Q_l)_0 Q_l + \cdots \tag{5.19}$$

Similarly, the first-order expansion of the μ^0 and α of Eq. (5.1) is, respectively, responsible for IR absorption and Raman scattering. According to the parity, one can easily understand that selection rules for hyper-Raman scattering are rather similar to those for IR [17, 18]. Moreover, some of the silent modes, which are IR- and Raman-inactive vibrational modes, can be allowed in hyper-Raman scattering because of the nonlinearity. Incidentally, hyper-Raman-active modes and Raman-active modes are mutually exclusive in centrosymmetric molecules. Similar to Raman spectroscopy, hyper-Raman spectroscopy is feasible by visible excitation. Therefore, hyper-Raman spectroscopy can, in principle, be used as an alternative for IR spectroscopy, especially in *IR-opaque media* such as an aqueous solution [103]. Moreover, its spatial resolution, caused by the diffraction limit, is expected to be much better than IR microscopy.

Hyper-Raman spectroscopy is not a surface-specific technique while SFG vibrational spectroscopy can selectively probe surfaces and interfaces, although both methods are based on the second-order nonlinear process. The vibrational SFG is a combination process of IR absorption and Raman scattering and, hence, only accessible to IR/Raman-active modes, which appear only in non-centrosymmetric molecules. Conversely, the hyper-Raman process does not require such broken centrosymmetry. Energy diagrams for IR, Raman, hyper-Raman, and vibrational SFG processes are summarized in Figure 5.17.

5.6.2
Enhancement of Hyper-Raman Scattering Intensity

The hyper-Raman scattering cross section is extremely small, typically of the order of 10^{-65} cm^4 per molecule [24]. Therefore, an enhancement of signal intensity is essential in order to utilize this phenomenon as a practical spectroscopic tool in the field of molecular science. In a similar manner to the enhancement of Raman scattering

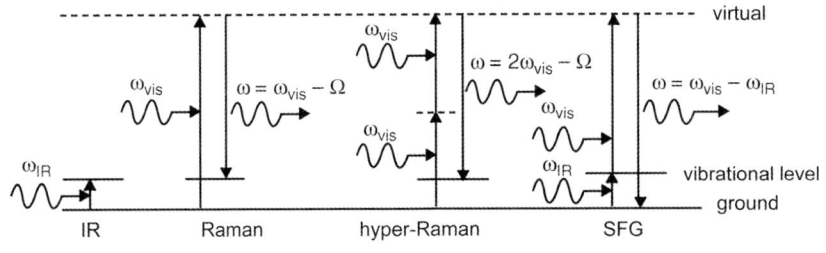

Figure 5.17 Energy diagrams of various vibrational spectroscopies.

intensities, there are two methods for the signal enhancement in hyper-Raman scattering: resonance hyper-Raman scattering [104, 105] and surface-enhanced hyper-Raman scattering [24]. The former gains intensities from resonances to electronic excited states of molecules. Therefore, this enhancement is practically useful only for centrosymmetric molecules [106]. (In the case of non-centrosymmetric molecules, Raman-active modes are dominantly enhanced via a Frank–Condon mechanism; both two-photon upward and one-photon downward transitions are allowed.) On the other hand, the latter gains intensities from plasmonic resonances at metal surfaces. Therefore, surface-enhanced hyper-Raman spectroscopy is surface-specific, although the enhancement is restricted by surface selection rules [102].

Figure 5.18 shows resonance hyper-Raman spectrum of fullerene C_{60} microcrystals [22] using pulsed excitation for which 2ω was near-resonant with the dipole-allowed $1T_{1u}$–$1A_g$ transition band (the input wavelength of 790 nm and the spectral-width of 14 cm^{-1} FWFM). According to group theory, C_{60} has 174

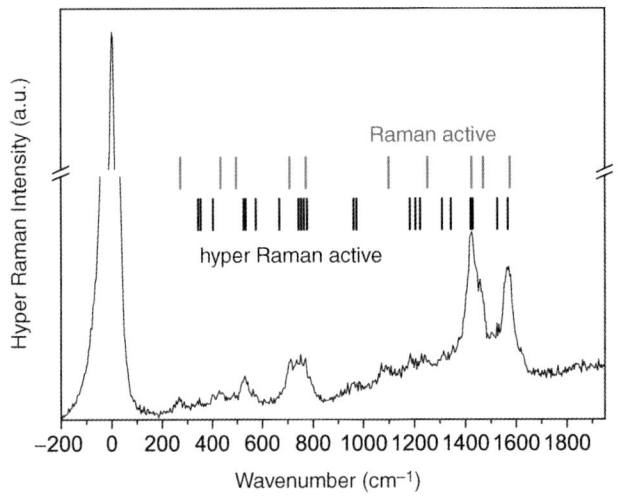

Figure 5.18 Resonance hyper-Raman spectrum of C_{60} microcrystals with theoretically calculated hyper-Raman-active modes (black bars) and Raman-active modes (gray bars).

internal degrees of freedom with 46 distinct vibrational coordinates having the representation.

$$\Gamma_{vib} = 2a_g + 1a_u + 3t_{1g} + 4t_{1u} + 4t_{2g} + 5t_{2u} + 6g_g + 6g_u + 8h_g + 7h_u \quad (5.20)$$

The numbers of IR- and Raman-active modes are 4 ($4t_{1u}$) and 10 ($2a_g + 8h_g$), respectively. On the other hand, hyper-Raman-active modes are all of the modes with u symmetry, including the silent modes. Compared with the theoretically calculated result, the expected modes are clearly seen in the spectrum. (The appearance of Raman-active modes is due to magnetic dipole contributions.)

In a similar manner to surface-enhanced Raman scattering, surface-enhancement of hyper-Raman scattering is a promising method to study adsorbed molecules on metal surfaces [24]. Based on recent developments in plasmonics, design and fabrication of metal substrates with high enhancement activities is now becoming possible [21]. Combination of the surface enhancement with the electronic resonances would also be helpful for the practical use of hyper-Raman spectroscopy. Development of enhanced hyper-Raman spectroscopy is awaited for the study of solid/liquid interfaces.

5.6.3
Conclusion

The background of hyper-Raman scattering has been described and several enhancement mechanisms presented. Unique features of hyper-Raman scattering have been demonstrated for C_{60}.

5.7
General Conclusion

Here we have described two second-order non-linear spectroscopies, SFG in detail and hyper-Raman scattering briefly.

SFG spectroscopy was applied to various solid/liquid interfaces and the following results were obtained. (i) SFG spectra in the OH-stretching region at a Pt/electrolyte solution interface showed two broad peaks at about 3200 and 3400 cm^{-1}, corresponding to "liquid-like" water and "ice-like" water molecules, respectively, in contrast to the Au thin film/0.05 M H_2SO_4 solution interface where the spectra were dominated by the peak corresponding to "liquid-like" water at all potentials, showing that water molecules are more highly oriented at the Pt electrode than at the Au electrode. (ii) SFG intensity in the OH-stretching region at a Pt/electrolyte solution interface exhibited parabolic behavior between -200 and 600 mV with a minimum around 200 mV, which is close to the pzc of a Pt electrode in $HClO_4$ electrolyte solution, and decreased again in the more positive potential region where Pt oxide was formed (more positive than 600 mV). (iii) The population of on-top CO at a Pt electrode/electrolyte interface instantly decreased, accompanied by an increase in multi-bonded CO, due to the transient temperature jump caused by the irradiation of picosecond visible pulses; the initial state was recovered within 100 ps. (iv) The relative intensity of the SFG signal at the PVA/quartz interface, where friction was low, due to the presence of "liquid-like" water rather than "ice-like" water, became higher when the PVA gel was pressed against the quartz surface while no such increase was observed when the PVA gel was contacted to the hydrophobic OTS-modified quartz surface where friction was high, suggesting the important role of water structure for low friction at polymer gel/solid interfaces.

The background of hyper-Raman scattering was described, several enhancement mechanisms were presented, and the unique features of hyper-Raman scattering were demonstrated for C_{60}.

Acknowledgments

This work was partially supported by Grants-in-Aid for Scientific Research (KAKEN-HI) in the Priority Area of "Molecular Nano Dynamics" (No. 16072202) from the Ministry of Education, Culture, Sports, Science, and Technology, Japan.

References

1 Shen, Y. R. (1989) Surface properties probed by second-harmonic and sum-frequency generation. *Nature*, **337**, 519–525.
2 Bain, C. D. (1995) Sum-frequency vibrational spectroscopy of the solid/liquid interface. *J. Chem. Soc. Faraday Trans.*, **91**, 1281–1296.
3 Vidal, F. and Tadjeddine, A. (2005) Sum-frequency generation spectroscopy of interfaces. *Rep. Prog. Phys.*, **68**, 1095–1127.
4 Buck, M. and Himmelhaus, M. (2001) Vibrational spectroscopy of interfaces by infrared-visible sum frequency generation. *J. Vac. Sci. Technol. A*, **19**, 2717–2736.
5 Richmond, G. L. (2002) Molecular bonding and interactions at aqueous surface as probed by vibrational sum frequency spectroscopy. *Chem. Rev.*, **102**, 2693–2724.
6 Gopalakrishnan, S., Liu, D., Allen, H. C., Kuo, M. and Shultz, M. J. (2006) Vibrational spectroscopic studies of aqueous interfaces: salts, acids, bases, and nanodrops. *Chem. Rev.*, **106**, 1155–1175.
7 Ataka, K. and Osawa, M. (1998) In situ infrared study of water-sulfate coadsorption on gold(111) in sulfuric acid solutions. *Langmuir*, **14**, 951–959.
8 Osawa, M., Tsushima, M., Mogami, H., Samjeske, G. and Yamakata, A. (2008) Structure of water at the electrified platinum-water interface: a study by surface-enhanced infrared absorption spectroscopy. *J. Phys. Chem. C*, **112**, 4248–4256.
9 Chen, Y. X. and Tian, Z. Q. (1997) Dependence of surface enhanced Raman scattering of water on the hydrogen evolution reaction. *Chem. Phys. Lett.*, **281**, 379–383.
10 Ye, S., Nihonyanagi, S. and Uosaki, K. (2001) Sum frequency generation (SFG) study of the pH-dependent water structure on a fused quartz surface modified by an octadecyltrichlorosilane (OTS) monolayer. *Phys. Chem. Chem. Phys.*, **3**, 3463–3469.

11 Uosaki, K., Yano, T. and Nihonyanagi, S. (2004) Interfacial water structure at as-prepared and UV-induced hydrophilic TiO_2 surfaces studied by sum frequency generation spectroscopy and quartz crystal microbalance. *J. Phys. Chem. B*, **108**, 19086–19088.

12 Nihonyanagi, S., Ye, S. and Uosaki, K. (2001) Sum frequency generation study on the molecular structures at the interfaces between quartz modified with amino-terminated self-assembled monolayer and electrolyte solutions of various pH and ionic strength. *Electrochim. Acta*, **46**, 3057–3061.

13 Nihonyanagi, S., Ye, S., Uosaki, K., Dreesen, L., Humbert, C., Thirty, P. and Peremans, A. (2004) Potential-dependent structure of the interfacial water on the gold electrode. *Surf. Sci.*, **573**, 11–16.

14 Noguchi, H., Okada, T. and Uosaki, K. (2008) SFG study on potential-dependent structure of water at Pt electrode/electrolyte solution interface. *Electrochim. Acta*, **53**, 6841–6844.

15 Noguchi, H., Okada, T. and Uosaki, K. (2008) Molecular structure at electrode/electrolyte solution interfaces related to electrocatalysis. *Faraday Discussion*, **140**, 125–137.

16 Noguchi, H., Minowa, H., Tominaga, T., Gong, J. P., Osada, Y. and Uosaki, K. (2008) Interfacial water structure at polymer gel/quartz interfaces investigated by sum frequency generation spectroscopy. *Phys. Chem. Chem. Phys.*, **10**, 4987–4993.

17 Cyvin, S. J., Rauch, J. E. and Decius, J. C. (1965) Theory of hyper-Raman effects (nonlinear inelastic light scattering): selection rules and depolarization ratios for the second-order polarizability. *J. Chem. Phys.*, **43**, 4083–4095.

18 Christie, J. H. and Lockwood, D. J. (1971) Selection rules for three- and four-photon Raman interactions. *J. Chem. Phys.*, **54**, 1141–1154.

19 Shimada, R., Kano, H. and Hamaguchi, H. (2008) Intensity enhancement and selective detection of proximate solvent molecules by molecular near-field effect in resonance hyper-Raman scattering. *J. Chem. Phys.*, **129**, 024505/1–024505/9.

20 Ikeda, K., Saito, Y., Hayazawa, N., Kawata, S. and Uosaki, K. (2007) Resonant hyper-Raman scattering from carbon nanotubes. *Chem. Phys. Lett.*, **438**, 109–112.

21 Ikeda, K., Takase, M., Sawai, Y., Nabika, H., Murakoshi, K. and Uosaki, K. (2007) Hyper-Raman scattering enhanced by anisotropic dimer plasmons on artificial nanostructures. *J. Chem. Phys.*, **127**, 111103/1–111103/4.

22 Ikeda, K. and Uosaki, K. (2008) Resonance hyper-Raman scattering of fullerene C_{60} microcrystals. *J. Phys. Chem. A.*, **112**, 790–793.

23 Leng, W., Woo, H. Y., Vak, D., Bazan, G. C. and Kelly, A. M. (2006) Surface-enhanced resonance Raman and hyper-Raman spectroscopy of water-soluble substituted stilbene and distyrylbenzene chromophores. *J. Raman Spectrosc.*, **37**, 132–141.

24 Kneipp, J., Kneipp, H. and Kneipp, K. (2006) Two-photon vibrational spectroscopy for biosciences based on surface-enhanced hyper-Raman scattering. *Proc. Natl. Acad. Sci. U.S.A.*, **103**, 17149–17153.

25 Bass, M., Franken, P. A., Hill, A. E., Peters, C. W. and Weinreich, G. (1962) Optical mixing. *Phys. Rev. Lett.*, **8**, 18.

26 Bloembergen, N. (1965) *Nonlinear Optics*, Benjamin, New York.

27 Simon, H. J., Mitchell, D. E. and Watson, J. G. (1974) Optical second-harmonic generation with surface plasmons in silver films. *Phys. Rev. Lett.*, **33**, 1531–1534.

28 Zhu, X. D., Shur, H. and Shen, Y. R. (1987) Surface vibrational spectroscopy by infrared-visible sum frequency generation. *Phys. Rev. B*, **35**, 3047–3050.

29 Hirose, C., Akamatsu, N. and Domen, K. (1992) Formulas for the analysis of the surface SFG spectrum and

transformation coefficients of cartesian SFG tensor components. *Appl. Spectrosc.*, **46**, 1051–1072.

30 Hirose, C., Akamatsu, N. and Domen, K. (1992) Formulas for the analysis of surface sum-frequency generation spectrum by CH stretching modes of methyl and methylene groups. *J. Chem. Phys.*, **96**, 997–1004.

31 Ward, R. N., Davies, P. B. and Bain, C. D. (1993) Orientation of surfactants adsorbed on a hydrophobic surface. *J. Phys. Chem.*, **97**, 7141–7143.

32 Bell, G. R., Bain, C. D. and Ward, R. N. (1996) Sum-frequency vibrational spectroscopy of soluble surfactants at the air/water interface. *J. Chem. Soc., Faraday Trans.*, **92**, 515–523.

33 G-Sionnest, P., Dumas, P., Chabal, Y. J. and Higashi, G. S. (1990) Lifetime of an adsorbate-substrate vibration: H on Si (111). *Phys. Rev. Lett.*, **64**, 2156–2159.

34 Peremans, A., Tadjeddine, A. and G-Sionnest, P. (1995) Vibrational dynamics of CO at the (100) platinum electrochemical interface. *Chem. Phys. Lett.*, **247**, 243–248.

35 Domen, K., Bandara, A., Kubota, J., Onda, K., Wada, A., Kano, S. S. and Hirose, C. (1999) SFG study of unstable surface species by picosecond pump–probe method. *Surf. Sci*, **427–428**, 349–357.

36 Backus, E. H. G., Eichler, A., Kleyn, A. W. and Bonn, M. (2005) Real-time obsevation of molecular motion on a surface. *Science*, **310**, 1790–1793.

37 Noguchi, H., Okada, T. and Uosaki, K. (2006) Photoinduced surface dynamics of CO adsorbed on a platinum electrode. *J. Phys. Chem. B*, **110**, 15055–15058.

38 Heinz, T. F. (1991) *Nonlinear Surface Electromagnetic Phenomena* (eds H-.E. Ponath and G. I. Stegeman), Elsevier Science Publishers B.V., Amsterdam, p. 353.

39 Bockris, J. O'M. and Potter, E. C. (1952) The mechanism of hydrogen evolution at nickel cathodes in aqueous solutions. *J. Chem. Phys.*, **20**, 614–628.

40 Mott, N. F. and Watts-Tobin, R. J. (1961) The interface between a metal and an electrolyte. *Electrochim. Acta*, **4**, 79–107.

41 Damaskin, B. B. and Frumkin, A. N. (1974) Potentials of zero charge, interaction of metals with water and adsorption of organic substances—III. The role of the water dipoles in the structure of the dense part of the electric double layer. *Electrochim. Acta*, **19**, 173–176.

42 Bockris, J. O'M., Gileadi, E. and Muller, K. (1967) A molecular theory of the charge dependence of competitive adsorption. *Electrochim. Acta*, **12**, 1301–1321.

43 Kunimatsu, K. and Bewick, A. (1986) Electrochemically modulated infrared spectroscopy of adsorbed water in the inner part of the double layer: part 1. Oxygen-hydrogen stretching spectra of water on gold in 1M perchloric acid. *Ind. J. Technol.*, **24**, 407–412.

44 Pettinger, B., Philpott, M. R. and Gordon, J. G. (1981) Contribution of specifically adsorbed ions, water, and impurities to the surface enhanced Raman spectroscopy (SERS) of Ag electrodes. *J. Chem. Phys.*, **74**, 934–940.

45 Zou, S. Z., Chen, Y. X., Mao, B. W., Ren, B. and Tian, Z. Q. (1997) SERS studies on electrode/electrolyte interfacial water I. Ion effects in the negative potential region. *J. Electroanal. Chem.*, **424**, 19–24.

46 Bewick, A. and Russell, J. W. (1982) Structural investigation by infra-red spectroscopy of adsorbed hydrogen on platinum. *J. Electroanal. Chem.*, **132**, 329–344.

47 Iwasita, T. and Xia, X. (1996) Adsorption of water at Pt(111) electrode in $HClO_4$ solutions. The potential of zero charge. *J. Electroanal. Chem.*, **411**, 95–102.

48 Toney, M. F., Howard, J. N., Richer, J., Borges, G. L., Gordon, J. G., Melroy, O. R., Wiesler, D. G., Yee, D. and Sorensen, L. B. (1994) Voltage-dependent ordering of water molecules at an electrode-electrolyte interface. *Nature*, **368**, 444–446.

49 Ito, M. and Yamazaki, M. (2006) A new structure of water layer on Cu(111) electrode surface during hydrogen evolution. *Phys. Chem. Chem. Phys.*, **8**, 3623–3626.

50 Buch, U. and Huisken, F. (2000) Infrared spectroscopy of size-selected water and methanol clusters. *Chem. Rev.*, **100**, 3863–3890.

51 Du, Q., Freysz, E. and Shen, Y. R. (1994) Vibrational spectra of water molecules at quartz/water interfaces. *Phys. Rev. Lett.*, **72**, 238–241.

52 Climent, V. et al. (1999) *Interfacial Electrochemistry* (ed. A. Wieckowski), Marcel Dekker, New York, chapter 26.

53 Cuesta, A. (2004) Measurement of the surface charge density of CO-saturated Pt(1 1 1) electrodes as a function of potential: the potential of zero charge of Pt(1 1 1). *Surf. Sci.*, **572**, 11–22.

54 Hamm, U. W., Kramer, D., Zhai, R. S. and Kolb, D. M. (1996) The pzc of Au(111) and Pt(111) in a perchloric acid solution: an ex situ approach to the immersion technique. *J. Electroanal. Chem.*, **414**, 85–89.

55 Nagy, G. and Heinzinger, K. (1992) A molecular dynamics study of water monolayers on charged platinum walls. *J. Electroanal. Chem.*, **327**, 25–30.

56 Akiyama, R. and Hirata, F. (1998) Theoretical study for water structure at highly ordered surface: Effect of surface structure. *J. Chem. Phys.*, **108**, 4904–4911.

57 Ward, R. N., Davis, P. B. and Bain, C. D. (1993) Orientation of surfactants adsorbed on a hydrophobic surface. *J. Phys. Chem.*, **97**, 7141–7143.

58 Sashikata, K., Furuya, N. and Itaya, K. (1991) In situ electrochemical scanning tunneling microscopy of single-crystal surfaces of Pt(111), Rh(111), and Pd(111) in aqueous sulfuric acid solution. *J. Vac. Sci. Technol. B.*, **9**, 457–464.

59 Damjanovic, A., Birss, V. I. and Boudreaux, D. S. (1991) Electron transfer through thin anodic oxide films during the oxygen evolution reactions at Pt electrodes. *J. Electrochem. Soc.*, **138**, 2549–2555.

60 Dai, H. L. and Ho, W. (eds) (1995) *Laser Spectroscopy and Photochemistry on Metal Surfaces*, World Scientific, Singapore.

61 Rubahn, H. G. (1999) *Laser Applications in Surface Science and Technology*, John Wiley & Sons, New York.

62 Noguchi, H., Yoda, E., Ishizawa, N., Kondo, J. N., Wada, A., Kobayashi, H. and Domen, K. (2005) Direct observation of unstable intermediate species in the reaction of trans-2-butene on ferrierite zeolite by picosecond infrared laser spectroscopy. *J. Phys. Chem B*, **109**, 17217–17223.

63 Yamakata, A., Uchida, T., Kubota, J. and Osawa, M. (2006) Laser-induced potential jump at the electrochemical interface probed by picosecond time-resolved surface-enhanced infrared absorption spectroscopy. *J. Phys. Chem. B*, **110**, 6423–6427.

64 Douhal, A. and Santamaria, J. (eds) (2002) *Femtochemistry and Femtobiology*, World Scientific, Singapore.

65 Schmidt, E. M. and G-Sionnest, P. (1996) Electrochemical tuning of the lifetime of the CO stretching vibration for CO/Pt (111). *J. Chem. Phys.*, **104**, 2438–2445.

66 Bonn, M., Funk, S., Hess, Ch., Denzler, D. N., Stampfl, C., Scheffler, M., Wolf, M. and Ertl, G. (1999) Phonon-versus electron-mediated desorption and oxidation of CO on Ru(0001). *Science*, **285**, 1042–1045.

67 Matranga, C. and G-Sionnest, P. (2000) Vibrational relaxation of cyanide at the metal/electrolyte interface. *J. Chem. Phys.*, **112**, 7615–7621.

68 Noguchi, H., Okada, T., Onda, K., Kano, S. S., Wada, A. and Domen, K. (2003) Time-resolved SFG study of formate on a Ni(111) surface under irradiation of picosecond laser pulses. *Surf. Sci.*, **528**, 183–188.

69 Batista, E. A., Malpass, G. R. P., Motheo, A. J. and Iwasita, A. T. (2004) New mechanistic aspects of methanol

oxidation. *J. Electroanal. Chem.*, **571**, 273–282.
70 Yajima, T., Uchida, H. and Watanabe, M. (2004) *In-situ* ATR-FTIR spectroscopic study of electro-oxidation of methanol and adsorbed CO at Pt-Ru alloy. *J. Phys. Chem. B*, **108**, 2654–2659.
71 Miki, A., Ye, S., Senzaki, T. and Osawa, M. (2004) Surface-enhanced infrared study of catalytic electrooxidation of formaldehyde, methyl formate, and dimethoxymethane on platinum electrodes in acidic solution. *J. Electroanal. Chem.*, **563**, 23–31.
72 Nakabayashi, S., Sugiyama, N., Yagi, I. and Uosaki, K. (1996) Dissociative adsorption dynamics of formaldehyde on a platinum electrode surface; one-dimensional domino? *Chem. Phys.*, **205**, 269–275.
73 Nakabayashi, S., Yagi, I., Sugiyama, N., Tamura, K. and Uosaki, K. (1997) Reaction pathway of four-electron oxidation of formaldehyde on platinum electrode as observed by *in situ* optical spectroscopy. *Surf. Sci.*, **386**, 82–88.
74 Mai, C. F., Shue, C.-H., Yang, Y.-C., OuYang, L.-Y., Yau, S.-L. and Itaya, K. (2005) Adsorption of formaldehyde on Pt (111) and Pt(100) electrodes: cyclic voltammetry and scanning tunneling microscopy. *Langmuir*, **21**, 4964–4970.
75 Nishimura, K., Ohnishi, R., Kunimatsu, K. and Enyo, M. (1989) Surface species produced on Pt electrodes during HCHO oxidation in sulfuric acid solution as studied by infrared reflection-absorption spectroscopy (IRRAS) and differential electrochemical mass spectroscopy (DEMS). *J. Electroanal. Chem.*, **258**, 219–225.
76 Dederichs, F., Friedrich, K. F. and Daum, W. (2000) Sum-frequency vibrational spectroscopy of CO adsorption on Pt(111) and Pt(110) electrode surfaces in perchloric acid solution: effects of thin-layer electrolytes in spectro-electrochemistry. *J. Phys. Chem. B*, **104**, 6626–6632.
77 Lu, G. Q., Lagutchev, A., Dlott, D. D. and Wieckowski, A. (2005) Quantitative vibrational sum-frequency generation spectroscopy of thin layer electrochemistry: CO on a Pt electrode. *Surf. Sci.*, **585**, 3–16.
78 Hess, Ch., Bonn, M., Funk, S. and Wolf, M. (2000) Hot-band excitation of CO chemisorbed on Ru(001) studied with broadband-IR sum-frequency generation. *Chem. Phys. Lett.*, **325**, 139–145.
79 Jakob, P. and Persson, B. N. J. (1998) Infrared spectroscopy of overtones and combination bands. *J. Chem. Phys.*, **109**, 8641–8651.
80 Peremans, A. and Tadjeddine, A. (1995) Spectroscopic investigation of electrochemical interfaces at overpotential by infrared-visible sum-frequency generation: Platinum in base and methanol-containing electrolyte. *J. Electroanal. Chem.*, **395**, 313–316.
81 Watanabe, S., Inukai, J. and Ito, M. (1993) Coverage and potential dependent CO adsorption on Pt(1111), (711) and (100) electrode surfaces studied by infrared reflection absorption spectroscopy. *Surf. Sci.*, **293**, 1–9.
82 Bonn, M., Funk, Ch. Hess. S., Miners, J. H., Persson, B. N. J., Wolf, M. and Ertl, G. (2000) Femtosecond surface vibrational spectroscopy of CO adsorbed on Ru(001) during desorption. *Phys. Rev. Lett.*, **84**, 4653–4656.
83 McCutchen, C. W. (1962) The frictional properties of animal joints. *Wear*, **5**, 1–17.
84 Oka, M., Noguchi, T., Kumar, P., Ikeuchi, K., Yamamuro, T., Hyon, S. H. and Ikeda, Y. (1990) Development of an artificial articular cartilage. *Clin. Mater.*, **6**, 361–381.
85 Ishikawa, Y., Hiratsuka, K. and Sasada, T. (2006) Role of water in the lubrication of hydrogel. *Wear*, **261**, 500–504.
86 Gong, J. P. (2006) Friction and lubrication of hydrogels—its richness and complexity. *Soft Matter*, **2**, 544–552.
87 Kagata, G. and Gong, J. P. (2007) Surface sliding friction of negatively charged

88 Gong, J. P., Higa, M., Iwasaki, Y., Katsuyama, Y. and Osada, Y. (1997) Friction of gels. *J. Phys. Chem. B*, **101**, 5487–5489.

89 Gong, J. P. and Osada, Y. (1998) Gel friction: A model based on surface repulsion and adsorption. *J. Chem. Phys.*, **109**, 8062–8068.

90 Persson, B. N. J. and Volokitin, A. I. (2002) Theory of rubber friction: Nonstationary sliding. *Phys. Rev. B*, **65**, 134106–134116.

91 Cisneros, S. E. Q. and Deiters, U. K. (2006) Generalization of the friction theory for viscosity modeling. *J. Phys. Chem. B*, **110**, 12820–12834.

92 Campbell, S. D. and Hillier, A. C. (1999) Nanometer-scale probing of potential-dependent electrostatic forces, adhesion, and interfacial friction at the electrode/electrolyte interface. *Langmuir*, **15**, 891–899.

93 Kagata, G., Gong, J. P. and Osada, Y. (2002) Friction of gels. 6. Effects of sliding velocity and viscoelastic responses of the network. *J. Phys. Chem. B*, **106**, 4596–4601.

94 McGuiggan, P. M. (2008) Stick slip contact mechanics between dissimilar materials: Effect of charging and large friction. *Langmuir*, **24**, 3970–3976.

95 Gong, J. P., Kagata, G. and Osada, Y. (1999) Friction of gels. 4. Friction on charged gels. *J. Phys. Chem. B*, **103**, 6007–6014.

96 Eisert, F., Gurka, M., Legant, A., Buck, M. and Grunze, M. (2000) Detection of molecular alignment in confined films. *Science*, **287**, 468–470.

97 Berg, O. and Kleneman, D. (2003) Vibrational spectroscopy of mechanically compressed monolayers. *J. Am. Chem. Soc.*, **125**, 5493–5500.

98 Briscoe, W. H., Titmuss, S., Tiberg, F., Thomas, R. K., McGillivray, D. J. and Klein, J. (2006) Boundary lubrication under water. *Nature*, **444**, 191–194.

99 Wang, J., Woodcock, S. E., Buck, S. M., Chen, C. and Chen, Z. (2001) Different surface-restructuring behaviors of poly(methacrylate)s detected by SFG in water. *J. Am. Chem. Soc.*, **123**, 9470–9471.

100 Tominaga, T., Takedomi, N., Biederman, H., Furukawa, H., Osada, Y. and Gong, J. P. (2008) Effect of substrate adhesion and hydrophobicity on hydrogel friction. *Soft Matter*, **4**, 1033–1040.

101 The friction coefficient at PVA gel/OTS modified quartz is 0.300 while that at PVA gel/unmodified quartz is 0.076. Frictions were measured by using a rheometer (ARES, TA instruments) as a function of sliding velocity in water and these values were calculated from the experimental results of lowest sliding velocity, 7.5×10^{-6} m/s.

102 Denisov, V. N., Mavrin, B. N. and Podobedov, V. B. (1987) Hyper-Raman scattering by vibrational excitations in crystals, glasses and liquids. *Phys. Rep.*, **151**, 1–92.

103 Shimada, R., Kano, H. and Hamaguchi, H. (2006) Hyper-Raman microspectroscopy: a new approach to completing vibrational spectral and imaging information under a microscope. *Opt. Lett.*, **31**, 320–322.

104 Mizuno, M., Hamaguchi, H. and Tahara, T. (2002) Observation of resonance hyper-Raman scattering of all-trans-retinal. *J. Phys. Chem. A*, **106**, 3599–3604.

105 Shoute, L. C. T., Bartholomew, G. P., Bazan, G. C. and Kelley, A. M. (2005) Resonance hyper-Raman excitation profiles of a donor-acceptor substituted distyrylbenzene: one-photon and two-photon states. *J. Chem. Phys.*, **122**, 184508.

106 Chung, Y. C. and Ziegler, L. D. (1988) The vibronic theory of resonance hyper-Raman scattering. *J. Chem. Phys.*, **88**, 7287–7294.

6
Fourth-Order Coherent Raman Scattering at Buried Interfaces
Hiroshi Onishi

6.1
Why Buried Interfaces?

Interfaces between two different media provide a place for conversion of energy and materials. Heterogeneous catalysts and photocatalysts act in vapor or liquid environments. Selective conversion and transport of materials occurs at membranes of biological tissues in water. Electron transport across solid/solid interfaces determines the efficiency of dye-sensitized solar cells or organic electroluminescence devices. There is hence an increasing need to apply molecular science to buried interfaces.

However, analyses of the interface surrounded by some medium are not easy. When an interface of interest is exposed to a vacuum, electron-based or ion-based methods are available to determine the chemical composition and molecular structure of the top layers. The charged particles with limited penetration range result in a good vertical resolution. Buried interfaces are beyond the range of penetration. Photons, an alternative class of probe particles, have better ability for penetration. When the linear response to the incident electric field is analyzed, the vertical resolution is limited to the order of the wavelength, which is greater than the thickness of the top layers.

Sum-frequency (SF) spectroscopy [1] has been used to achive vertical resolutions much better than the wavelength. Sum-frequency light is generated at an interface irradiated with infrared (IR) and visible light. The probability of sum-frequency generation is governed by a second-order susceptibility $\chi^{(2)}$ to be zero in any medium with inversion symmetry. The second-order transition is allowed at interfaces where the symmetry breaks. The generation probability, when allowed, is enhanced by the vibrational resonance of the IR light, as shown in Figure 6.1a. The SF light intensity as a function of the IR wavenumber provides a vibrational spectrum at the interface. Vibrational SF spectroscopy is successful in probing interfaces exposed to vapor [2–5]. Infrared light is still sensitive to absorption by condensed media. Access to interfaces buried in liquids is achieved by irradiating the probe light from the side of a weak IR-absorptive material [6–9].

Molecular Nano Dynamics, Volume I: Spectroscopic Methods and Nanostructures
Edited by H. Fukumura, M. Irie, Y. Iwasawa, H. Masuhara, and K. Uosaki
Copyright © 2009 WILEY-VCH Verlag GmbH & Co. KGaA, Weinheim
ISBN: 978-3-527-32017-2

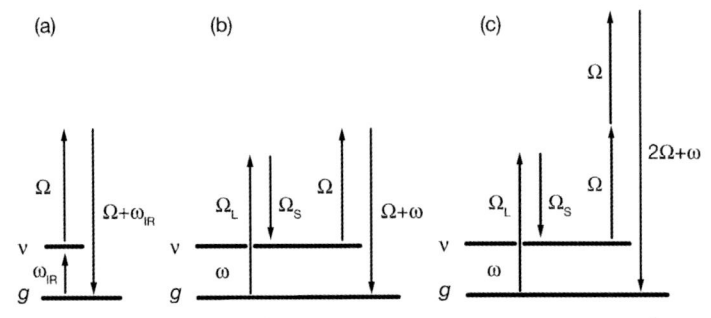

Figure 6.1 Nonlinear optical responses. (a) Second-order SF generation, the transition probability is enhanced when the IR light is resonant to the transition from the ground state g to a vibrational excited state v. ω is the angular frequency of the vibration. (b) Third-order coherent Raman scheme, the vibrational coherence is generated via impulsive stimulated Raman excitation. Ω_L and Ω_S are the high-frequency and low-frequency components of the pump light pulse. A probe pulse of frequency Ω interacts with the coherence to present the optical response of the fundamental frequency $\Omega + \omega \approx \Omega$. (c) Fourth-order coherent Raman scattering, the optical response of the second harmonic frequency $2\Omega + \omega \approx 2\Omega$ is modulated by the vibrational coherence.

The author has proposed application of fourth-order coherent Raman scatterring to provide IR-free observation of vibrations at buried interfaces. This method was originally developed in 1997 by Chang, Xu and Tom to observe coherent vibrations of the GaAs lattice [10]. The vibrations are coherently pumped with near-infrared light via a stimulated Raman transition instead of the IR resonant transition used in the SF scheme. Another ability of this Raman-based method is to access low-frequency vibrations. It is difficult to prepare low-frequency IR light compatible with SF generation. A specifically modified tabletop laser light source [11] or a free-electron laser facility [12] is required to observe an SF spectrum of wavenumbers below 1000 cm^{-1}. Low-frequency vibrations are important in dissipating the excess energy released in endothermic reactions. Lateral and vertical transport of molecules at an interface is initiated by multiple excitation of low-frequency modes, including frustrated rotations and frustrated translations.

In the current chapter, the principles of Raman excitation and interface-selective detection of vibrational coherence are described, including applications to air/liquid, liquid/liquid, air/solid interfaces, and an organic submonolayer.

6.2
Optical Transitions

A light pulse of a center frequency Ω impinges on an interface. Raman-active modes of nuclear motion are coherently excited via impulsive stimulated Raman scattering, when the time width of the pulse is shorter than the period of the vibration. The ultrashort light pulse has a finite frequency width related to the Fourier transformation of the time width, according to the energy–time uncertainty relation.

When the full width at half maximum (fwhm) of a Gaussian pulse is 20 fs, its frequency width is 740 cm^{-1} as the fwhm. Frequency components Ω_L and Ω_S are present in the pulse and are used to generate the vibrational coherence, where $\Omega_L - \Omega_S$ is equal to the vibration frequency ω.

Another light pulse of frequency Ω comes at a time delay t_d and interacts with the vibrationally excited molecules. The intensity of the probe light transmitted through the interface is modulated as a function of the delay. The modulation is Fourier-transformed to provide the frequency and phase of the vibrational coherence. The decay of the coherence is traced with the modulation amplitude. This series of optical transitions shown in Figure 6.1b contains three incident electric fields (Ω_L, Ω_S, and Ω) and offers bulk-sensitive, time-domain detection of vibrational coherence. The whole portion of the material having interacted with the pump pulse contributes to the generation of the response. This method is known as third-order coherent Raman spectroscopy in the time domain and has been successfully applied to bulk liquids [13], bulk solids [14, 15], and molecular submonolayers [16, 17].

To ensure interface-selective detection of the Raman-pumped vibrational coherence, one more incident electric field is required. A fourth-order optical response is thereby generated. The requirement is fulfilled by observing the second harmonic (SH) light generated at the interface, instead of the transmitted fundamental light. The SH probe scheme contains four incident electric fields (Ω_L, Ω_S, Ω, and Ω) as shown in Figure 6.1c. The two-photon transition probing the coherence is equivalent to a hyper-Raman scattering from v to g [18]. The cross section of the pump-and-probe transitions is therefore proportional to the product of the Raman tensor for the pump from g to v, and the hyper-Raman tensor for the probe from v to g.

When two or more vibrational modes are excited, the fourth-order response field E_{fourth} is presented as the sum of exponentially decayed modulations,

$$E_{fourth}(t_d, 2\Omega) \propto \sum_v A_v \cos(\omega_v t_d + \varphi_v) \exp(-t_d/T_v) \tag{6.1}$$

where A_v, ω_v, φ_v, and T_v are the amplitude, frequency, phase, and dephasing time of each mode. The time-domain response is Fourier-transformed to a frequency spectrum of the fourth-order susceptibility, $\chi^{(4)}(\omega_v)$. The efficiency of the impulsive stimulated Raman excitation is enhanced when an electronic transition from g to an electronic state e is resonant with the photon energy. In the on-resonance extreme a cosine-like coherence is generated with $\varphi_v = 0$ or π. In the out-of-resonance extreme a sine-like coherence is expected with $\varphi_v = \pi/2$ or $3\pi/2$. This presentation is analogous to what has been established in third-order Raman spectroscopy [19].

In addition to the fourth-order response field E_{fourth}, the probe light generates two SH fields of the same frequency 2Ω, the pump-free SH field $E_0(2\Omega)$, and the pump-induced non-modulated SH field $E_{non}(t_d, 2\Omega)$. The ground-state population is reduced by the pump irradiation and the SH field is thereby weakened. The latter term $E_{non}(t_d, 2\Omega)$ is a virtual electric field to represent the weakened SH field. Time-resolved second harmonic generation (TRSHG) has been applied to observe $E_{non}(t_d, 2\Omega)$ with a picosecond time resolution [20–25]. The fourth-order field interferes with the two SH fields to be detected in a heterodyned form.

The pump-affected intensity of SH light $I(t_d, 2\Omega)$ is given by,

$$\frac{I(t_d, 2\Omega)}{I_0(2\Omega)} = \frac{|E_0(2\Omega) + E_{non}(t_d, 2\Omega)e^{i\Phi} + E_{fourth}(t_d, 2\Omega)e^{i\phi}|^2}{|E_0(2\Omega)|^2} \quad (6.2)$$

with respect to the pump-free intensity $I_0(2\Omega)$. Φ is the phase shift of $E_{non}(t_d, 2\Omega)$ relative to $E_0(2\Omega)$ and assumed to be zero, since E_{non} and E_0 are generated in a common, second-order transition. ϕ represents the phase shift of E_{fourth} relative to $E_0(2\Omega)$. With E_{fourth} being much smaller than the other terms, Eq. (6.2) is simplified to,

$$E_{fourth}(t_d, 2\Omega) E_{second}(t_d, 2\Omega) \propto I(t_d, 2\Omega) - I_{second}(t_d, 2\Omega) \quad (6.3)$$

where $E_{second}(t_d, 2\Omega) = E_0(2\Omega) + E_{non}(t_d, 2\Omega)$. $I_{second}(t_d, 2\Omega)$ is the SH light intensity relevant to E_{second} and determined with a non-oscillatory numerical function fitted to the observed $I(t_d, 2\Omega)$. The right-hand side of Eq. (6.3) represents the intensity modulation of the SH light, $I_{fourth}(t_d, 2\Omega)$, which is the experimetally determined quantity. On the left-hand side of the equation the fourth-order field E_{fourth} is multiplied by E_{second}. The intensity modulation is proportional to the amplitude of E_{fourth} and hence to the number of coherently vibrating molecules or atoms.

Hirose et al. [26] proposed a homodyne scheme to achieve the background-free detection of the fourth-order field. With pump irradiation in a transient grating configuration, the fourth-order field propagates in a direction different from that of the second-order field because of different phase match conditions. The fourth-order field is homodyned to make $I_{fourth}(t_d, 2\Omega)$ and spatially filtered from the second-order response $I_{second}(t_d, 2\Omega)$.

6.3
Experimental Scheme

Figure 6.2 illustrates the spectrometer used in our laboratory. A noncollinear optical parametric amplifier (TOPAS-white, Quantronix) is pumped by a Ti:sapphire regenerative amplifier (Hurricane, Spectra Physics, 800 nm, 90 fs, 1 kHz). The time width of the amplified light pulse is less than 20 fs, and the wavelength is tunable at 500–750 nm. A p-polarized pump and p-polarized probe pulses are focused at an interface with an incident angle θ of $50°$. The spot diameter of the focused beams is 0.1 mm. The p-polarized SH light beam emitted to the reflected direction is conducted to a photomultiplier tube. The multiplier output is gated with a boxcar integrator and sent to a PC on a pulse-to-pulse basis. The pump pulse is chopped at 500 Hz. The pump-on signal and pump-off signal are separately accumulated and the former is divided by the latter. The time origin is determined by monitoring the SH light intensity. A more detailed description is available [27].

The time resolution of the instrument determines the wavenumber-dependent sensitivity of the Fourier-transformed, frequency-domain spectrum. A typical response of our spectrometer is 23 fs, and a Gaussian function having a half width

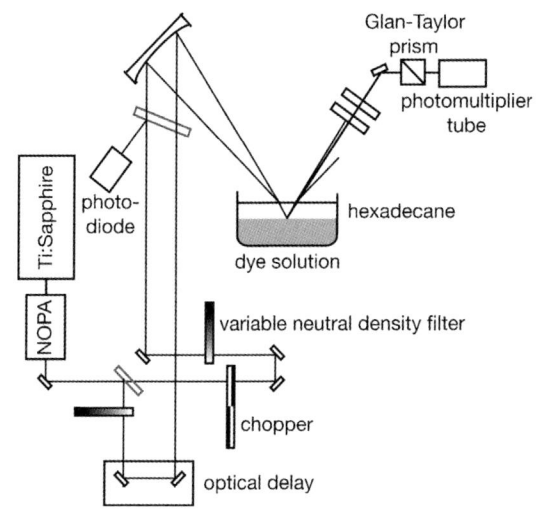

Figure 6.2 A fourth-order coherent Raman spectrometer constructed with a Ti:sapphire regenerative amplifier (Ti:sapphire) and noncollinear optical parametric amplifier (NOPA).

at half maximum (hwhm) of 640 cm^{-1} centered at 0 cm^{-1} represents the wavenumber-dependent sensitivity. Frequency spectra are presented in the following sections without correcting the sensitivity.

6.4
Application to a Liquid Surface

The fourth-order coherent Raman spectrum of a liquid surface was observed by Fujiyoshi et al. [28]. The same authors later reported a spectrum with an improved signal-to-noise ratio and different angle of incidence [27]. A water solution of oxazine 170 dye was placed in air and irradiated with light pulses. The SH generation at the oxazine solution was extensively studied by Steinhurst and Owrutsky [24]. The pump and probe wavelength was tuned at 630 nm to be resonant with the one-photon electronic transition of the dye. The probability of the Raman transition to generate the vibrational coherence is enhanced by the resonance. The efficiency of SH generation is also enhanced.

With the resonance to the electronic transition, the ground-state population is partially depleted by the pump irradiation and restored with the time delay. The raw intensity of SH light was accordingly damped at $t_d = 0$ and recovered in picoseconds, as seen in Figure 6.3a. Intensity modulation due to the vibrational coherence was superimposed on the non-modulated evolution as expected from Eq. (6.3). The coherence continued for picoseconds on this solution surface. The non-modulated component $I_{second}(t_d, 2\Omega)$ was fitted with a multiexponential

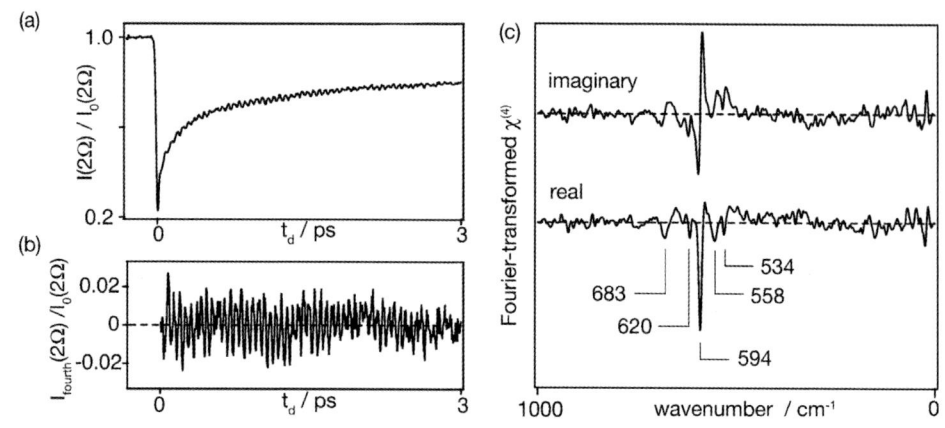

Figure 6.3 Vibrational coherence at a solution surface.
(a) The raw SH intensity, (b) the modulated component, and (c) the Fourier-transformed spectrum. The surface was irradiated with p-polarized pump (5 mJ cm^{-2}) and p-polarized probe (2.5 mJ cm^{-2}) pulses.

function. The modulated component $I_{\text{fourth}}(t_d, 2\Omega)$ was determined by subtracting the fitted $I_{\text{second}}(t_d, 2\Omega)$ from the raw intensity.

$I_{\text{fourth}}(t_d, 2\Omega)$ was multiplied with a window function and then converted to a frequency-domain spectrum via Fourier transformation. The window function determined the wavenumber resolution of the transformed spectrum. Figure 6.3c presents the spectrum transformed with a resolution of 6 cm^{-1} as the fwhm. Negative, symmetrically shaped bands are present at 534, 558, 594, 620, and 683 cm^{-1} in the real part, together with dispersive shaped bands in the imaginary part at the corresponding wavenumbers. The band shapes indicate the phase of the fourth-order field ϕ to be π. Cosine-like coherence was generated in the five vibrational modes by an impulsive stimulated Raman transition resonant to an electronic excitation.

The wavenumbers of the observed bands are identical with those of the spontaneous Raman spectrum of the solution and oxazine solid [27]. The impulsive stimulated Raman transition may initiate coherent vibrations in the electronic excited state. However, there was no sign of the excited-state vibrations superimposed on the ground-state bands in the spectrum of Figure 6.3.

6.5
Application to a Liquid/Liquid Interface

A 0.2-mm thick hexadecane layer was placed on the oxazine solution. The vibrational coherence at the hexadecane/solution interface was pump–probed in a similar manner [27]. The light pulses traveled in the hexadecane layer and experienced group velocity dispersion before arriving at the interface. This undesired dispersion

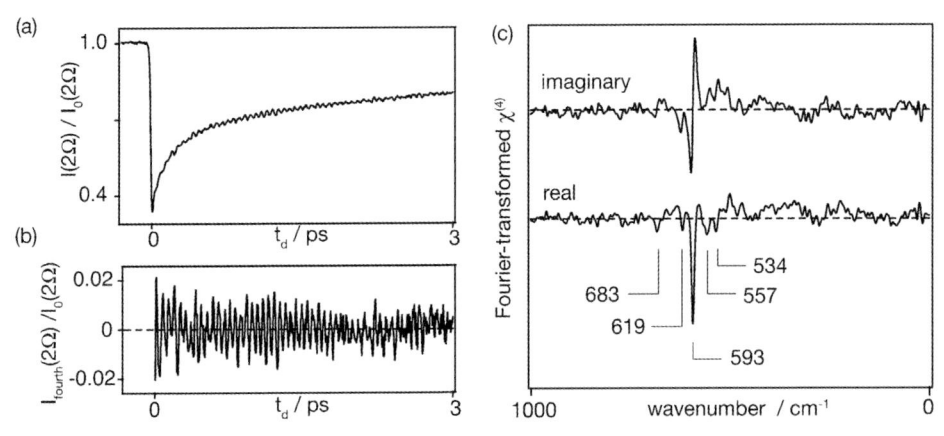

Figure 6.4 Vibrational coherence at a liquid/liquid interface. (a) The raw SH intensity, (b) the modulated component, and (c) the Fourier-transformed spectrum. A 0.2-mm thick hexadecane layer was placed on the oxazine solution of Figure 6.3. The interface was irradiated with p-polarized pump (5 mJ cm^{-2}) and p-polarized probe (2.5 mJ cm^{-2}) pulses of a 630-nm wavelength.

was compensated by adding an extra negative chirp in the noncollinear amplifier. As shown in Figure 6.4, the raw SH intensity and the Fourier-transformed spectrum were identical to those observed at the air/solution interface, being insensitive to the hexadecane overlayer. Vibrations in a chromophore are thought to be sensitive to the solvent structure around the chromophore [6, 7, 23]. This was not the case for this particular chromophore. The spontaneous Raman spectrum of oxazine was observed in the bulk solution and bulk solid. Dye monomers are equilibrated with dimerized species in the solution [24]. The solvated monomer, solvated dimer, and solid dye are in different dielectric environments. The observed Raman bands are, nevertheless, identical with the two spectra shown in Figures 6.3 and 6.4. It is, hence, not surprising that the oxazine vibrations are insensitive to the interface composition.

6.6
Applications to Solid Surfaces

A number of solid compounds have been examined with this time-domain method since the first report of coherent phonons in GaAs [10]. Coherent phonons were created at the metal/semiconductor interface of a GaP photodiode [29] and stacked GaInP/GaAs/GaInP layers [30]. Cesium-deposited [31–33] and potassium-deposited [34] Pt surfaces were extensively studied. Manipulation of vibrational coherence was further demonstrated on Cs/Pt using pump pulse trains [35–37]. Magnetic properties were studied on Gd films [38, 39].

Vibrational coherence at a TiO$_2$ surface covered with different organic molecules is described in this section. Titanium oxide is a wide-bandgap semiconductor. When buried in some medium a wide range of applications are known including photocatalysts, solar cells, hydrophilic coatings. The (1 1 0) plane of rutile polymorph is the most extensively studied single-crystalline surface of metal oxide [40]. Atomically flat surfaces of (1 1 0) orientation are routinely prepared and modified with chemisorbed formate [41], acetate [42], benzoate [43], fluorescein [44], N3 dye [45, 46] and so on. The (1 1 0) surface with and without adsorbates is characterized with various techniques including probe microscopes [47]. The vacuum-prepared surface may be contaminated when taken out of the vacuum chamber and exposed to laboratory air. White et al. [48] found that a trimethylacetate (TMA) monolayer with hydrophobic *tert*-butyl groups exposed to the environment passivates the surface in air.

The vibrational coherence at the TMA-covered surface was observed in air with pump and probe pulses of a 630-nm wavelength [49]. This red light is out of resonance with the bandgap excitation of TiO$_2$ (3 eV). The coherent response presented in panels Figure 6.5a and b decayed much more rapidly than the modulation at the solution interfaces described in the preceding sections. Four major bands are recognized at 180, 357, 444, and 826 cm^{-1} in the Fourier-transformed spectrum of Figure 6.5c. The wavenumbers of the first, third, and fourth bands agreed with those of Raman-active bulk phonon bands. In the bulk rutile having D_{4h} symmetry a Raman-active mode is inactive for hyper-Raman transitions [18]. Hence, the fourth-order transition of the three bands is forbidden in the

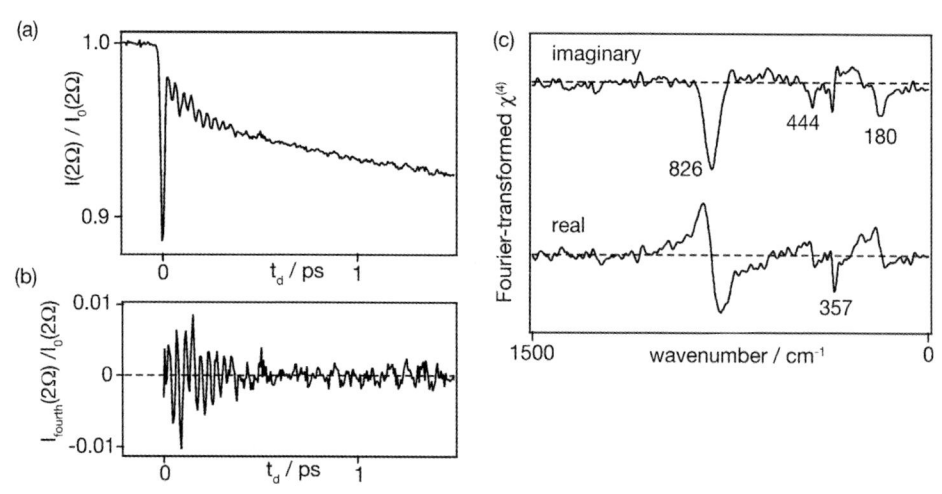

Figure 6.5 Vibrational coherence at a TiO$_2$(110) surface covered with TMA monolayer. (a) The raw SH intensity, (b) the modulated component, and (c) the Fourier-transformed spectrum. The TMA-covered surface was irradiated in air with p-polarized pump (14 mJ cm^{-2}) and p-polarized probe (6 mJ cm^{-2}) pulses.

bulk and should occur at the surface. The shape of the three bands is negative and symmetric in the imaginary part, while dispersive in the real part. The phase of the fourth-order field ϕ relative to E_{second} is $3\pi/2$.

The most intense 826-cm^{-1} band is broader than the other bands. The broadened band suggests a frequency distribution in the observed portion of the surface. Indeed, the symmetric peak in the imaginary part of the spectrum is fitted with a Gaussian function rather than with a Lorenz function. The bandwidth was estimated to be 56 cm^{-1} by considering the instrumental resolution, 15 cm^{-1} in this particular spectrum. This number is larger than the intrinsic bandwidth of the bulk modes [50]. Lanz and Corn [51] proposed a 20-nm thick space charge layer on the TiO_2 surface. When the fourth-order response with our TMA-covered surface is generated in the space charge layer, the broad width of the 826-cm^{-1} band is understood as a depth-dependent wavenumber of the lattice vibration.

Bulk phonon modes are absent in wave numbers near 357 cm^{-1}, the center-frequency of the second band. According to electron energy loss studies done in a vacuum [52, 53], TMA-free $TiO_2(110)$ surfaces exhibit surface optical phonons at 370–353 cm^{-1}. The 357-cm^{-1} band is related to the surface optical phonons.

The four bands observed on the TMA-covered TiO_2 surface are lattice vibrations of the substrate. Molecular vibration of TMA was not observed, probably because TMA has no resonance with the fundamental wavelength (630 nm) and SH wavelength (315 nm) of the pump and probe pulses. By putting p-nitrobenzoate ($NO_2C_6H_4COO$, pNB) on the TiO_2 surface, vibrations of molecular adsorbates were first observed on a solid [54]. pNB has more Raman-active vibrational freedoms than TMA. A TMA-covered $TiO_2(1\,1\,0)$ surface was immersed in dichloromethane containing p-nitrobenzoic acid. The acid in the solution is expected to be dissociatively adsorbed as pNB by exchanging the preadsorbed TMA. Similar exchange reactions are known in formate–acetate exchange [55], TMA–retinoate exchange [56], TMA–fluorescein exchange [44], and TMA–N3 exchange [45]. The immersed surface was characterized with XPS and a number density of 0.35 nitrogen atom nm^{-2} was determined.

As shown in Figure 6.6, the raw SH intensity and modulated component were almost identical to what was observed on the TMA-covered surface. A slight bump appeared at 570 cm^{-1} in addition to the four major bands of lattice vibrations. The Forurier-transformed spectrum was simulated using the sum of Lorentzian functions to properly identify the weak band. As a result, five bands are definitely identified at 825, 572, 439, 363, 180 cm^{-1}. Corrugations at 100 cm^{-1} or below are artifacts caused by fitting residue. There is no corresponding bulk and surface phonon mode of TiO_2 in this wavenumber region. The 572 cm^{-1} band should have originated from adsorption of pNB on TiO_2. Assignment of the band is not easy, unfortunately. According to a normal mode analysis [57] skeletal vibrations are predicted at 550–500 cm^{-1} and related to the vibrations observed in surface-enhanced Raman scattering of p-nitrobenzoic acid adsorbed on metal surfaces [58]. On the other hand, photoinduced coupling of p-nitrobenzoic acid to 4,4′-azodibenzoate

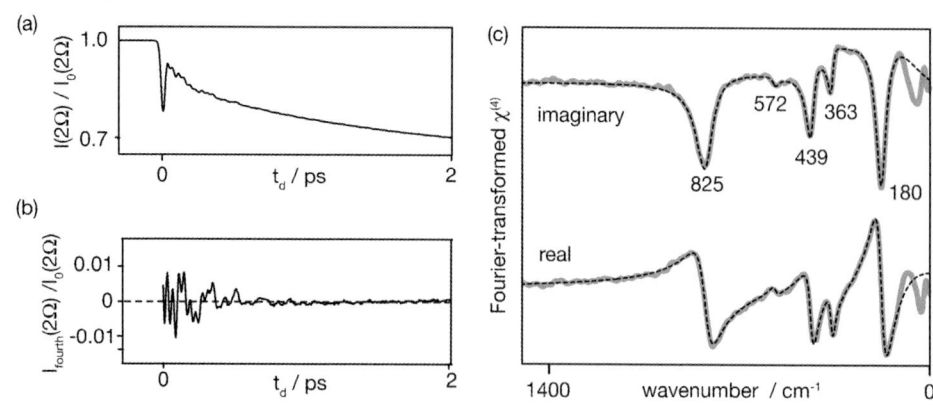

Figure 6.6 Vibrational coherence on a pNB-adsorbed TiO$_2$(110) surface. (a) The raw SH intensity, (b) the modulated component, (c) the Fourier-transformed spectrum, the gray lines show the transformed spectrum. The spectrum simulated with Lorentzian functions is overlaid with broken lines. The pNB-adsorbed surface was irradiated in air with p-polarized pump (8 mJ cm^{-2}) and p-polarized probe (8 mJ cm^{-2}) pulses of a 550-nm wavelength.

(OOCC$_6$H$_4$−N=N−C$_6$H$_4$COO) was found on metal surfaces. The coupled product also presents vibrations at around 550 cm^{-1} [58].

6.7
Frequency Domain Detection

One latest technical development is the direct observation of $\chi^{(4)}$ in the frequency domain. Yamaguchi and Tahara [59, 60] irradiated a Rhodamine solution surface with a white light continuum pulse and a narrow-bandwidth picosecond pulse. Vibrational coherence having different frequencies is generated in a stimulated Raman transition caused by the continuum and the narrow-bandwidth light. The energy distribution of the vibrational excited state v is projected on the narrow-bandwidth SH light by the hyper-Raman transition from v to g. The frequency-domain spectrum of the coherence is observed by using a polychrometer with a multichannel CCD. On the solution surface intra-molecular bands at 1200, 1350, 1500, 1650, and 2220 cm^{-1} were observed.

In the time-domain detection of the vibrational coherence, the high-wavenumber limit of the spectral range is determined by the time width of the pump and probe pulses. Actually, the highest-wavenumber band identified in the time-domain fourth-order coherent Raman spectrum is the phonon band of TiO$_2$ at 826 cm^{-1}. Direct observation of a frequency-domain spectrum is free from the high-wavenumber limit. On the other hand, the narrow-bandwidth, picosecond light pulse will be less intense than the femtosecond pulse that is used in the time-domain method and may cause a problem in detecting weak fourth-order responses.

6.8
Concluding Remarks

Successful applications of fourth-order coherent Raman scattering are presented. Interface-selective detection of Raman-active vibrations is now definitely possible at buried interfaces. It can be recognized as a Raman spectroscopy with interface selectivity. Vibrational sum-frequency spectroscopy provides an interface-selective IR spectroscopy in which the vibrational coherence is created in the IR resonant transition. The two interface-selective methods are complementary, as has been experienced with Raman and IR spectroscopy in the bulk.

On the other hand, we cannot ignore drawbacks in observing fourth-order responses. The desired response is always weak due to the high optical order. The damage threshold of the interface to be analyzed is severe with intense irradiation. The difficulty has been overridden by one-photon resonant enhancement of Raman-pump efficiency. The observable range of materials is somewhat limited as a result. There is still much room for technical improvements and the author is optimistic for the future.

Acknowledgments

The work reviewed in this chapter has been carried out in collaboration with Satoru Fujiyoshi, Tomonori Nomoto, and Taka-aki Ishibashi. We started this series of studies in Kanagawa Academy of Science and Technology and continued in Kobe University with support from Core Research for Evolutional Science and Technology of the Japan Science and Technology Agency, and from a Grant-in-Aid for Scientific Research in Priority Area "Molecular Nano Dynamics".

References

1 Zhu, X. D., Suhr, H. and Shen, Y. R. (1987) Surface vibrational spectroscopy by infared-visible sum frequency generation. *Phys. Rev. B*, **35**, 3047–3050.

2 Bandara, A., Kano, S. S., Onda, K., Katano, S., Kubota, J., Domen, K., Hirose, C. and Wada, A. (2002) SFG spectroscopy of CO/Ni(111): UV pumping and the transient hot band transition of adsorbed CO. *Bull. Chem. Soc. Jpn.*, **75**, 1125–1132.

3 Ishibashi, T. and Onishi, H. (2004) Multiplex sum-frequency spectroscopy with electronic resonance enhancement. *Chem. Lett.*, **33**, 1404–1407.

4 Schaller, R. D., Johnson, J. C., Wilson, K. R., Lee, L. F., Haber, L. H. and Saykally, R. J. (2002) Nonlinear chemical imaging nanomicroscopy: from second and third harmonic generation to multiplex (broad-bandwidth) sum frequency generation near-field scanning optical microscopy. *J. Phys. Chem. B*, **106**, 5143–5154.

5 Somorjai, G. A. and Rupprechter, G. (1999) Molecular studies of catalytic reactions on crystal surfaces at high pressures and high temperatures by infrared-visible sum frequency generation (SFG) surface vibrational spectroscopy. *J. Phys. Chem.*, **103**, 1623–1638.

6 Knock, M. M., Bell, G. R., Hill, E. K., Turner, H. J. and Bain, C. D. (2003) Sum-frequency spectroscopy of surfactant monolayers at the oil-water interface. *J. Phys. Chem. B*, **107**, 10801–10814.

7 Scatena, L. F., Brown, M. G. and Richmond, G. L. (2001) Water at hydrophobic surfaces: weak hydrogen bonding and strong orientation effects. *Science*, **292**, 908–912.

8 Wang, J., Even, M., Chen, X., Schmaier, A. H., Waite, J. H. and Chen, Z. (2003) Detection of amide I signals of interfacial proteins *in situ* using SFG. *J. Am. Chem. Soc.*, **125**, 9914–9915.

9 Ye, S., Nihonyanagi, S. and Uosaki, K. (2001) Sum frequency generation (SFG) study of the pH-dependent water structure on a fused quartz surface modified by an octadecyltrichlorosilane (OTS) monolayer. *Phys. Chem. Chem. Phys.*, **3**, 3463–3469.

10 Chang, Y. M., Xu, L. and Tom, H. W. K. (1997) Observation of coherent surface optical phonon oscillations by time-resolved surface second-harmonic generation. *Phys. Rev. Lett.*, **78**, 4649–4652.

11 Mani, A. A., Shultz, Z. D., Caudano, Y., Champagne, B., Humbert, C., Dreesen, L., Gewirth, A. A., White, J. O., Thiry, P. A. and Peremans, A. (2004) Orientation of thiophenol adsorbed on silver determined by nonlinear vibrational spectroscopy of the carbon skeleton. *J. Phys. Chem. B*, **108**, 16135–16138.

12 Braun, R., Casson, B. D., Bain, C. D., van der Ham, E. W. M., Vrehen, Q. H. F., Eliel, E. R., Brigg, A. M. and Davies, P. B. (1999) Sum-frequency generation from thiophenol on silver in the mid and far-IR. *J. Chem. Phys.*, **110**, 4634–4640.

13 Dhar, L., Rogers, J. A. and Nelson, K. A. (1994) Time-resolved vibrational spectroscopy in the impulsive limit. *Chem. Rev.*, **94**, 157–193.

14 Dekorsy, T., Cho, G. C. and Kurz, H. (2000) *Light Scattering in Solids VIII* (eds M. Cardona and G. Güntherodt), Springer, Berlin, Germany, Chapter 4.

15 Stevens, T. E., Kuhl, J. and Merlin, R. (2002) Coherent phonon generation and the two stimulated Raman tensors. *Phys. Rev. B*, **65**, 144304.

16 Fujiyoshi, S., Ishibashi, T. and Onishi, H. (2004) Time-domain Raman measurement of molecular submonolayers by time-resolved reflection spectroscopy. *J. Phys. Chem. B*, **108**, 1525–1528.

17 Fujiyoshi, S., Ishibashi, T. and Onishi, H. (2005) Low-frequency vibrations of molecular submonolayers detected by time-domain raman spectroscopy. *J. Mol. Struct.*, **735–736**, 169–177.

18 Long, D. A. (1977) *Raman Spectroscopy*, McGraw-Hill, New York, USA.

19 Ziegler, L. D., Fan, R., Desrosiers, A. E. and Sherer, N. F. (1994) Femtosecond polarization spectroscopy: A density matrix description. *J. Chem. Phys.*, **100**, 1823–1839.

20 Antoine, R., Tamburello-Luca, A. A., Hebért, P., Brevet, P. F. and Girault, H. H. (1998) Picosecond dynamics of Eosin B at the air/water interface by time-resolved second harmonic generation: orientational randomization and rotational relaxation. *Chem. Phys. Lett.*, **288**, 138–146.

21 Castro, A., Sitzmann, V., Zhang, D. and Eisenthal, K. B. (1991) Rotational relaxation at the air/water interface by time-resolved second harmonic generation. *J. Phys. Chem.*, **95**, 6752–6753.

22 Meech, S. R. and Yoshihara, K. (1990) Time-resolved surface second harmonic generation: a test of the method and its application to picosecond isomerization in adsorbates. *J. Phys. Chem.*, **94**, 4913–4920.

23 Shi, X., Borguet, B., Tarnovsky, A. N. and Eisenthal, K. B. (1996) Ultrafast dynamics and structure at aqueous interfaces by second harmonic generation. *Chem. Phys.*, **205**, 167–178.

24 Steinhurst, D. A. and Owrutsky, J. C. (2001) Second harmonic generation from oxazine dyes at the air/water interface. *J. Phys. Chem.*, **105**, 3062–3072.

25 Zimdars, D., Dadap, J. I., Eisenthal, K. B. and Heinz, T. F. (1999) Femtosecond

dynamics of solvation at the air/water interface. *Chem. Phys. Lett.*, **301**, 112–120.

26 Hirose, Y., Yui, H. and Sawada, T. (2005) Second harmonic generation-based coherent vibrational spectroscopy for a liquid interface under the nonresonant pump condition. *J. Phys. Chem. B*, **109**, 13063–13066.

27 Fujiyoshi, S., Ishibashi, T. and Onishi, H. (2006) Molecular vibrations at a liquid–liquid interface observed by fourth-order Raman spectroscopy. *J. Phys. Chem. B*, **110**, 9571–9578.

28 Fujiyoshi, S., Ishibashi, T. and Onishi, H. (2004) Interface-specific vibrational spectroscopy of molecules with visible lights. *J. Phys. Chem. B*, **108**, 10636–10639.

29 Chang, Y.-M. (2003) Coherent phonon spectroscopy of GaP Schottky diode. *Appl. Phys. Lett.*, **82**, 1781–1783.

30 Chang, Y.-M., Lin, H. H., Chia, C. T. and Chen, Y. F. (2004) Observation of coherent interfacial optical phonons in GaInP/GaAs/GaInP single quantum wells. *Appl. Phys. Lett.*, **84**, 2548–2550.

31 Watanabe, K., Takagi, N. and Matsumoto, Y. (2002) Impulsive excitation of a vibrational mode of Cs on Pt(111). *Chem. Phys. Lett.*, **366**, 606–610.

32 Watanabe, K., Takagi, N. and Matsumoto, Y. (2004) Direct time-domain observation of ultrafast dephasing in adsorbate-substrate vibration under the influence of a hot electron bath:Cs adatoms on Pt(111). *Phys. Rev. Lett.*, **92**, 057401.

33 Watanabe, K., Takagi, N. and Matsumoto, Y. (2005) Femtosecond wavepacket dynamics of Cs adsorbates on Pt(111): Coverage and temperature dependences. *Phys. Rev. B*, **71**, 085414.

34 Fuyukui, M., Watanabe, K. and Matsumoto, Y. (2006) Coherent surface phonon dynamics at K-coverd Pt(111) surface investigated by time-resolved second harmonic generation. *Phys. Rev. B*, **74**, 195412.

35 Watanabe, K., Takagi, N. and Matsumoto, Y. (2005) Mode-selective excitation of coherent surface phonons on alkali-covered metal surafaces. *Phys. Chem. Chem. Phys.*, **7**, 2697–2700.

36 Matsumoto, Y. (2007) Photochemistry and photo-induced ultrafast dynamics at metal surfaces. *Bull. Chem. Soc. Jpn.*, **80**, 842–855.

37 Matsumoto, Y. and Watanabe, K. (2006) Coherent vibrations of adsorbates induced by femtosecond laser excitation. *Chem. Rev.*, **106**, 4234–4260.

38 Melnikov, A. V., Radu, I., Bovensiepen, U., Krupin, O., Starke, K., Matthias, E. and Wolf, M. (2003) Coherent optical phonons and parametrically coupled magnons induced by femtosecond laser excitation of the Gd(0001) surface. *Phys. Rev. Lett.*, **91**, 227403.

39 Melnikov, A. V., Radu, I., Bovensiepen, U., Starke, K., Wolf, M. and Matthias, E. (2005) Spectral dependence of time-resolved coherent and incoherent second-harmonic response of ferromagnetic Gd(0001). *J. Opt. Soc. Am. B*, **22**, 204–210.

40 Diebold, U. (2003) The surface science of titanium dioxide. *Surf. Sci. Rep.*, **48**, 53–229.

41 Onishi, H. and Iwasawa, Y. (1994) STM-imaging of formate intermediates adsorbed on a TiO_2(110) surface. *Chem. Phys. Lett.*, **226**, 111–114.

42 Onishi, H., Yamaguchi, Y., Fukui, K. and Iwasawa, Y. (1996) Temperature-jump STM observation of reaction intermediate on metal-oxide surface. *J. Phys. Chem.*, **100**, 9582–9584.

43 Guo, Q. and Williams, E. M. (1999) The effect of adsorbate–adsorbate interaction on the structure of chemisorbed overlayers on TiO_2(110). *Surf. Sci.*, **433–435**, 322–326.

44 Pang, C., Ishibashi, T. and Onishi, H. (2005) Adsorption of fluorescein isothiocyanate isomer-I (FITC-I) Dye on TiO_2(110) from an acetone solution. *Jpn. J. Appl. Phys.*, **44**, 5438–5442.

45 Sasahara, A., Pang, C. and Onishi, H. (2006) STM observation of a ruthenium dye adsorbed on a TiO_2(110) surface. *J. Phys. Chem. B*, **110**, 4751–4755.

46 Ikeda, M., Koide, N., Han, L., Sasahara, A. and Onishi, H. (2008) Work function on

dye-adsorbed TiO$_2$ surfaces measured by using a Kelvin probe force microscope. *J. Phys. Chem. C*, **112**, 6961–6967.

47 Onishi, H., Fukui, K. and Iwasawa, Y. (1995) Atomic-scale surface structures of TiO$_2$(110) determined by scanning tunneling microscopy: A new surface-limited phase of titanium oxide. *Bull. Chem. Soc. Jpn.*, **68**, 2447–2458.

48 White, J. M., Szanyi, J. and Henderson, M. A. (2003) The photon-driven hydrophilicity of titania: A model study using TiO$_2$(110) and adsorbed trimethyl acetate. *J. Phys. Chem. B*, **107**, 9029–9033.

49 Fujiyoshi, S., Ishibashi, T. and Onishi, H. (2005) Fourth-order Raman spectroscopy of wide-bandgap materials. *J. Phys. Chem. B*, **109**, 8557–8561.

50 Proto, S. P. S., Fleury, P. A. and Damen, T. C. (1967) Raman spectra of TiO$_2$, MgF$_2$, ZnF$_2$, FeF$_2$, and MnF$_2$. *Phys. Rev.*, **154**, 522–526.

51 Lantz, J. M. and Corn, R. M. (1994) Electrostatic field measurements and band flattening during electron-transfer processes at single-crystal TiO$_2$ electrodes by electric field-induced optical second harmonic generation. *J. Phys. Chem.*, **98**, 4899–4905.

52 Cox, P. A., Egdell, R. G., Eriksen, S. and Flavell, W. R. (1986) The high-resolution electron-energy-loss spectrum of TiO$_2$(110). *J. Electron Spectrosc. Relat. Phenom.*, **39**, 117–126.

53 Rocker, G., Schaefer, J. A. and Göpel, W. (1984) Localized and delocalized vibrations on TiO$_2$(110) studied by high-resolution electron-energy-loss spectroscopy. *Phys. Rev. B*, **30**, 3704–3708.

54 Nomoto, T. and Onishi, H. (2008) Fourth-order Raman spectroscopy of adsorbed organic species on TiO$_2$ surface. *Chem. Phys. Lett.*, **455**, 343–347.

55 Uetsuka, H., Sasahara, A., Yamakata, A. and Onishi, H. (2002) Microscopic identification of a bimolecular reaction intermediate. *J. Phys. Chem. B*, **106**, 11549–11552.

56 Ishibashi, T., Uetsuka, H. and Onishi, H. (2004) An ordered retinoate monolayer prepared on rutile TiO$_2$(110). *J. Phys. Chem. B*, **108**, 17166–17170.

57 Ernstbrunner, E. E., Girling, R. B. and Hester, R. E. (1978) Free radical studies by resonance Raman spectroscopy. Part 3–4-Nitrobenzoate radical Dianion. *J. Chem. Soc. Faraday Trans. 2*, 1540–1549.

58 Venkatachalam, R. S., Boerio, F. J. and Roth, P. G. (1988) Formation of p,p'-azodibenzoate from p-aminobenzoic acid on silver island films during surface-enhanced Raman scattering. *J. Raman Spectrosc.*, **19**, 281–287.

59 Yamaguchi, S. and Tahara, T. (2005) Interface-specific $\chi^{(4)}$ coherent raman spectroscopy in the frequency domain. *J. Phys. Chem. B*, **109**, 24211–24214.

60 Yamaguchi, S. and Tahara, T. (2007) $\chi^{(4)}$ Raman spectroscopy for buried water interfaces. *Angew. Chem. Int. Ed.*, **46**, 7609–7612.

7
Dynamic Analysis Using Photon Force Measurement
Hideki Fujiwara and Keiji Sasaki

7.1
Introduction

7.1.1
Weak Force Measurements

Micro- and nano-particles dispersed in a liquid experience various forces [1–5]. Although gravity, viscous drag, Van der Waals interactions, surface double layer forces, and Brownian motion exerted on a single particle or between particles and those between a particle and a solid surface are only of the order of femto-newtons in strength, these forces play an important role in the adhesion mechanism. Short-range forces due to hydrophobic and hydrophilic interactions, hydration and solvation energy, and hydrogen bonding networks sometimes govern the behavior of microparticles in solution. In addition, depending on local environmental conditions (temperature, concentration gradient of gases, solvents, and molecules) around the single particles, the chemical and physical characteristics of particles and cells are also changed due to the changes in molecular conformation and reaction yields. Thus, various phenomena (aggregation, adhesion, sedimentation, chemical reactions, and transition processes) of organic, metallic, semiconductor colloidal particles, surfactant micelles, macromolecules, and so on, should be strongly affected by the strength of the balances of those forces and environmental conditions around the nanoparticles [1].

To measure the strength of the forces exerted on particles, various analytical techniques have been developed [6, 7]. Unfortunately, since most of these techniques are based on hydrodynamics, assumption of the potential profiles is required and the viscosities of the fluid and the particle sizes must be precisely determined in separate experiments, for example, using the viscous flow technique [8, 9] and power spectrum analysis of position fluctuation [10]. Furthermore, these methods provide information on ensemble averages for a mass of many particles. The sizes, shapes, and physical and chemical properties of individual particles may be different from each other, which will result in a variety of force strengths. Thus, single-particle measurement is indispensable for the analysis of van der Waals interactions and electric double layer forces and for observation of ionic dissociation and adsorption processes. Recently, a technique for measuring the adhesion force between single particles and a solid surface has been developed by bringing a single particle attached to the cantilever of an atomic force microscope (AFM) close to the solid surface and by monitoring the displacement of the cantilever [11]. However, it is technically difficult to attach the single particle with a sub-micrometer diameter to the cantilever and to eliminate the influence of the cantilever on the physical and chemical characteristics of a sample. Furthermore, since the detectable force using the AFM cantilever is of the order of pico-newtons, the sensitivity of the AFM is insufficient for the measurements of very weak forces of the order of femto-newtons.

7.1.2
Potential Analysis Method Using Photon Force Measurement

For this purpose, we have proposed a method that makes it possible to precisely and instantaneously observe arbitrary profiles of three-dimensional potentials exerted on a single microparticle [12]. This method utilizes the photon force as an optical spring to characterize the physical and chemical forces exerted on single particles [13]. The force measurement technique we used is based on thermodynamic analyses of the Brownian motion, evaluated with nanometer position sensing [14]. Neither the assumption of potential profiles nor knowledge of media viscosities or particle diameters is necessary. The only physical parameter that needs to be known is the temperature of the sample [15]. The principle is shown in Figure 7.1. When a microparticle is irradiated with a focused laser beam, the particle can be three-dimensionally trapped at the focal spot [16, 17] and fluctuate within the trapping potential well due to the Brownian motion [14]. In the method, this fluctuation of particle position is sequentially measured for a sufficiently long time by a position sensing system and a histogram is obtained as a function of the three-dimensional position. This histogram gives a probability density function of the particle position. Since the particle motion is caused by thermal energy, the three-dimensional potential profile can be determined from the position histogram by a simple logarithmic transformation of the Boltzmann distribution. By measuring the

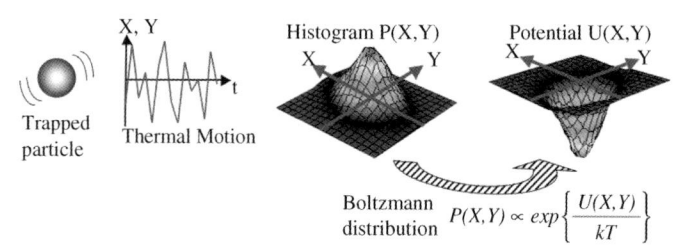

$$P(X,Y) \propto \exp\left\{\frac{U(X,Y)}{kT}\right\}$$

Figure 7.1 The principle of thermodynamic analysis for the measurement of three-dimensional potentials exerted on a single microparticle.

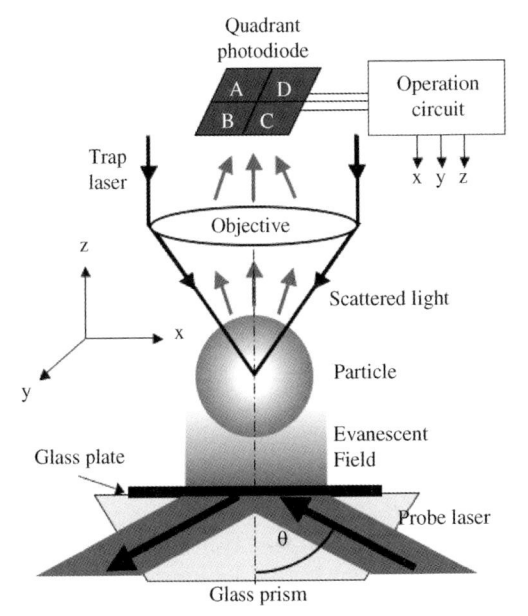

Figure 7.2 A schematic diagram of nanometer position sensing. Light from the evanescent field scattered by the microparticle is measured with a quadrant photodiode detector, whose differential outputs correspond to the x and y displacements and the total intensity depends exponentially on the distance z between the particle and the glass plate.

potentials with and without the additional weak force acting on the particle, the subtraction of these potentials gives the potential of the weak force, resulting in the determination of the strength of the additional weak force.

The fluctuation of the particle position is measured by the nanometer position sensing system based on total internal reflection microscopy [12, 14, 18] and quadrant photodiode detection. A schematic diagram of the system is shown in Figure 7.2. A microparticle dispersed in a liquid is optically trapped by a focused near-infrared trapping laser beam under an optical microscope. The particle is positioned in the close vicinity of the surface of a microscope glass slide. A weak S-polarized illumination laser beam is introduced into a prism, which is optically coupled to the glass slide through matching oil, with the incident angle on the glass slide surface larger than the critical angle of the total internal reflection condition. Thus the laser beam induces an evanescent field, which illuminates the trapped particle. The scattered light from the microparticle is collected by the objective lens, passed through an interference filter, and then the particle position is detected by a quadrant photodiode (QPD).

Using this photon force measurement technique, radiation pressure induced by a focused laser beam and an evanescent field [12, 14, 19, 20] was investigated for polymer latexes and metallic particles. Electrostatic forces of charged particles in solutions have also been precisely observed using this method [21]. This photon force measurement can be extended to the analysis of radiation pressure exerted on a single absorbing particle and to the estimation of the imaginary part of the refractive index [22]. Furthermore, using this technique, we have clarified that the trapped particle position in the vicinity of an interface is discretely shifted by the interference effect of the incident trapping laser beam and reflected beams from a substrate and the focal point of the trapping laser beam [23–25]. Thus, by using the photon force measurement technique, we have succeeded in analyzing the potential profiles and precisely evaluating femto-newton order forces acting on single particles. Since this technique measures the thermal fluctuation of single particles and analyzes the potentials acting on the particle on the basis of the thermodynamic analysis, the only required physical parameter for the highly precise weak force measurement is the temperature of the sample and a priori information, such as medium viscosity, particulate mass and size.

However, although various weak force measurements using the potential analysis method have been reported, these measurements have been mainly of the forces acting on single particles, not those between two particles. It is well known that various forces, such as the Van der Waals force, the electric double layer effect, the electrostatic force, the hydrodynamic interaction force, and so on, are also exerted between two adjacent particles dispersed in liquid [1]. Therefore, the weak force acting between particles should also be of importance for a wide range of fields, such as colloidal science, biology, medical science, and so on. For the measurement of the interaction forces acting between two particles, the interaction force has been mainly investigated by use of a cross-correlation analysis of the temporal position fluctuations of two particles [26, 27]. This method typically requires a priori knowledge such as particle size, mass, and media viscosity, and so on, in order to evaluate the interaction. Since the photon force measurement technique we used is based on the thermodynamic analysis of the Brownian motion evaluated with the nanometer position sensing system, the only physical parameter needed is the temperature of the sample; neither *a priori* knowledge of media viscosity and particle size nor the assumption of potential profiles is necessary [14]. However, in order to investigate the interaction force by the use of potential analysis, it is necessary to know the effect of the interaction forces on the potentials and how the trapping or kinetic potential profiles are changed by the interaction. This is still unclear.

In this chapter, we introduce a two-beam photon-force measurement system using a coaxial illumination for sensing the particle position as a new technique for dynamically analyzing the interaction forces exerted between two adjacent objects. By utilizing this potential analysis method for the interaction forces, especially, we focused on the hydrodynamic interaction force between adjacent trapped particles and the potentials at different distances between two particles were measured. Then, we discussed the validity of the proposed method, comparing with the typically used correlation method and/or the theoretical calculation. From the results, we found that the profiles of the two-dimensional plots of the trapping or kinetic potentials of two particles, were collapsed on decreasing the distance between two particles, indicating the increase in the interaction force between two particles.

7.2 Measurement of the Hydrodynamic Interaction Force Acting between Two Trapped Particles Using the Potential Analysis Method

From the changes in the obtained potential profiles, we evaluated the interaction coefficient β and, compared it with the theoretical calculation of β, we also confirmed that the proposed method could evaluate the interaction force without any fitting parameters, which have usually been required in the typical correlation method.

7.2.1 Two-Beam Photon Force Measurement System

Figure 7.3 shows the two-beam photon-force measurement system using a coaxial illumination photon force measurement system. Two microparticles dispersed in a liquid are optically trapped by two focused near-infrared beams (~1 μm spot size) of a CW Nd:YAG laser under an optical microscope (1064 nm, 1.2 MW cm^{-2}, 100X oil-immersion objective, NA = 1.4). The particles are positioned sufficiently far from the surface of a glass slide in order to neglect the interaction between the particles and the substrate. Green and red beams from a green LD laser (532 nm, 21 kW cm^{-2}) and a He–Ne laser (632.8 nm, 21 kW cm^{-2}) are introduced coaxially into the microscope and slightly focused onto each microparticle as an illumination light (the irradiated area was about 3 μm in diameter). The sizes of the illumination areas for the green and red beams are almost the same as the diameter of the microparticles (see Figure 7.4). The back scattered light from the surface of each microparticle is

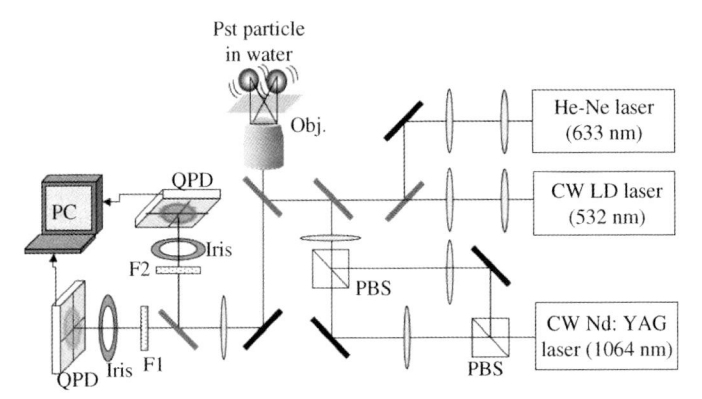

Figure 7.3 A schematic of a two-beam photon-force measurement system. Obj: objective lens (100× oil immersion, N.A. 1.4), PBS: polarization beam splitter, F1: color filter for eliminating red illumination laser beam, F2: color filter for eliminating green illumination laser beam DM: dichroic mirror, QPD: quadrant photo diode. Two polystyrene latex particles dispersed in water were trapped by a focused CW Nd:YAG laser beam. Two particles were illuminated by red and green lasers, respectively.

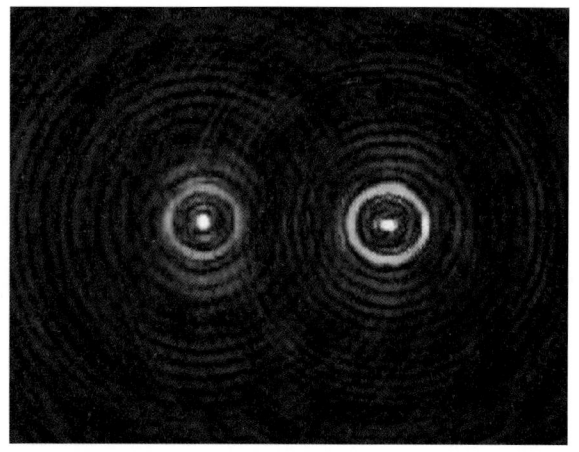

Figure 7.4 A microscope image of two trapped microparticles in water, illuminated by a green and red laser beam, respectively. The diameter of the particles is 3 μm.

collected by the same objective lens. Then, the green and red scattered lights are divided by a dichroic mirror and imaged on QPDs set in the individual paths of the green and red scattered lights. The sum of the four output signals (z-signal) and two differential outputs normalized by the sum (x- and y-signals) from each QPD are transferred, with analog-to-digital conversion, to a computer for data acquisition and analysis of each position of the two trapped particles. Calibrations of the x- and y- position were performed by analyzing the variation of long-time-averaged signals while changing the focal position of the trapping laser beam. The sampling rate for the output signals of the QPDs was set to be 10 kHz, that is, shorter than the mechanical response time of the particle. The fluctuations of the particle positions are recorded sequentially for a sufficiently long time (60 s) by position sensing systems in each path. In the experiments, polystyrene particles (diameter 3 μm) dispersed in water were used as a sample. On changing the distance between the surfaces of two trapped particles by manipulating the focused positions of the trapping laser beams (from 1 to 20 μm), we measured the position fluctuations of each particle in the x- and y-directions.

7.2.2
Potential Analysis Method for Hydrodynamic Force Measurement

For the analysis, we developed a new method that makes it possible to observe correlated potentials between two trapped particles. The principle is shown in Figure 7.5. From the recorded position fluctuations of individual particles (indicated by the subscripts 1 and 2), histograms are obtained as a function of the three-dimensional position. Since the particle motion is caused by thermal energy, the three-dimensional potential profile can be determined from the position histogram by a simple logarithmic transformation of the Boltzmann distribution. Similarly, the

7.2 Measurement of the Hydrodynamic Interaction Force Acting between Two Trapped Particles

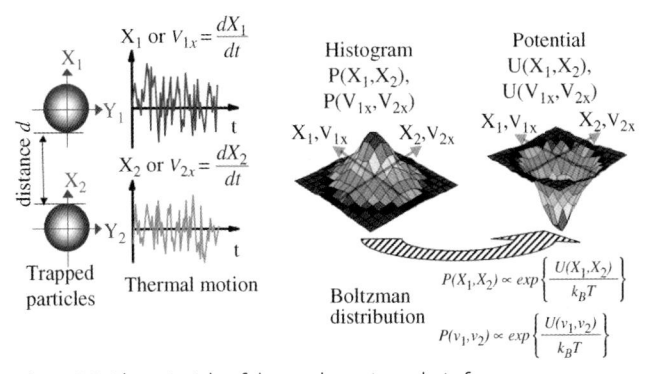

Figure 7.5 The principle of thermodynamic analysis for measuring trapping or kinetic potentials exerted between two trapped particles.

velocity of each particle motion was also derived from the differentiation of the recorded position fluctuations, resulting in the kinetic potentials.

In order to discuss the correlation between the position fluctuations, the potentials are two-dimensionally plotted against the positions or velocities of individual particles (indicated as x_1 and x_2 or v_1 and v_2). From the Langevin equations, assuming that the mass and diameter of the particles and the irradiation power of the trapping laser beams were the same, the motions of each trapped particle can be given by [28]

$$m\frac{dv_{1j}}{dt} + \gamma v_{1j} + k_j \int v_{1j}\,dt - \beta_j v_{2j} = n_{1j} \qquad (7.1)$$

and

$$m\frac{dv_{2j}}{dt} + \gamma v_{2j} + k_j \int v_{2j}\,dt - \beta_j v_{1j} = n_{2j} \qquad (7.2)$$

where $j = x$, y, and z. In the equations, m, γv_{ij}, $k_j \int v_{ij}\,dt$, $\beta_j v_{ij}$, and n_{ij} ($i = 1, 2$) indicate the particle mass, the viscosity, the restoring force of the optical trap, the interaction force, and the random force in the j-direction ($j = x$, y, z), respectively. Note that β_j is the interaction coefficient and the function of the distance between two particles. From the sum and difference of these equations, we can derive the following equations,

$$m\frac{d(v_{1j} + v_{2j})}{dt} + (\gamma - \beta_j)(v_{1j} + v_{2j}) + k_j \int (v_{1j} + v_{2j})\,dt = n_{1j} + n_{2j} \qquad (7.3)$$

and

$$m\frac{d(v_{1j} - v_{2j})}{dt} + (\gamma + \beta_j)(v_{1j} - v_{2j}) + k_j \int (v_{1j} - v_{2j})\,dt = n_{1j} - n_{2j} \qquad (7.4)$$

Since the histogram gives a probability density function of the particle position, the correlation in the velocities v_{1j} and v_{2j} in the j-direction causes the change in the shape of the histogram plotted against v_{1j} and v_{2j}, due to the different coefficient $\gamma - \beta_j$ in the $(v_{1j} + v_{2j})$ direction and $\gamma + \beta_j$ in the $(v_{1j} - v_{2j})$ direction in the second terms of the Eqs. (7.4) and (7.5). If the interaction becomes large on decreasing the distance between the particles, the shape of the histogram is squeezed in the $(v_{1j} - v_{2j})$ direction and extended in the $(v_{1j} + v_{2j})$ direction, depending on the strength of the interaction. However, if the interaction can be neglected ($\beta_j = 0$), as the coefficients in the second terms in Eqs. (7.4) and (7.5) become the same, the histogram shape does not change in the $(v_{1j} + v_{2j})$ and $(v_{1j} - v_{2j})$ directions.

For example, let consider the case in the direction between the central axes of the two trapped particles, which is indexed as v_{1x} and v_{2x}. The correlation of the kinetic potential in the v_x-directions changes depending on the coefficients of $\gamma - \beta_x$ in the $(v_{1x} + v_{2x})$ direction and $\gamma + \beta_x$ in $(v_{1x} - v_{2x})$ direction. Thus, as the interaction coefficient β_x changes with the distance between the two particles, the shape of the kinetic potential plotted against v_{1x} and v_{2x} should change in the $(v_{1x} + v_{2x})$ and $(v_{1x} - v_{2x})$ directions.

7.2.3
Trapping Potential Analysis

Figure 7.6 shows the two-dimensional plots of trapping potentials in the x_1- and x_2-directions, when the distances between adjacent particles were (a) 1, (b) 5, and (c) 20 µm. In the figure, the vertical and horizontal axes indicate the x positions of the particles 1 and 2, respectively, and the depth of the trapping potential is presented by the gray-scale image. From the results, although the collapse of the potential shape should appear with decreasing distance (increasing interaction), the observed potentials were round-shaped and it was difficult to identify the change in the shape. We considered that this is due to the fluctuation of the system, such as fluctuation in the trapping laser and noise of electronic circuits, and the change in the trapping potential would be hidden in the noise.

We also calculated the cross-correlations of the position fluctuations in the x_1- and x_2-directions at each distance, which method has been utilized for analyzing the

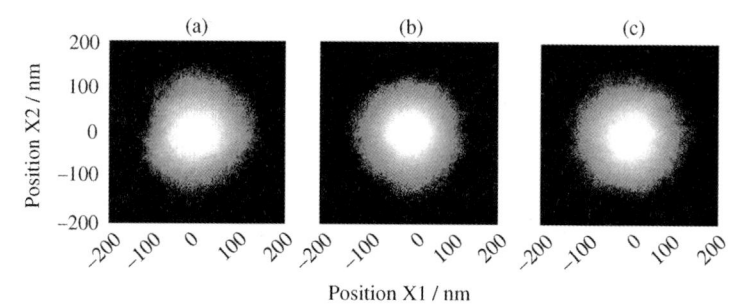

Figure 7.6 Two-dimensional plots of x-directional trapping potentials. The particles were trapped in water by two focused laser beams with a power of 10 mW. The distances between the surfaces of the two trapped particles were (a) 1, (b) 5, and (c) 20 µm.

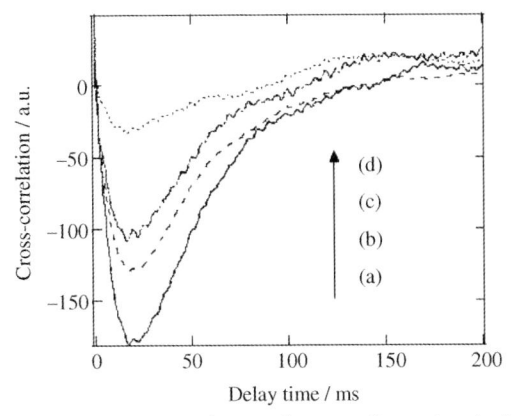

Figure 7.7 Cross-correlations of position fluctuations in the x_1- and x_2-directions. The distances between the surfaces of the two trapped particles were (a) 1, (b) 3, (c) 5, and (d) 20 µm.

interaction. Figure 7.7 shows the results when the distances were (a) 1, (b) 3, (c) 5, and (d) 20 µm. From the results, we can surely confirm the typical cross-correlation curves and found the correlation dip around 20 ms. On decreasing the distance, the correlation dip becomes deeper and this behavior coincided well with the experimental data on the hydrodynamic force in ref. [26, 27]. From the correlation data, the observed position fluctuation includes the influence of the interaction, however, it is difficult to confirm the influence of the interaction from the potential analysis. We considered that since the position fluctuation induced by the interaction was smaller than the amplitude of the system noise and was buried in the noise, the collapse of the trapping potential could not be clearly observed.

7.3
Kinetic Potential Analysis

On the other hand, in order to observe the influence of the interaction, we also measured the kinetic potentials, which are plotted against the velocities of the particles 1 and 2, when changing the distance from 1 to 20 µm. Since the velocities of the x- and y-directions were obtained by the derivation of the position fluctuations of each particle, the low-frequency components were suppressed and the interaction of two particles would be clearly obtained rather than the position fluctuation. Surely, from the cross-correlation of the velocities v_{1x} and v_{2x} (Figure 7.8), we can observe much clearer correlation peaks with the change in the distance from 1 to 20 µm, in which the cross-correlation of the derivation of the position should be identical to the second-order derivation of the position cross-correlation. From the results, different from the position cross-correlations, the system noise components were eliminated from the results and the correlation peak was increased with decreasing distance. Thus, we were able to observe the interaction clearly.

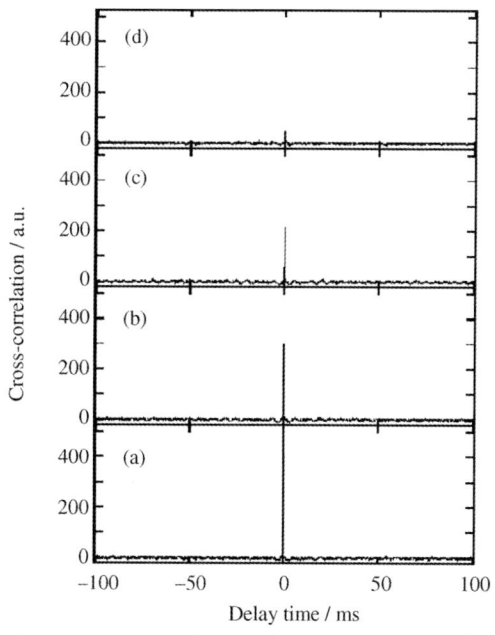

Figure 7.8 Cross-correlations of velocity fluctuations in the x_1- and x_2-directions. The distances between the surfaces of the two trapped particles were (a) 1, (b) 3, (c) 5, and (d) 20 µm.

Figure 7.9 shows the two-dimensional plots of kinetic potentials in the v_{1x}- and v_{2x}-directions, when the distances between adjacent particles were (a) 1, (b) 5, and (c) 20 µm. In the figure, the vertical and horizontal axes indicate the velocity of the particles 1 and 2 in the x-direction, and the depth of the kinetic potential is shown by the gray-scale image. From the results, when the distance was sufficiently large (see Figure 7.9c), the profile was almost round-shaped and the profiles in the

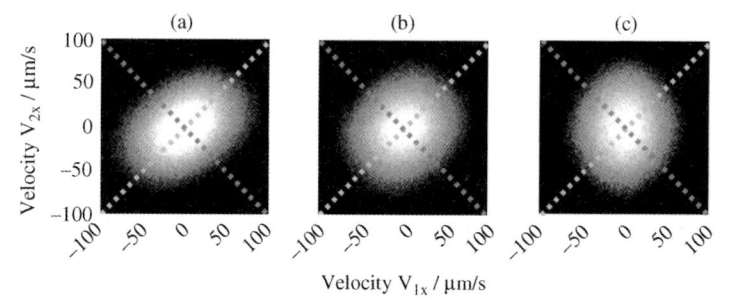

Figure 7.9 Two-dimensional plots of x-directional kinetic potentials exerted on two trapped particles. The particle was trapped in water by two focused laser beams with a power of 10 mW. The distances between the surfaces of the two trapped particles were (a) 1, (b) 5, and (c) 20 µm.

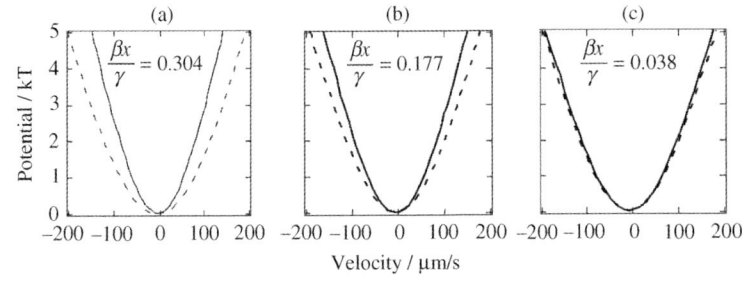

Figure 7.10 Cross-section of the kinetic potentials at the dashed lines in Figure 7.9. The distances between the surfaces of two trapped particles were (a) 1, (b) 5, and (c) 20 μm. Solid and dashed lines indicate the cross-section of the $(v_{1x} - v_{2x})$- and $(v_{1x} + v_{2x})$-directions, respectively.

$(v_{1x} + v_{2x})$- and $(v_{1x} - v_{2x})$-directions (dashed lines in the figure) almost coincided with each other, although the potential profile was slightly distorted. However, when the distance became shorter (see Figure 7.9a and b), we found that the profile was squeezed in the $(v_{1x} - v_{2x})$-direction and extended in the $(v_{1x} + v_{2x})$-direction.

In order to confirm the collapse of the potential profiles, Figure 7.10 shows the cross-section of the kinetic potentials in the $(v_{1x} - v_{2x})$- and $(v_{1x} + v_{2x})$-directions, when the distances were (a) 1, (b) 5, and (c) 20 μm. In the figure, black and dashed lines indicate the cross-sections in the $(v_{1x} - v_{2x})$- and $(v_{1x} + v_{2x})$-directions, respectively. The profiles of the kinetic potentials collapsed in the $(v_{1x} - v_{2x})$-direction on decreasing the distance. Since the collapse was caused by the increase in the influence of the interaction via the coefficient β in the x-direction, which should be the function of the distance, we estimated the value of β_x/γ at each distance by fitting the curves with hyperbolic functions (αv^2) and taking the ratio of $\alpha_{(v1x-v2x)}$ and $\alpha_{(v1x+v2x)}$. Since the ratio of $\alpha_{(v1x-v2x)}$ and $\alpha_{(v1x+v2x)}$ should be proportional to the ratio of the coefficients in the second term in Eqs. (7.3) and (7.4), β_x/γ can be estimated from the results as 0.304, 0.177, and 0.038 at the distances of (a) 1, (b) 5, and (c) 20 μm, respectively. Thus, we confirmed that the interaction coefficient was changed with the change in the distance.

Similarly, we measured the kinetic potentials in the v_y-direction, which was perpendicular to the direction between the central axes of the two trapped particles. Since the hydrodynamic force exerted similarly on the y-direction as on the x-direction, we expected similar results to Figures 7.9 and 7.10. Figure 7.11 shows the kinetic potentials in the v_{1y}- and v_{2y}-directions, when the distances were (a) 1, (b) 5, and (c) 20 μm. Comparing Figure 7.11b and c at the distances of 5 and 20 μm, the clear difference was not found in the $(v_{1y} - v_{2y})$- and $(v_{1y} + v_{2y})$-directions. However, at a distance of 1 μm (Figure 7.11a), the potential profile was modified and the changes in the $(v_{1y} - v_{2y})$- and $(v_{1y} + v_{2y})$-directions were observed. From the plots of the cross-sections in Figure 7.12, the interaction coefficients β_y/γ in the v_y-direction were estimated to be 0.207, 0.037, and 0.004, when the distances were (a) 1, (b) 5, and (c) 20 μm, respectively. From the results, the tendency was similar to the case of the v_x-direction, but the strength of the interaction was smaller.

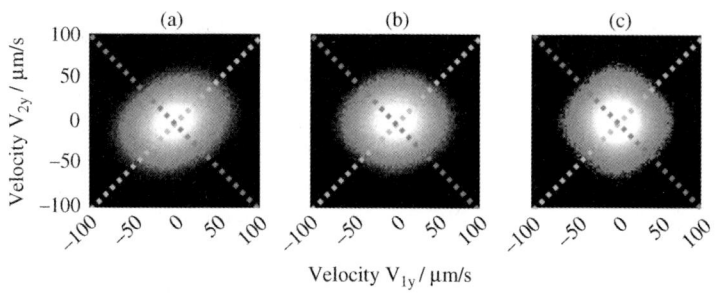

Figure 7.11 Two-dimensional plots of y-directional kinetic potentials exerted on two trapped particles. The particle was trapped in water by two focused laser beams with a power of 10 mW. The distances between the surfaces of the two trapped particles were (a) 1, (b) 5, and (c) 20 μm.

From the theoretical calculations of the hydrodynamic force, we can estimate the distance dependences of the interaction coefficients in the x- and y-directions. According to ref. [29], the interaction coefficients β_x/γ and β_y/γ are calculated from the following approximated equations, assuming that the two particles are spherical in shape with the same diameter.

$$\beta_x/\gamma = \frac{\frac{3}{2}k + \frac{19}{8}k^3}{1 + \frac{9}{4}k^2} \tag{7.5}$$

and

$$\beta_y/\gamma = \frac{\frac{3}{4}k + \frac{59}{64}k^3}{1 + \frac{9}{16}k^2} \tag{7.6}$$

where

$$k = \left(\frac{a}{d + 2a}\right) \tag{7.7}$$

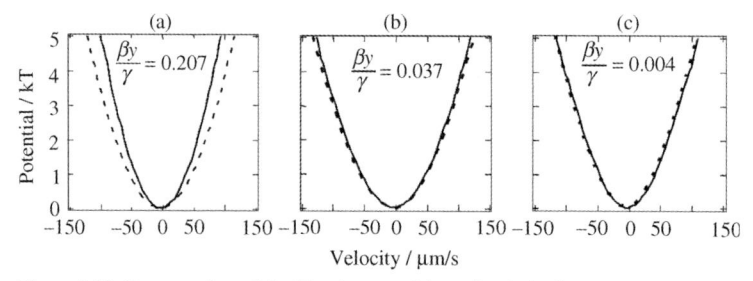

Figure 7.12 Cross-section of the kinetic potentials at the dashed lines in Figure 7.11. The distances between the surfaces of the two trapped particles were (a) 1, (b) 5, and (c) 20 μm. Solid and dashed lines indicate the cross-section of the $(v_{1x} - v_{2x})$- and $(v_{1x} + v_{2x})$-directions, respectively.

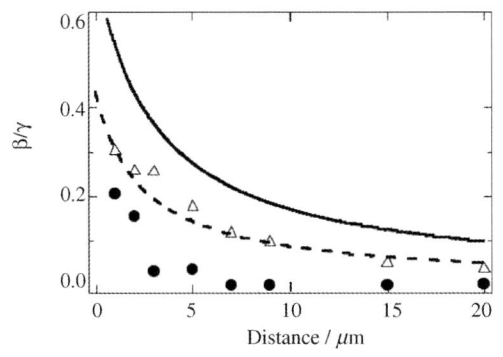

Figure 7.13 Distance dependence of β/γ. Solid and dashed lines indicate the theoretical calculation of β/γ in the x- and y-directions. Triangles and circles indicate the experimentally obtained data for β/γ in the x- and y-directions.

a and d indicate the radius of the trapped particle and the distance between the surfaces of the two trapped particles. Figure 7.13 shows the plots of the calculated results from Eqs. (7.5) and (7.6) against the distance d, in which the solid and dashed lines indicate the calculation of β_x/γ and β_y/γ, respectively. In addition, the experimentally obtained data of β_x/γ and β_y/γ were also plotted in the same figure as triangles and circles. From the results, the absolute values were about half the values obtained from the calculation. We consider one possible origin of the difference to be that the electronic noise from the electronic circuits with high-frequency components, which could not be eliminated by the derivation procedure, distorted the potential profiles, resulting in the poor estimation of β/γ. However, we would like to emphasize here that the tendencies of the distance dependence of both experimentally obtained β_x/γ and β_y/γ coincided well with the calculated results and the data were obtained only by measuring the potentials with no fitting parameters.

7.4
Summary

We introduced the technique for measuring the weak interaction forces acting between two particles using the photon force measurement method. Compared with the previous typically used methods, such as cross-correlation analysis, this technique makes it possible to evaluate the interaction forces without a priori information, such as media viscosity, particle mass and size. In this chapter, we focused especially on the hydrodynamic force as the interaction between particles and measured the interaction force by the potential analysis method when changing the distance between particles. As a result, when the particles were close to each other, the two-dimensional plots of the kinetic potentials for each particle were distorted in the diagonal direction due to the increase in the interaction force. From the results, we evaluated the interaction coefficients and confirmed that the dependence of the obtained values on the distance coincided well with the theoretical calculations and the experimental results from the cross-correlation analysis method. Thus, we have developed the two-beam photon-force measurement system and the potential analysis method for evaluating the weak interaction force acting between two trapped particles using the photon force measurement technique. Although in the experiments introduced here we used micrometer-sized particles, if the scattered light from particles is sufficiently intense to observe with a quadrant photodiode, it can be applicable to particles of any size and shape (metal nanoparticles, semiconductor powders, swelled micelles, organic nanocrystals, etc.).

Acknowledgments

This work was supported in part by a Grant-in-Aid for Scientific Research on Priority Area from MEXT.

References

1 Israelachvili, J. N. (1985) *Intermolecular and Surface Forces*, Academic Press, London.
2 Hamaker, H. C. (1937) The London-van der Waals attraction between spherical particles. *Physica*, **4**, 1058–1072.
3 Derjaguin, B. V. (1934) Interaction forces between hydrophobic and hydrophilic self-assembled monolayers. *Kolloid Zeits.*, **69**, 155–164.
4 Derjaguin, B. V. and Landau, L. (1941) Theory of the stability of strongly charged particles in solutions of electrolytes. *ActaPhysiochim., URSS.*, **14**, 633–662.
5 Verway, E. J. W. and Overbeek, J. Th. G. (1948) *Theory of Stability of Lyophobic Colloids*, Elsevier, Amsterdam.
6 Shaw, D. J. (1992) *Introduction to Colloid and Surface Chemistry*, 4th edn, Butterworth-Heinemann, Oxford.
7 Shaw, D. J. (1969) *Electrophoresis*, Academic Press, New York.
8 Kuo, S. C. and Sheetz, M. P. (1993) Force of single kinesin molecules measured with optical tweezers. *Science*, **260**, 232–234.
9 Finer, J. T., Simmons, R. M. and Spudich, J. A. (1994) Single myosin molecule mechanics: piconewton forces and nanometre steps. *Nature*, **368**, 113–119.
10 Svoboda, K., Schmidt, C. F., Schnapp, B. J. and Block, S. M. (1993) Direct observation of kinesin stepping by optical trapping interferometry. *Nature*, **365**, 721–727.
11 Ducker, W. A., Senden, T. J. and Pashley, R. M. (1991) Direct measurement of colloidal forces using an atomic force microscope. *Nature*, **353**, 239–241.
12 Sasaki, K., Tsukima, M. and Masuhara, H. (1997) Three-dimensional potential analysis of radiation pressure exerted on a single microparticle. *Appl. Phys. Lett.*, **71**, 37–39.
13 Masuhara, H., De Schryver, F. C., Kitamura, N. and Tamai, N. (eds) (1994) *Microchemistry*, Elsevier, Amsterdam.
14 Sasaki, K. (2003) *Single Organic Nanoparticles* (eds H. Masuhara, H. Nakanishi and K. Sasaki), Springer, Berlin, Chapter 9.
15 Hofkens, J., Hotta, J., Sasaki, K., Masuhara, H. and Iwai, K. (1997) Molecular assembling by the radiation pressure of a focused laser beam: Poly (N-isopropylacrylamide) in aqueous solution. *Langmuir*, **13**, 414–419.
16 Ashkin, A. (1992) Forces of a single-beam gradient laser trap on a dielectric sphere in

the ray optics regime. *Biophys. J.*, **61**, 569–582.
17 Ashkin, A., Dziedzic, J. M., Bjorkholm, J. E. and Chu, S. (1986) Observation of a single-beam gradient force optical trap for dielectric particles. *Opt. Lett.*, **11**, 288–290.
18 Prieve, D. C. (1986) Hydrodynamic measurement of double-layer repulsion between colloidal particle and flat plate. *Science*, **231**, 1269–1270.
19 Wada, K., Sasaki, K. and Masuhara, H. (2000) Optical measurement of interaction potentials between a single microparticle and an evanescent field. *Appl. Phys. Lett.*, **76**, 2815–2817.
20 Sasaki, K., Hotta, J., Wada, K. and Masuhara, H. (2000) Analysis of radiation pressure exerted on a metallic particle within an evanescent field. *Opt. Lett.*, **25**, 1385–1387.
21 Wada, K., Sasaki, K. and Masuhara, H. (2002) Electric charge measurement on a single microparticle using thermodynamic analysis of electrostatic forces. *Appl. Phys. Lett.*, **81**, 1768–1770.
22 Matsuo, Y., Takasaki, H., Hotta, J. and Sasaki, K. (2001) Absorption analysis of a single microparticle by optical force measurement. *J. Appl. Phys.*, **89**, 5438–5441.
23 Inami, W. and Kawata, Y. (2001) Analysis of the scattered light distribution of a tightly focused laser beam by a particle near a substrate. *J. Appl. Phys.*, **89**, 5876–5880.
24 Fujiwara, H., Takasaki, H., Hotta, J. and Sasaki, K. (2004) Observation of the discrete transition of optically trapped particle position in the vicinity of an interface. *Appl. Phys. Lett.*, **84**, 13–15.
25 Hotta, J., Takasaki, H., Fujiwara, H. and Sasaki, K. (2002) Precise analysis of optically trapped particle position and interaction forces in the vicinity of an interface. *Int. J. Nanosci.*, **1**, 645–649.
26 Meiners, J.-C. and Quake, S. R. (1999) Direct measurement of hydrodynamic cross correlations between two particles in an external potential. *Phys. Rev. Lett.*, **82**, 2211–2214.
27 Bartlett, P., Henderson, S. I. and Mitchell, S. J. (2001) Measurement of the hydrodynamic forces between two polymer-coated spheres. *Philos. Trans. R. Soc. London, Ser.A*, **359**, 883–895.
28 Doi, M. and Edwards, S. F. (1986) *The Theory of Polymer Dynamics*, Clarendon Press, Oxford, Chapter 3.
29 Happel, J. and Brenner, H. (1983) *Low Reynolds Number Hydrodynamics with Special Applications to Particulate Media*, Martinus Nijhoff Publishers, The Hague, Chapter 6.

8
Construction of Micro-Spectroscopic Systems and their Application to the Detection of Molecular Dynamics in a Small Domain

Syoji Ito, Hirohisa Matsuda, Takashi Sugiyama, Naoki Toitani, Yutaka Nagasawa, and Hiroshi Miyasaka

8.1
Introduction

The elucidation of molecular dynamics in a small domain, especially intermolecular interaction via rotational/translational diffusion, is of significant importance for a rational understanding of cooperative/hierarchical phenomena of molecular systems, such as crystallization from the solution phase, crosslinking of polymers, and segregation processes. In order to comprehensively elucidate the dynamics in these complex phenomena taking place in a small domain, an apparatus with spatial and temporal resolution is indispensable. In this chapter, we will introduce several microscopic systems that have been developed by the combination of a confocal optical set-up and appropriate laser light sources. First, we will show a laser microscope with 35 fs excitation pulse under objectives and its application to higher order multiphoton excitation and fluorescence imaging. Then we will introduce the measurement of temperature in micro-space under a laser trapping condition. In this measurement, we have utilized a fluorescence correlation technique. In the third section, we will discuss the single-particle-level relaxation dynamics of quantum dots diffusing freely in water, as revealed by fluorescence correlation detection under the microscope.

8.2
Development of a Near-Infrared 35 fs Laser Microscope and its Application to Higher Order Multiphoton Excitation

Spatially resolved measurements, based on the confocal laser microscope and related techniques, have recently enabled direct detection of individual molecules, single nanoparticles, and molecular assemblies, leading to elucidation of the heterogeneous nature of these systems and its dependence on the individual environments.

Molecular Nano Dynamics, Volume I: Spectroscopic Methods and Nanostructures
Edited by H. Fukumura, M. Irie, Y. Iwasawa, H. Masuhara, and K. Uosaki
Copyright © 2009 WILEY-VCH Verlag GmbH & Co. KGaA, Weinheim
ISBN: 978-3-527-32017-2

In addition, combining the microscope with the use of a pulsed laser light source provides temporal information on these systems in a small domain. The dispersion of refractive index, however, strongly affects the temporal resolution in the measurements of dynamics under the microscope and typical resolution stays around 100 fs when a Ti:Sapphire laser is used as an excitation source.

The dispersion of the refractive indices is small in the near-infrared (NIR) region >1 μm, compared to that in the visible region. Hence, we can expect ultrashort pulse duration attainable under the microscope in the NIR region. The ultrashort NIR pulse can also provide several advantages: (i) the low probability of light scattering leading to deeper penetration depth than that of visible light can improve imaging capability for opaque materials, such as biological specimens; (ii) the NIR pulse causes less damage to samples; and (iii) the quite high photon intensity in the short pulse duration can easily induce higher order multiphoton processes, leading to higher spatial resolution in three dimensions.

From these viewpoints, we have developed a femtosecond NIR laser microscope with a home-built cavity dumped chromium: forsterite (Cr:F) laser as an excitation light source whose output wavelength is centered at 1260 nm. In the following the set-up of the NIR laser microscope and its application to multiphoton imaging are presented.

8.2.1
Confocal Microscope with a Chromium: Forsterite Ultrafast Laser as an Excitation Source

A schematic diagram of the confocal microscope system with the Cr:F laser is shown in Figure 8.1a. The spectrum of the laser pulse covers the wavelength region from 1.2 to 1.35 μm with an FWHM of 77 nm, as shown in Figure 8.1b. The output of the Cr:F laser was guided into an optical microscope (IX 71, Olympus) after passing through a prism pair for the optimal compensation of the pulse duration at the sample plane of the microscope. In the optical set-up, a Michelson interferometer was introduced for the time-resolved measurements; the NIR pulse was divided into two pulses with the same intensity. The optical delay unit (nanomover, Melles Griot) with a minimum step of 10 nm enabled us to measure ultrafast excitation dynamics at 66.7 as (attosecond) intervals. A dichroic mirror (800DCSX, Chroma Technology) attached to the optical microscope realized the selective reflection of the NIR laser pulse toward an objective (MPlan 100 × IR, Olympus, NA = 0.95) and selective transmission of visible light to the detection system of the microscope. The emission in the visible region from samples was guided into an avalanche photodiode (C5460-01, Hamamatsu) or an optical fiber. For the effective detection of the fluorescence signal, a lock-in amplifier (Model 5210, EG&G Instruments) was introduced into the detection system. A fiber-coupled spectrometer (SD2000, Ocean Optics) was employed for the spectral measurements.

Figure 8.1c shows the interferometric autocorrelation signal of second harmonic generation (SHG) from a BBO crystal positioned at the sample plane of the microscope. The shape of the SHG trace was symmetrical with respect to the time origin; the ratio of the maxima to the background was 8 : 1, indicating that nearly ideal

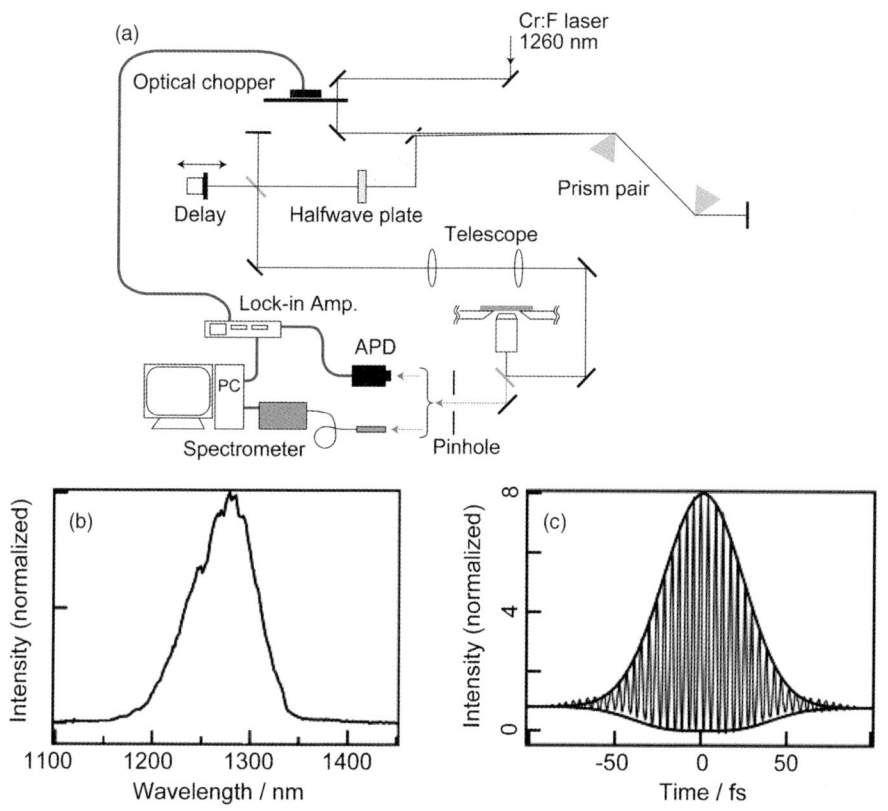

Figure 8.1 (a) Block diagram of the femtosecond near-infrared laser microscope system. (b) Spectrum of the light pulse from the Cr:F laser. (c) Interferometric autocorrelation trace of SHG signal with envelope curve calculated assuming a chirp-free Gaussian pulse with 35 fs fwhm.

irradiation conditions were preserved even under the objective with high refractive index dispersion. The envelope of the SHG signal in Figure 8.1c is the result calculated on the assumption of a Gaussian pulse with 35 fs duration under the objective. From the reproduction of the SHG autocorrelation signal by the envelope thus calculated, we can safely conclude that the dispersion of the ultrashort laser pulse in the NIR region could be well compensated to almost its transform limit.

8.2.2
Detection of Higher Order Multiphoton Fluorescence from Organic Crystals

Figure 8.2a–c shows optical transmission images of organic microcrystals of perylene, anthracene, and pyrene, excited at a laser power of 1.7 nJ pulse^{-1} under similar excitation conditions as in Figure 8.1c. The bright spots <2 μm in diameter at the center of each microcrystal were areas irradiated with the NIR laser pulse.

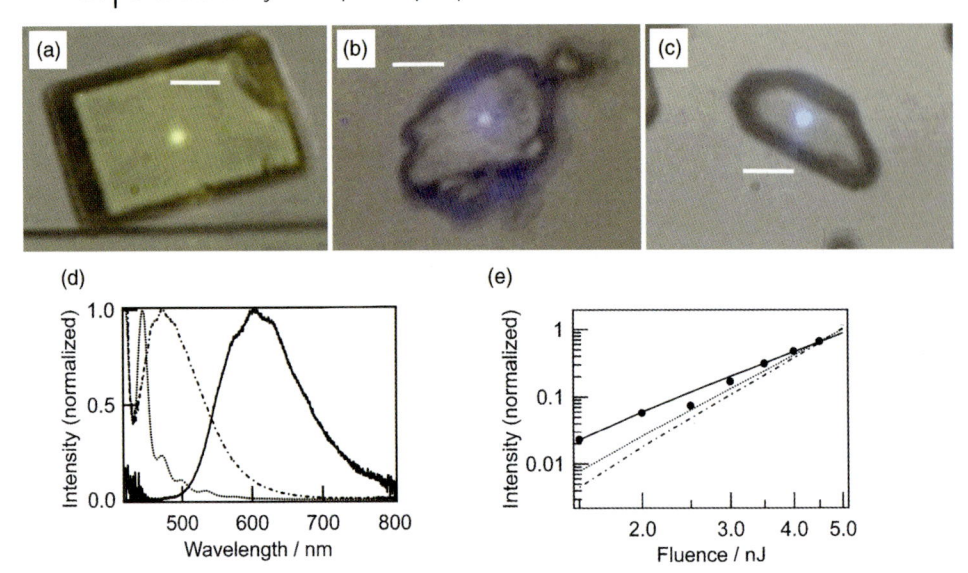

Figure 8.2 Optical transmission images of perylene (a), anthracene (b), and pyrene (c) microcrystals irradiated by the NIR laser; scale bar ~5 μm. (d) Emission spectra of fluorescence spots in the microcrystals of anthracene (dotted line), pyrene (broken line), and perylene (smooth line). (e) The dependence of the fluorescence intensities of the pyrene (closed circle), anthracene (open circle), and perylene (open square) microcrystals on the incident NIR Cr:F laser intensity with lines calculated by the least-squares method. The slopes were 2.8, 3.9, and 4.3 for perylene, anthracene, and pyrene, respectively.

Emission spectra at these points are shown in Figure 8.2d. The band shapes were independent of the excitation intensity from 0.1 to 2.0 nJ pulse^{-1}. The spectrum of the anthracene crystal with vibronic structures is ascribed to the fluorescence originating from the free exciton in the crystalline phase [1, 2], while the broad emission spectra of the pyrene microcrystal centered at 470 nm and that of the perylene microcrystal centered at 605 nm are, respectively, ascribed to the self-trapped exciton in the crystalline phase of pyrene and that of the α-type perylene crystal. These spectra clearly show that the femtosecond NIR pulse can produce excited singlet states in these microcrystals.

Figure 8.2e shows the dependence of the fluorescence intensity on the excitation power of the NIR light for the microcrystals measured with a 20× objective. In this plot, both axes are given in logarithmic scales. The slope of the dependence for the perylene crystal is 2.8, indicating that three-photon absorption is responsible for the florescence. On the other hand, slopes for the perylene and anthracene crystals are 3.9 for anthracene and 4.3 for pyrene, respectively. In these cases, four-photon absorption resulted in the formation of emissive excited states in the crystals. These orders of the multiphoton absorption are consistent with the absorption-band edges for each crystal. The four-photon absorption cross section for the anthracene crystal was estimated to be 4.0×10^{-115} cm^8 s^3 photons^{-3} by comparing the four-photon induced fluorescence intensity of the crystal with the two-photon induced fluorescence intensity of the reference system (see ref. [3] for more detailed information).

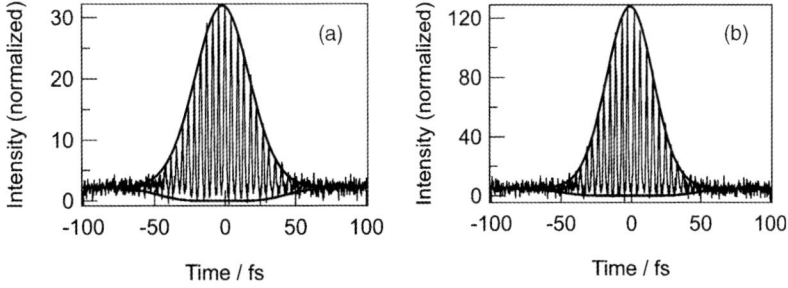

Figure 8.3 Interferometric autocorrelation traces of the fluorescence intensities of perylene (a) and anthracene (b) microcrystals irradiated by two NIR Cr:F laser pulses centered at 1.26 μm with the same intensity.

The four-photon absorption cross section thus obtained is overall consistent with those of other aromatic molecules in solution described in literatures [4, 5].

To confirm the mode of the multiphoton excitation process more directly, a simultaneous process or stepwise process via metastable states attained by lower order multiphoton processes such as S–T absorption, we applied the interferometric detection of the fluorescence from the microcrystals using a pair of NIR pulses with the same intensity. Figure 8.3 shows interferometric autocorrelation signals from perylene and anthracene microcrystals on the sample plane of the confocal microscope under irradiation with NIR light at about 100 pJ pulse^{-1}. The peak-to-background ratios of the autocorrelation curves were ~32:1 for perylene and 128:1 for anthracene, respectively. These results clearly exhibit the distinctive features of three- and four-photon autocorrelation signals [6, 7]. In addition, envelopes of the autocorrelation curves in Figure 8.3 were well reproduced by curves calculated with the parameter of 35 fs pulse duration, indicating that the multiphoton absorption was completed within the pulse duration. From these results we conclude that the excited states of the microcrystals were produced via simultaneous multiphoton absorption.

8.2.3
Multiphoton Fluorescence Imaging with the Near-Infrared 35 fs Laser Microscope

Figure 8.4a and b shows a scanning three-photon fluorescence image of perylene microcrystals obtained at an excitation power of 70 pJ pulse^{-1} and the corresponding optical transmission image. The width of a microcrystal along the line PQ in the fluorescence image is about 2.5 μm, while the corresponding length in the transmission image is about 1.8 μm. The difference in size between the two images is consistent with the accessible size of a three-photon reaction (400 nm, FWHM) based on the diffraction limit. A scanning four-photon fluorescence image of anthracene microcrystal obtained at an excitation power of 400 pJ pulse^{-1} is also shown in Figure 8.4c with the corresponding optical transmission image (Figure 8.4d). The characteristic structure of the anthracene microcrystal observed in the transmission

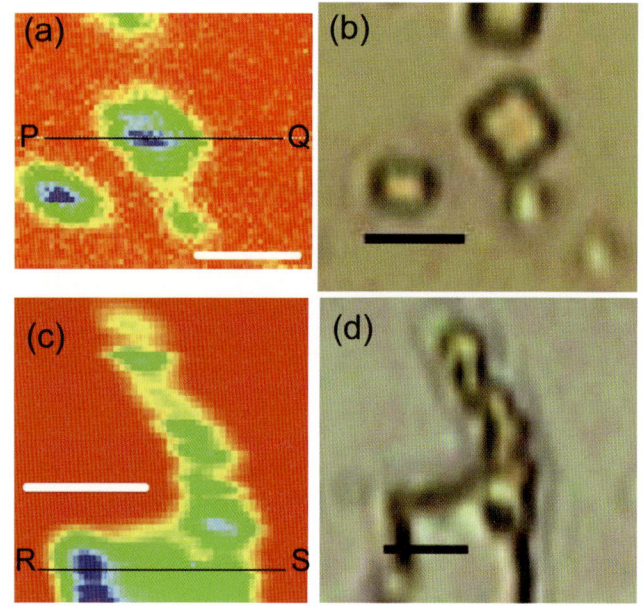

Figure 8.4 (a) Scanning three-photon fluorescence image of perylene microcrystals obtained by irradiation of the NIR pulse of 1260 nm with power 70 pJ pulse^{-1}; scanning step ~100 nm. (b) Corresponding optical transmission image of the perylene crystals. (c) Scanning four-photon fluorescence image of an anthracene microcrystal obtained by irradiation of the NIR pulse with power 400 pJ pulse^{-1}; scanning step ~100 nm. (d) Corresponding optical transmission image of the anthracene microcrystal. Scale bar ~2 m.

image was well reproduced in the scanning fluorescence image. The width of the microcrystal along the line RS in Figure 8.4c is about 4.0 μm in the four-photon fluorescence image, while it is about 3.7 μm in the optical transmission image. The difference in crystal size between the two images is smaller than that in the three-photon imaging. This can be explained by improvement in the spatial resolution due to higher-order multiphoton fluorescence. During the imaging process, no photo-damage of the sample was observed even through the three- or four-photon absorption cross section is very small.

As stated in the introductory part of this section, the multiphoton excitation by NIR laser pulse has an advantage in the excitation of opaque materials with less damage. This advantage seems to be effective in the measurement of biological systems. However, water has an absorption band originating from the overtone of OH-stretching in the wavelength region of around 1.25 μm. The one-photon absorption may decrease the penetration depth of the excitation light into aqueous samples, and consequently this absorption may prevent clear imaging. With the aim to check the applicability of the multiphoton microscope to such samples with water, we obtained a two-photon fluorescence image of a kind of algae in water. As shown in Figure 8.5b, *Zygnema* in water was clearly visualized by plotting two-photon induced fluorescence intensity from the sample, indicating that the NIR multiphoton microscope can be used for imaging micro systems in an aqueous environment.

(a)   (b)

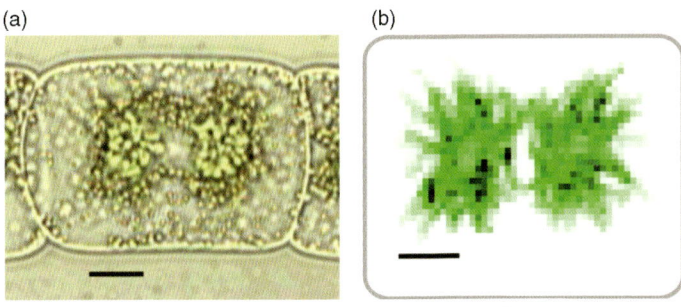

Figure 8.5 Transmitting (a) and two-photon fluorescence (b) images of a part of Zygnema in water. Scale bar is 5 μm in length.

8.3
Application of Fluorescence Correlation Spectroscopy to the Measurement of Local Temperature at a Small Area in Solution

Fluorescence intensity detected with a confocal microscope for the small area of diluted solution temporally fluctuates in sync with (i) motions of solute molecules going in/out of the confocal volume, (ii) intersystem crossing in the solute, and (iii) quenching by molecular interactions. The degree of fluctuation is also dependent on the number of dye molecules in the confocal area (concentration); with an increase in the concentration of the dye, the degree of fluctuation decreases. The autocorrelation function (ACF) of the time profile of the fluorescence fluctuation provides quantitative information on the dynamics of molecules. This method of measurement is well known as fluorescence correlation spectroscopy (FCS) [8, 9].

We have applied FCS to the measurement of local temperature in a small area in solution under laser trapping conditions. The translational diffusion coefficient of a solute molecule is dependent on the temperature of the solution. The diffusion coefficient determined by FCS can provide the temperature in the small area. This method needs no contact of the solution and the extremely dilute concentration of dye does not disturb the sample. In addition, the FCS optical set-up allows spatial resolution less than 400 nm in a plane orthogonal to the optical axis. In the following, we will present the experimental set-up, principle of the measurement, and one of the applications of this method to the quantitative evaluation of temperature elevation accompanying optical tweezers.

8.3.1
Experimental System of FCS

The experimental set-up for the FCS measurement is illustrated schematically in Figure 8.6. A CW Ar^+ laser (LGK7872M, LASOS lasertechnik GmbH) at 488 nm was coupled to a single mode optical fiber to isolate the laser device from an experimental table on which the confocal microscope system was constructed. This excitation laser light transmitted through the optical fiber was collimated with a pair of lenses, and then was guided into a microscope objective (100X, NA: 1.35, Olympus).

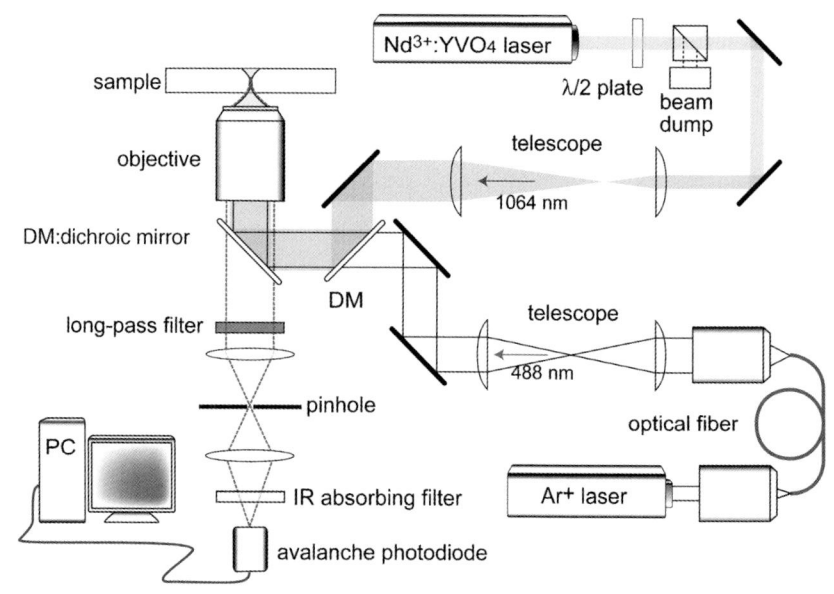

Figure 8.6 Schematic illustration of the FCS system with optical tweezers.

The detection volume (confocal volume) of the FCS measurement was controlled by a pinhole (typically 25–50 μm) attached to an optical microscope (IX70, Olympus). The photons emitted from dye molecules in the confocal volume were guided to an avalanche photodiode (SPCM-AQR-14, Perkin Elmer), which was connected to a counting board (M9003, Hamamatsu photonics K.K.). An edge (long-pass) filter (Semrock, LP01-488RU) blocked the light of 488 nm scattered from the sampling area toward the photodetector. The autocorrelation function of the fluorescent intensity was obtained using FCS software (U9451, Hamamatsu photonics K.K.) that presents autocorrelation function from the raw temporal data integrated for ∼3 s. In the present study, the autocorrelation functions were accumulated 20–30 times to improve the signal to noise (S/N) ratio. A CW beam of 1064 nm from a NIR laser (J20-BF-106W, Spectra-Physics) was overlapped with the 488-nm beam coaxially and was focused to the same point as the blue light, realizing the FCS measurement at the optical trapping point. The stray light of the NIR beam propagating to the detection system was also eliminated with an IR-absorbing filter (Sigma Koki, HAF-50S-30H).

8.3.2
The Principle of the Method of Measurement of Local Temperature Using FCS

The autocorrelation function, $G(\tau)$, of the temporal fluctuation of the fluorescence intensity at the confocal volume is analytically represented by the following equation [8, 9]:

$$G(\tau) = 1 + \frac{1}{N}\left(1 + \frac{p}{1-p}\exp\left(-\frac{\tau}{\tau_T}\right)\right)\left(1 + \frac{\tau}{\tau_D}\right)^{-1}\left(1 + \frac{\tau}{w^2\tau_D}\right)^{-1/2} \quad (8.1)$$

8.3 Application of Fluorescence Correlation Spectroscopy

where, N is the average number of molecules in the confocal volume, V_{conf}, with cylindrical shape, p is the fraction of molecules in the triplet state, τ_T is the lifetime of the triplet state, and w is the structure parameter defined by $w = w_z/w_{xy}$. Here, w_z and w_{xy} are, respectively, the axial length and radial radii of the cylindrical confocal volume ($V_{conf} = 2\pi w_z w_{xy}^2$). τ_D is the average time for a molecule to go across the confocal volume; this characteristic time is called the "mean residence time" or "diffusion time". This diffusion time, τ_D, is related to the translational diffusion coefficient, D, as represented by Eq. (8.2).

$$\tau_D = \frac{w_{xy}^2}{4D} \tag{8.2}$$

Under the condition that the Stokes–Einstein model holds, the translational diffusion coefficient, D, can be represented by Eq. (8.3). the diffusion time, τ_D, obtained through the analysis is given by Eq. (8.4).

$$D = \frac{kT}{6\pi\eta(T)a} \tag{8.3}$$

Here, T is the absolute temperature, k the Boltzmann constant, $\eta(T)$ the viscosity of the solution at temperature, T, and a is the hydrodynamic radius of a probe molecule in the sample solution. From Eqs. (8.2) and (8.3), we obtain Eq. (8.4).

$$\frac{\gamma}{\tau_D} = \frac{T}{\eta(T)} \quad \left(\gamma = \frac{6\pi a w_{XY}^2}{4k}\right) \tag{8.4}$$

The $\eta(T)$ can be independently measured by a viscometer and the value of γ is determined by the FCS measurement at a certain temperature (typically 21~22 °C). Under the condition that the hydrodynamic diameter of the probe molecule is constant in the temperature range examined, we can obtain the temperature of the confocal area. It is worth noting that the present method estimates "average" temperature inside the confocal volume of the microscopic system because FCS provides the average value of the translational diffusion velocity over multiple fluorescent molecules passing through the sampling area.

8.3.3
Measurement of Local Temperature for Several Organic Solvents

As described in the above section, information on the relation between the temperature and the viscosity of sample solutions is indispensable for determining the temperature from the result of FCS measurement. Examples of $\eta(T)$ obtained by a conventional method [10] are shown in Figure 8.7.

Figure 8.8a and b, respectively, show fluorescence autocorrelation curves of R6G in ethylene glycol and R123 in water at 294.4 K. The solid lines in these traces are curves analyzed by the nonlinear least square method with Eq. (8.1). Residuals plotted on top of the traces clearly indicate that the experimental results were well reproduced by the

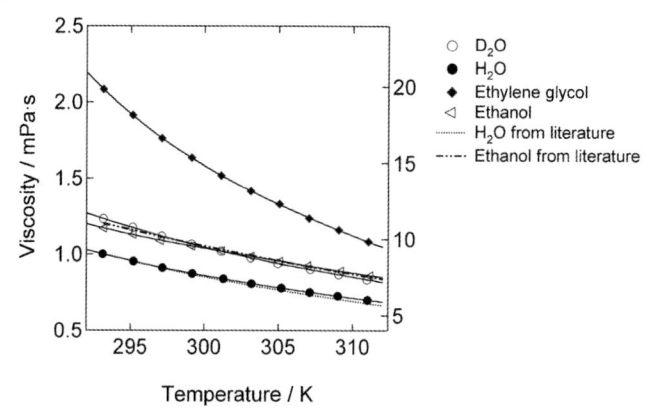

Figure 8.7 Temperature dependences of viscosity for several solvents measured with conventional Ostwald viscometers. Markers exhibit experimental results. Data points were interpolated by polynomial function; the calculated curves are drawn with lines.

calculated curves. FCS signals for other solutions examined here were also well reproduced by Eq. (8.1). The lateral size of the sampling volume, w_{xy}, under the present experimental condition was determined to be 350 nm in diameter on the basis of the diffusion constants of R6G [11, 12] (2.8×10^{-10} m^2 s^{-1}) and R123 [13] (3.0×10^{-10} m^2 s^{-1}) in water as references.

The fluorescence autocorrelation curves of R6G in ethylene glycol and R123 in water in the presence of the focused NIR laser light are plotted in Figure 8.8c and d, respectively, indicating that decay of the correlation curves becomes fast with an increase in the incident NIR laser power. These correlation curves obtained under irradiation of the NIR laser light were also well reproduced by Eq. (8.1), as was shown in Figure 8.6 without incident NIR laser light.

On the change in the translational motion of the molecule, it is worth noting the optical trapping effect by the radiation force of the tightly focused NIR laser light. The optical force that a particle experiences under the optical trapping condition is dependent on its size and the intensity of the incident laser light. From a simple model using the Rayleigh approximation [14], it is predicted that the optical force potential depth for a particle of about 10 nm in diameter under irradiation of several hundred mW was equal to the averaged energy of thermal motion at room temperature ($\sim 4.0 \times 10^{-21}$ J). This indicates that a particle size of at least 10–20 nm in diameter is necessary for effective optical trapping. The optical force for the dye molecules examined here (about 1 nm diameter, estimated from their diffusion coefficient in water) is estimated to be $\leq 10^{-3}$ kT, indicating that the effect of optical trapping is negligible. Hence we can safely attribute the increase in diffusion velocity to the elevation in temperature. Actually, the mean residence time for the R6G in heavy water, that has no significant absorption at the wavelength of 1064 nm, was not influenced by irradiation with the NIR laser light. Summarizing these results and discussion, we can safely conclude that estimation of the local temperature based on Eq. (8.4) can provide reliable values.

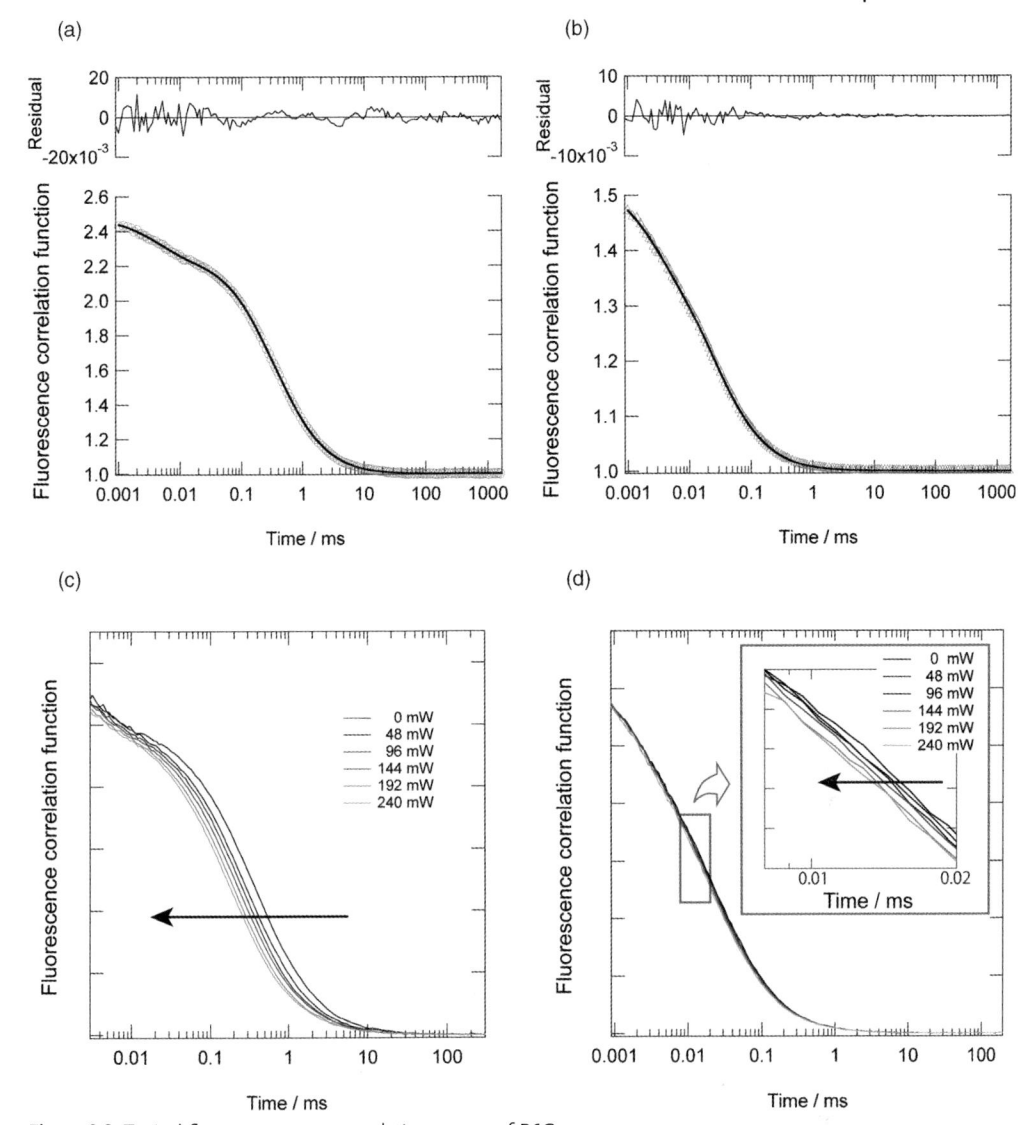

Figure 8.8 Typical fluorescence autocorrelation curves of R6G in ethylene glycol (a) and R123 in water (b) without the NIR laser light with calculated curves (solid line) based on Eq. (8.1) and residuals. Fluorescence autocorrelation curves of R6G in ethylene glycol (c) and R123 in water (d) under irradiation of the NIR laser at several powers up to 240 mW. The inset of Figure 8.8d shows a magnified view of a part of the figure enclosed by a rectangle.

Figure 8.9 shows the temperature at the focusing point of the NIR light for several solvents, measured by the present method, as a function of incident laser power. These plots show that the temperature increases linearly with an increase in the NIR

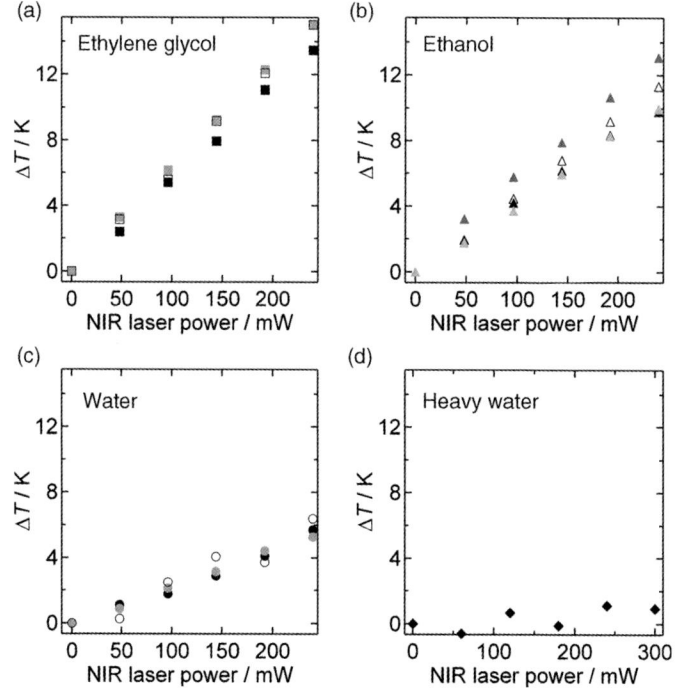

Figure 8.9 Temperature at the focusing point of the NIR laser light measured by the present method for ethylene glycol (a), ethanol (b), water (c), and heavy water (d). The temperature elevation coefficients for these solutions are summarized in Table 8.1

power within the power range examined here, except for heavy water that has no effective absorption at the wavelength of the NIR light. The coefficient of the temperature rise versus the incident laser power, $\Delta T/\Delta P$, is summarized in Table 8.1. No remarkable temperature elevation in D_2O solution again confirms that the temperature elevation is due to the absorption of the NIR laser light by the solvents.

As was discussed in the previous part, the temperature elevation in the solutions can be ascribed to the absorption of the NIR light by the solvents. In order to quantitatively explain the temperature elevation coefficient, $\Delta T/\Delta P$, for other solvents, we proposed a simple model that can parametrize the temperature elevation. As easily predicted, the $\Delta T/\Delta P$ value is closely related to the extinction coefficient of light absorption, α, and the thermal conductivity, λ. Heat generated at the focal point of the NIR beam Q_{abs} is proportional to the extinction coefficient, α, and the incident laser power, P, as represented by Eq. (8.5).

$$Q_{abs} \propto \alpha P \tag{8.5}$$

Thermal conduction is assumed to take place from the small spherical heat source with a radius, r_1. This approximation leads to the one-dimensional heat conduction

Table 8.1 Local temperature deviation, extinction coefficient, thermal conductivity.

Solvent	Extinction coefficient, α [m^{-1}]	Thermal conductivity, λ [W m^{-1} K^{-1}]	α/λ [W^{-1} K]	Molecule	Distance from the bottom of sample cell [μm]	$\Delta T/\Delta P$ [K W^{-1}] (experimental data)
Ethylene glycol	19.9	0.26	75	R6G	30	56 ± 0.5
				R6G	30	63 ± 0.6
				R6G	30	64 ± 0.4
				R6G	20	66 ± 1
				R6G	5	67 ± 3
Ethanol	11.2	0.17	69	R123	30	47 ± 0.5
				R6G	30	56 ± 0.9
				R123	30	42 ± 0.5
				R6G	30	42 ± 0.6
H$_2$O	14.5	0.59	24	R123	30	22 ± 0.8
				R123	30	24 ± 1.9
				R123	30	22 ± 0.3
D$_2$O	~0		~0	R6G	30	2.6 ± 0.1

equation represented by Eq. (8.6).

$$q = -\lambda \frac{dT}{dr} \tag{8.6}$$

Here, q is the flux of heat (W m^{-2}), λ is the thermal conductivity (W m^{-1} K^{-1}), T is temperature (K), and r is the distance from the center of the spherical heat source. Under the steady state approximation, the heat generated in the small sphere, Q_{in}, is equal to the heat flow, Q_{flow}, from the surface of the small sphere to the surrounding medium, as expressed by Eq. (8.7).

$$Q_{in} = Q_{flow} = 4\pi r^2 q \tag{8.7}$$

Substitution of Eq. (8.6) into Eq. (8.7) gives the differential equation:

$$\frac{dT}{dr} = -\frac{Q_{in}}{4\pi\lambda} \frac{1}{r^2} \tag{8.8}$$

Under the boundary condition, $T = T_1$ at $r = r_1$, $T = T_{Room}$ at $r_2 = \infty$, and $Q_{in} \propto Q_{abs}$, we can obtain Eq. (8.9) from Eq. (8.8).

$$\Delta T \propto \frac{\alpha P}{\lambda} \left(\frac{1}{r_1}\right) \tag{8.9}$$

Finally, the relation between the temperature elevation coefficient, $\Delta T/\Delta P$, is given by Eq. (8.10).

$$\frac{\Delta T}{\Delta P} \propto \frac{\alpha}{\lambda} \left(\frac{1}{r_1}\right) \tag{8.10}$$

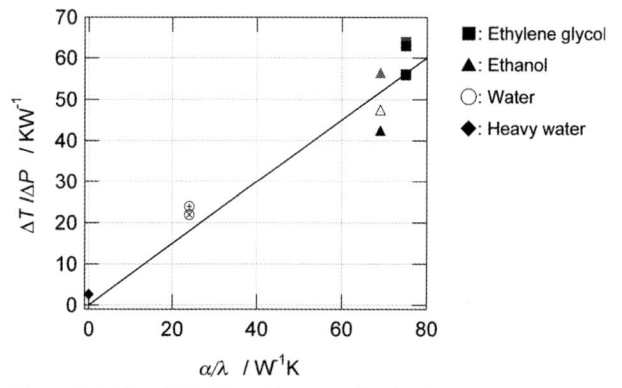

Figure 8.10 Plot of $\Delta T/\Delta P$ as a function of α/λ. The temperature elevation coefficients are proportional to the ratio of the extinction coefficient of the solvents, α, and the thermal conductivity of the solvents, λ.

Eq. (8.9) predicts that the temperature at the focusing point of the NIR light increases in proportion to the incident laser power; this was confirmed experimentally, as shown in Figure 8.9. The simple model expressed by Eq. (8.10) also predicts a linear relation between $\Delta T/\Delta P$ and α/λ. As shown in Figure 8.10, the experimental results obtained in the present study well reproduced this prediction. From these results, it can be concluded that the temperature elevation coefficient is qualitatively determined by these two parameters of solvents, α and λ; we can predict this coefficient for other solvents.

8.3.4
Summary

We have developed a method for measuring temperature in a small domain using FCS and applied the technique to a quantitative evaluation of temperature elevation at the focusing point of NIR laser light in solution under optical trapping conditions. The temperature at the focus of the NIR beam increased in proportion to the incident laser power for solvents with slight absorption at the wavelength. On the other hand, such a temperature rise did not take place in heavy water that has no significant absorption of the NIR light. In order to parametrize the temperature elevation, we proposed a simple model that predicts a linear relation between the temperature elevation coefficient, $\Delta T/\Delta P$, and the value of α/λ; here, α and λ are, respectively, the extinction coefficient and the thermal conductivity of the solvent. We foresee that the present measurement method with high spatial selectivity will be useful for monitoring temperature in small systems and elucidating these dynamics in terms of temperature distribution.

8.4
Relaxation Dynamics of Non-Emissive State for Water-Soluble CdTe Quantum Dots Measured by Using FCS

Since the development of a stabilizing method for semiconductor nanocrystals in solution [15, 16], radiative colloidal particles ranging from 1.5 to 8 nm in diameter, so-called quantum dots, have been attracting considerable attention owing to their several advantages: extreme brightness, higher photostability than that of organic fluorescent molecules, and easy color tunability by control of particle size.

In order to utilize these excellent advantages of the new emissive nanomaterials, it is necessary to comprehensively understand the relaxation dynamics from excited states, which is completely different from that in fluorescence molecules. Since the first report on the intermittency for single quantum dots [17], single particle detection (SPD) methods using scanning confocal microscopy and a wide field imaging technique have been widely applied to the investigation of the blinking behavior. One of the most fruitful results of the investigations is that the on/off time distribution of a quantum dot follows power-law statistics [18, 19] in the time region from several tens of milliseconds to hundreds of seconds. In other words, the average on/off time of quantum dots is dependent on the measurement (integration) time, that is, the time scale of the blinking depends on the observation time.

Although the studies with SPD techniques have provided significant results on the intermittency in quantum dots, the systems of observation were limited to immobile quantum dots in solids, such as polymer films and glass matrices. The immobilization results in intrinsic heterogeneity of the local environment around each quantum dot; the SPD cannot cover the photophysical kinetics in quantum dots in solution of a more homogeneous environment. In addition, the SPD approaches needed conventional "bin-time" longer than ~10 ms for reliable determination of on and off states. This also limits the elucidation of relaxation dynamics for shorter time scales.

As briefly mentioned in the previous section, FCS provides quantitative information on the lifetime of the non-radiative state for molecules in solution in the time range from sub-microseconds to seconds. This method can, potentially, be applied to the characterization of the photophysical properties of quantum dots freely diffusing in solution with higher temporal resolution than the previous SPD. In spite of the potential advantage, however, only a handful studies on quantum dots by FCS have been published up to now, for CdSe [20–23] and for CdTe [24]. In the following, the results of FCS measurement for CdTe quantum dots in water will be presented and the time scale of the "off" time will be characterized on the basis of the autocorrelation function obtained.

8.4.1
Samples and Analysis of Experimental Data Obtained with FCS

CdTe nanocrystals of various sizes were synthesized according to the procedure in the report of Weller et al. [25] The sizes of the CdTe nanocrystals were estimated from their steady state absorption/emission spectrum. The dilute colloidal suspension (typically of the order of 10^{-9} M) of the CdTe quantum dots in water was injected into plastic vessels with cover slips on the bottom for FCS measurement.

As mentioned in the introductory part of this section, quantum dots exhibit quite complex non-radiative relaxation dynamics. The non-radiative decay is not reproduced by a single exponential function, in contrast to triplet states of fluorescent organic molecules that exhibit monophasic exponential decay. In order to quantitatively analyze fluorescence correlation signals of quantum dots including such complex non-radiative decay, we adopted a fluorescence autocorrelation function including the decay component of a stretched exponential as represented by Eq. (8.11).

$$G(\tau) = 1 + \frac{1}{N}\left(1 + \frac{p}{1-p}\exp\left(-\left(\frac{\tau}{\tau_{dark}}\right)^{\alpha}\right)\right)\left(1+\frac{\tau}{\tau_D}\right)^{-1}\left(1+\frac{\tau}{w^2\tau_D}\right)^{-1/2}$$

(8.11)

Here, α is a stretched factor, and τ_{dark} is the characteristic time of the non-radiative relaxation.

8.4.2
Non-Emissive Relaxation Dynamics in CdTe Quantum dots

Figure 8.11a shows steady-state absorption spectra of the CdTe quantum dots in water. Each spectrum in the figure exhibits a distinct peak at a different band corresponding to its size, indicating that all of these suspensions include monodispersed nanocrystals. This mono-dispersibility is also supported by their emission spectra with different peak bands corresponding to particle size, as in Figure 8.11b.

FCS curves for the quantum dots with mono-dispersity were obtained at various excitation laser powers ranging from 10 to 200 μW. Figure 8.12 shows an autocor-

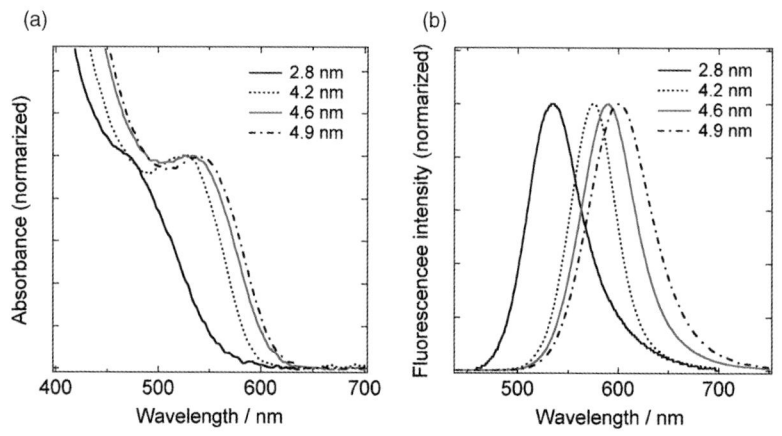

Figure 8.11 Absorption (a) and emission (b) spectra for the water-soluble CdTe quantum dots examined. The sizes for the CdTe dots are indicated in the figures.

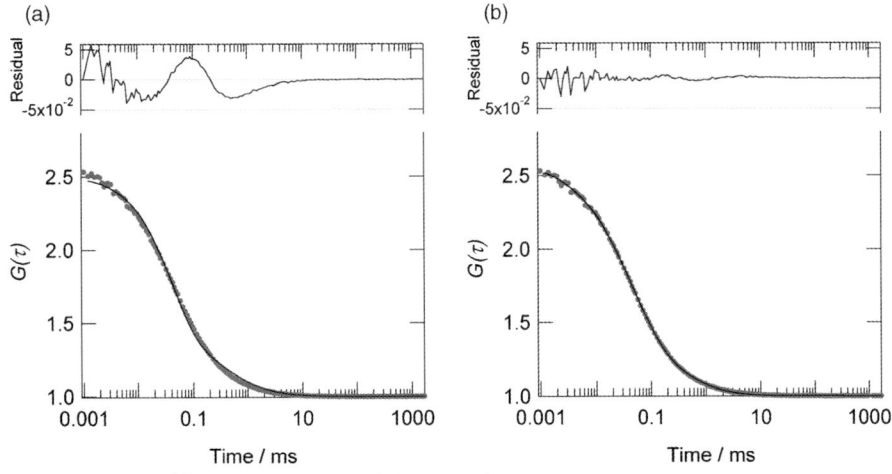

Figure 8.12 Typical fluorescence autocorrelation curve (gray closed circles) of the CdTe quantum dots with 4.6 nm diameter in water with a calculated curve (solid line) based on Eq. (8.1) (a) and based on Eq. (8.3) (b). Residuals are also indicated at the top of each trace.

relation curve for the CdTe nanocrystal with diameter 4.6 nm at the excitation power of 100 μW as a typical example. Figure 8.12a shows that the analytical model with a monophasic decay component expressed by Eq. (8.1) did not reproduce the autocorrelation curve obtained. On the other hand, the analytical model including stretched exponential decay, Eq. (8.11), well reproduced the experimental results, as shown in Figure 8.12b. Similar results were confirmed for other samples with different particle size. These results indicate that the non-emissive state in the monodispersed quantum dots exhibit dispersive kinetics.

The autocorrelation curve for the CdTe quantum dots exhibited excitation intensity dependence; representative results for the CdTe nanocrystals 4.9 nm in diameter are shown in Figure 8.13. The value of the autocorrelation curve at the temporal origin, $G(0)$, decreased with increasing excitation laser power (Figure 8.13a), and the decay of the autocorrelation curves became faster with increasing excitation laser power (Figure 8.13b). Because the medium (water) does not absorb the excitation laser light, the temperature at the measurement area of the FCS measurement stayed constant [26]. Hence, the Brownian motion of the quantum dots was not affected by irradiation of the excitation laser beam. We can therefore attribute the change in the autocorrelation curve in the temporal region of μs–ms to the blinking of the CdTe quantum dots. The decrease in $G(0)$ with increasing excitation power was previously reported for CdSe quantum dots [21]; this was attributed to the convolution between excitation saturation and blinking. However, the saturation effect predicts that the diffusion time becomes slower with an increase in excitation power [27]. Contrary to the prediction, the decay of the autocorrelation curve showed the opposite dependence on the excitation intensity, as in Figure 8.13b, indicating that the simple saturation effect cannot be the origin of the anomalous change in the autocorrelation traces.

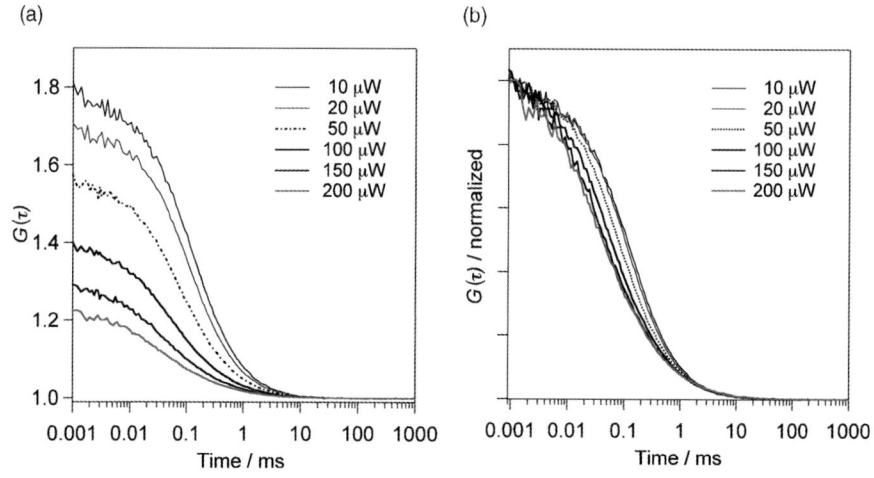

Figure 8.13 Autocorrelation curves for the CdTe quantum dots with diameter 4.9 nm at the excitation laser power from 10–200 μW (a). Comparison of the shapes of these autocorrelation curves by normalization (b).

In addition to the apparent change in the autocorrelation curves for the quantum dots, analysis with Eq. (8.11) provided more quantitative information on the parameter, τ_{dark}, which is a representative of time scale for nonradiative relaxation dynamics in the quantum dots. In this data analysis, the values of τ_{dark} were obtained by the nonlinear least-square method with constant τ_D values, that were determined from the diameters of the quantum dots and the viscosity of water, with the assumption that the diffusion time of the quantum dots is not seriously affected by the intensity change of the excitation laser light. The main two results obtained from the quantitative analysis are summarized in Figure 8.14. First, the values of τ_{dark} were linearly dependent on the diffusion time for each sample, as shown in Figure 8.14a. Second, the parameter τ_{dark} was also dependent on the excitation laser power; the values of τ_{dark} decreased with increasing excitation laser power in the range 10–200 μW, as in Figure 8.14b. Integration of the above results and discussion strongly suggest that the detrapping process of a carrier from trap sites is accelerated by the additional absorption of the excitation light, and the depth of the trap sites may have a distribution and affect the dispersive kinetics.

8.5
Summary

With the aim of elucidating molecular dynamics in a small domain, we have constructed several microspectroscopic systems, that is, (i) the confocal microscope with the excitation light source being a femtosecond NIR laser emitting a 35 fs pulse, and (ii) the fluorescence correlation spectroscopic system with optical tweezers.

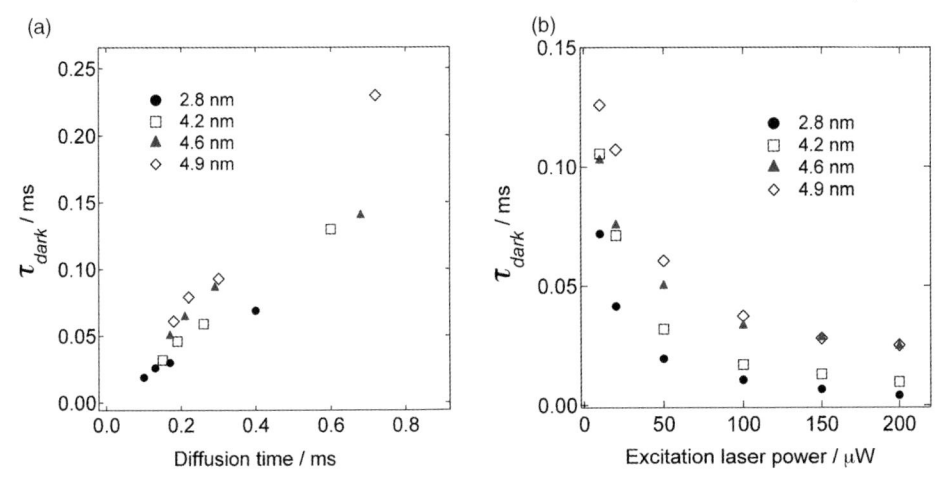

Figure 8.14 (a) Values of τ_{dark} as a function of corresponding diffusion times (observation times) for each CdTe quantum dot. Four sets of measurements for one sample were conducted with different sized pinholes and different solvent as follows: (1) 25 μm pinhole, in water; (2) 25 μm pinhole, in deuterated water; (3) 50 μm pinhole, in water; (4) 100 μm pinhole, in water. All fluorescence correlation curves were obtained at the same excitation laser power <50 μW. (b) Value of the apparent lifetime of the non-emissive states, τ_{dark}, as a function of the excitation laser power for each CdTe quantum dot in water.

The NIR femtosecond laser microscope realized higher order multiphoton excitation for aromatic compounds; interferometric autocorrelation detection of the fluorescence from the microcrystals of the aromatic molecules confirmed that their excited states were produced not via stepwise multiphoton absorption but by simultaneous absorption of several photons. The microscope enabled us to obtain three-dimensional multiphoton fluorescence images with higher spatial resolution than that limited by the diffraction theory for one-photon excitation.

We have also developed a method of measurement for local temperature in microspace with a fluorescence correlation technique. Using this method, the temperature elevation at the optical trapping point due to absorption of the NIR trapping beam by solvent was quantitatively evaluated; the temperature at the trapping point increased linearly with increase in the incident NIR light, and the temperature elevation coefficient was mainly dependent on two physical parameters of the solvent: the absorption coefficient at 1064 nm and the thermal conductivity.

Using the FCS system, we have also examined the relaxation dynamics of non-emissive states in water-soluble CdTe quantum dots in the time region from μs to ms. The autocorrelation curves for the quantum dots were not reproduced by the analytical model including a single diffusion term and a monophasic dark-state contribution, but were well reproduced by the analytical model taking a stretched exponential type relaxation into consideration. The result strongly suggested the widely distributed relaxation pathway of the dark state in the timescale from μs to ms. The apparent lifetime of the non-emissive state determined through the present study exhibited excitation power dependence, suggesting the relaxation process of the non-emissive state promoted by additional absorption of photons. These results could not be obtained by the previous approaches using single molecule detection techniques for immobile particles in solid matrices.

As shown above, we have demonstrated that the microspectroscopic systems can potentially reveal dynamics in heterogeneous small systems. We foresee that the study of microspectroscopy, including the development of new measurement methods and new applications, can pave the way to a rational understanding of complex, heterogeneous, and hierarchical molecular dynamics taking place in small spaces and then give new significant insights to science and technology.

Acknowledgment

The authors thank Professor Hiroshi Masuhara, Professor Tsuyoshi Asahi, Professor Naoto Tamai, and Dr Lingyun Pan for collaboration and helpful discussion. This work was partly supported by Grant-in-Aids for Research in Priority Area (No. 432), from the Ministry of Education, Culture, Sports, Science, and Technology (MEXT) of the Japanese Government and CREST JST.

References

1 Nishimura, H., Yamaoka, T., Hattori, K., Matsui, A. and Mizuno, K. (1985) Wavelength-dependent decay times and time-dependent spectra of the singlet-exciton luminescence in anthracene crystals. *J. Phys. Soc. Jpn.*, **54**, 4370–4381.

2 Matsui, A. and Nishimura, H. (1980) Luminescence of free and self trapped excitons in pyrene. *J. Phys. Soc. Jpn.*, **49**, 657–663.

3 Matsuda, H., Fujimoto, Y., Ito, S., Nagasawa, Y., Miyasaka, H., Asahi, T. and Masuhara, H. (2006) Development of near-infrared 35 fs laser microscope and its application to the detection of three- and four-photon fluorescence of organic microcrystals. *J. Phys. Chem. B*, **110**, 1091.

4 Pantell, R., Pradere, F., Hanus, J., Schott, M. and Puthoff, H. (1967) Theoretical and experimental values for two-, three-, and four-photon absorptions. *J. Chem. Phys.*, **46**, 3507–3511.

5 Hernández, F. E., Belfield, K. D., Cohanoschi, I., Balu, M. and Schafer, K. J. (2004) Three- and four-photon absorption of a multiphoton absorbing fluorescent probe. *Appl. Opt.*, **43**, 5394–5398.

6 Meshulach, D., Barad, Y. and Silberberg, Y. (1997) Measurement of ultrashort optical pulses by third-harmonic generation. *J. Opt. Soc. Am. B*, **14**, 2122–2125.

7 Squier, J. A., Fittinghoff, D. N., Barty, C. P. J., Wilson, K. R., Müller, M. and Brakenhoff, G. J. (1998) Characterization of femtosecond pulses focused with high numerical aperture optics using interferometric surface-third-harmonic generation. *Opt. Commun.*, **147**, 153–156.

8 Rigler, R. and Elson, E. S. (eds) (2001) *Fluorescence Correlation Spectroscopy*, Springer Series in Chemical Physics, **65**, Springer, Berlin.

9 Krichevsky, O. and Bonnet, G. (2002) Fluorescence correlation spectroscopy: the technique and its applications. *Rep. Prog. Phys.*, **65**, 251–297.

10 Riddick, J. A. and Bunger, W. B. (1970) Organic solvents, in *Techniques of Chemistry*, vol. **II**, 3rd edn (ed. A.

Weissberger), Wiley-Interscience, New York, pp. 28–34.
11 Elson, E. L. and Magde, D. (1974) Fluorescence correlation spectroscopy. I. Conceptual basis and theory. *Biopolymers*, **13**, 1–27; Elson, E. L. and Webb, W. W. (1974) Fluorescence correlation spectroscopy. II. An experimental realization. *Biopolymers*, **13**, 29–61.
12 Dorn, I. T., Neumaier, K. R. and Tampe, R. (1998) Molecular recognition of histidine-tagged molecules by metal-chelating lipids monitored by fluorescence energy transfer and correlation spectroscopy. *J. Am. Chem. Soc.*, **120**, 2753.
13 Masuda, A., Ushdia, K. and Okamoto, T. (2005) New fluorescence correlation spectroscopy enabling direct observation of spatiotemporal dependence of diffusion constants as an evidence of anomalous transport in extracellular matrices. *Biophys. J.*, **88**, 3584–3591.
14 Ashkin, A., Dziedzic, J. M., Bjorkholm, J. E. and Chu, S. (1986) Observation of a single-beam gradient force optical trap for dielectric particles. *Opt. Lett.*, **11**, 288–290.
15 Peng, X., Schlamp, M. C., Kadavanich, A. V. and Alivisatos, A. P. (1997) Epitaxial growth of highly luminescent CdSe/CdS core/shell nanocrystals with photostability and electronic accessibility. *J. Am. Chem. Soc.*, **119**, 7019–7029.
16 Dabbousi, B. O., Rodriguez-Viejo, J., Mikulec, F. V., Heine, J. R., Mattoussi, H., Ober, R., Jensen, K. F. and Bawendi, M. G. (1997) (CdSe)ZnS core-shell quantum dots: synthesis and characterization of a size series of highly luminescent nanocrystallites. *J. Phys. Chem. B*, **101**, 9463–9475.
17 Nirmal, M., Dabbousi, B. O., Bawendi, M. G., Macklin, J. J., Trautman, K., Harris, T. D. and Brus, L. E. (1996) Fluorescence intermittency in single cadmium selenide nanocrystals. *Nature*, **383**, 802–804.
18 Kuno, M., Fromm, D. P., Hamann, H. F., Gallagher, A. and Nesbitt, D. J. (2000) Nonexponential "blinking" kinetics of single CdSe quantum dots: A universal power law behavior. *J. Chem. Phys.*, **112**, 3117–3120.
19 Shimizu, K. T., Neuhauser, R. G., Leatherdale, C. A., Empedocles, S. A., Woo, W. K. and Bawendi, M. G. (2001) Blinking statistics in single semiconductor nanocrystal quantum dots. *Phys. Rev. B*, **63**, 205316.
20 Larson, D. R., Zipfel, W. R., Williams, R. M., Clark, S. W., Bruchez, M. P., Wise, F. W. and Webb, W. W. (2003) Water-soluble quantum dots for multiphoton fluorescence imaging *in vivo*. *Science*, **300**, 1434–1436.
21 Doose, S., Tsay, J. M., Pinaud, F. and Weiss, S. (2005) Comparison of photophysical and colloidal properties of biocompatible semiconductor nanocrystals using fluorescence correlation spectroscopy. *Anal. Chem.*, **77**, 2235–2242.
22 Yao, J., Larson, D. R., Vishwasrao, H. D., Zipfel, W. R. and Webb, W. W. (2005) Blinking and nonradiant dark fraction of water-soluble quantum dots in aqueous solution. *Proc. Natl. Acad. Sci.*, **102**, 14284–14289.
23 Swift, J. L., Heuff, R. F. and Cramb, D. T. (2006) A two-photon excitation fluorescence cross-correlation assay for a model ligand-receptor binding system using quantum dots. *Biophys. J.*, **90**, 1396–1410.
24 Ito, S., Toitani, N., Pan, L., Tamai, N. and Miyasaka, H. (2007) Fluorescence correlation spectroscopic study on water-soluble cadmium telluride nanocrystals: fast blinking dynamics in the µs–ms region. *J. Phys.: Condens. Matter*, **19**, 486208.
25 Gaponik, N., Talapin, D. V., Rogach, A. L., Hoppe, K., Shevchenko, E. V., Kornowski, A., Eychmu1ller, A. and Weller, H. (2002) Thiol-capping of CdTe Nanocrystals: An alternative to organometallic synthetic routes. *J. Phys. Chem. B*, **106**, 7177–7185.
26 Ito, S., Sugiyama, T., Toitani, N., Katayama, G. and Miyasaka, H. (2007)

Application of fluorescence correlation spectroscopy to the measurement of local temperature in solutions under optical trapping condition. *J. Phys. Chem. B*, **111**, 2365–2371.

27 Gregor, I., Patra, D. and Enderlein, J. (2005) Optical saturation in fluorescence correlation spectroscopy under continuous-wave and pulsed excitation. *ChemPhysChem*, **6**, 164–170.

9
Nonlinear Optical Properties and Single Particle Spectroscopy of CdTe Quantum Dots

Lingyun Pan, Yoichi Kobayashi, and Naoto Tamai

9.1
Introduction

It is well known that semiconductor quantum dots (QDs) consisting of several hundreds to thousands of atoms show optical properties different from the bulk, in which the discrete electronic states are formed by a quantum confinement effect [1–3]. The preparation of CdSe QDs as a family of II–VI semiconductor nanomaterials has shown enormous development over the past two decades since the new synthetic method using organic ligands was developed in the early 1990s by Bawendi *et al.* [4]. Various investigations have been carried out to improve the synthetic methods and luminescence quantum yields by core–shell methods, and to analyze single particle properties and conductivities in thin films [5–7]. Semiconductor QDs prepared by such colloidal chemistry have various new possibilities that were not considered for QDs prepared by the chemical vapor method, sputtering, or molecular beam epitaxy. One such application is luminescent labeling of biological materials, as suggested by Alivisatos *et al.* and Nie *et al.* [8, 9]. QDs have various advantages such as, (i) high resistance to light as compared to organic materials, (ii) narrow bandwidth of luminescence spectra, (iii) high tunability of absorption and luminescence spectra by the quantum size effect, (iv) large molar extinction coefficient ($>10^5\,\mathrm{M}^{-1}\,\mathrm{cm}^{-1}$), and (v) high luminescence quantum yields and long lifetime, and so on [10]. These advantages make possible superior imaging of proteins or cells and long-term observation as compared with organic molecules.

On the other hand, the nonlinear optical properties of nanometer-sized materials are also known to be different from the bulk, and such properties are strongly dependent on size and shape [11]. In 1992, Wang and Herron reported that the third-order nonlinear susceptibility, $\chi^{(3)}$, of silicon nanocrystals increased with decreasing size [12]. In contrast to silicon nanocrystals, $\chi^{(3)}$ of CdS nanocrystals decreased with decreasing size [13]. These results stimulated the investigation of the nonlinear optical properties of other semiconductor QDs. For the CdTe QDs that we are concentrating on, there have been few studies of nonresonant third-order nonlinear parameters.

Molecular Nano Dynamics, Volume I: Spectroscopic Methods and Nanostructures
Edited by H. Fukumura, M. Irie, Y. Iwasawa, H. Masuhara, and K. Uosaki
Copyright © 2009 WILEY-VCH Verlag GmbH & Co. KGaA, Weinheim
ISBN: 978-3-527-32017-2

As for the size dependence of nonlinear optical properties of semiconductor nanomaterials, detailed investigations are required from both the theoretical and experimental points of view.

Here we summarize our investigations on the nonlinear optical properties of CdTe QDs in the nonresonant wavelength region, in which the size dependence of $\chi^{(3)}$ will be emphasized. In addition, these nonlinear optical properties can be used to monitor the optical manipulation process of nanometer-sized CdTe QDs under a microscope, which may be useful for various arrangements of QDs suitable for information science and technology. Single particle spectroscopy of CdTe QDs is also important to understand the optical and dynamical properties of single QDs. Blinking dynamics of CdTe QDs characteristic for single particles will be explained. We will stress the important role of the Auger process and the environmental conditions for blinking dynamics.

9.2
Nonlinear Optical Properties of CdTe QDs

For the application of QDs to three-dimensional biological imaging, a large two-photon absorption cross section is required to avoid cell damage by light irradiation. For application to optoelectronics, QDs should have a large nonlinear refractive index as well as fast response. Two-photon absorption and the optical Kerr effect of QDs are third-order nonlinear optical effects, which can be evaluated from the third-order nonlinear susceptibility, $\chi^{(3)}$, or the nonlinear refractive index, γ, and the nonlinear absorption coefficient, β. Experimentally, third-order nonlinear optical parameters have been examined by four-wave mixing and Z-scan experiments.

The Z-scan is a sensitive technique for the measurement of nonlinear parameters developed by Sheik-Bahae at the end of the 1980s [14, 15]. Figure 9.1 illustrates a

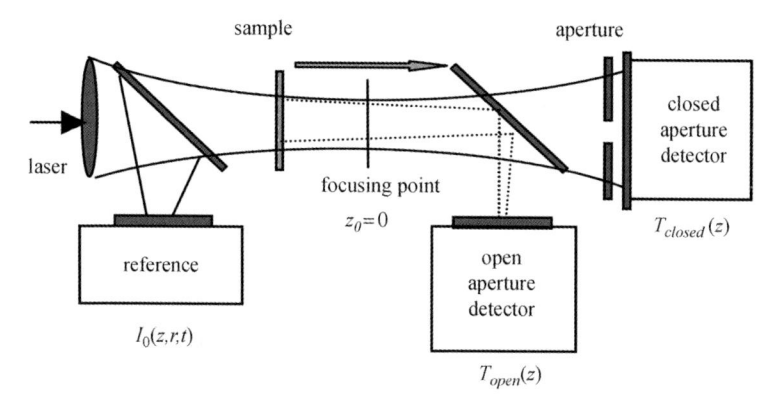

Figure 9.1 A typical Z-scan set-up. A sample is scanned near the focusing point z_0 within ±30 mm. Transmitted light after the sample is separated into two beams; one beam is detected by an open aperture detector to get the open aperture signal $T_{open}(z)$ that includes information on the nonlinear absorption coefficient, β. Another beam is detected by a closed aperture detector after passing through a small aperture to get the closed aperture signal $T_{close}(z)$ that includes information on the nonlinear refractive index, γ.

typical schematic diagram of a Z-scan experiment. A Gaussian beam is focused into a sample, and the sample is scanned around ±30 mm of the focusing point z_0 along the z direction (the propagating direction of the laser) by electronically driving the stage to observe the transmission change. Transmittance change induced by the Kerr effect could be detected by the open and closed aperture. The transmittance detected by the open aperture ($T_{open}(z,\beta)$) and the closed aperture ($T_{close}(z,\gamma)$) contains information on the nonlinear absorption β and nonlinear refractive index γ. β and γ values can be decided from the simulation of transmittance by using Huygens–Fresnel integration. It should be noted that the high repetition rate (~80 MHz) laser may give a large deviation of nonlinear parameters from the expected values since the thermal effect may be superimposed on the Z-scan signals [16].

Figure 9.2 illustrates a typical example of normalized transmittance, $T(z)$, of CdTe QDs against the sample position z from the focusing point with and without aperture [17]. Since the peak of the normalized transmittance for the closed aperture precedes the valley, the sign of the nonlinear refractive index of CdTe QDs is negative. This result suggests that CdTe QDs are self-defocusing materials. Nonlinear absorption was examined by the open aperture experiments. β and γ can be obtained from the intensity dependence on $T_{open}(z,\beta)$ and $T_{close}(z,\gamma)$. These nonlinear parameters, were corrected for volume fraction f_v and local-field effect ϕ to get normalized β_{norm} and γ_{norm}. The real and imaginary parts of $\chi^{(3)}$, Re $\chi^{(3)}$ and Im $\chi^{(3)}$, were also obtained by using β and γ with the following equations, Re $\chi^{(3)} = 2n_{QD}^2 \varepsilon_0 c \gamma$ and Im $\chi^{(3)} = n_{QD}^2 \varepsilon_0 c \lambda \beta / 2\pi$, where n_{QD} is the linear refractive index of the QDs, and ε_0 is the vacuum dielectric constant. The nonlinear parameters of CdTe QDs are summarized in Table 9.1. The two-photon absorption cross section, σ, is calculated

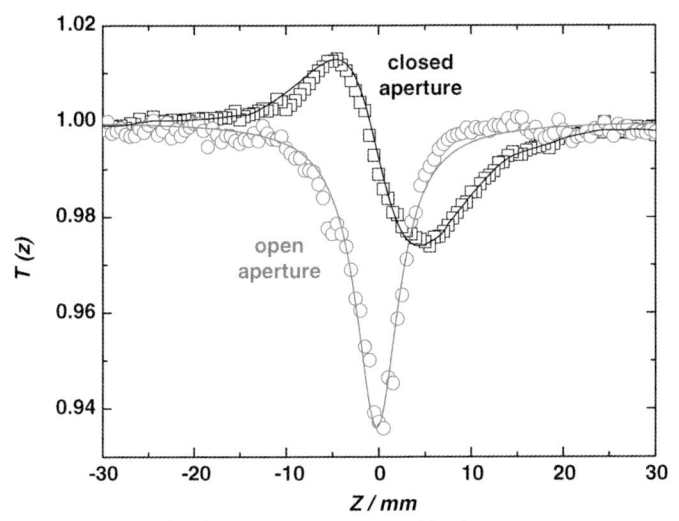

Figure 9.2 Normalized transmittance measured by the Z-scan with and without the collecting aperture for CdTe QDs with the diameter of 4.1 nm excited at 803 nm (0.4 μJ pulse^{-1}). The open aperture Z-scan corresponds to the nonlinear absorption and the closed aperture Z-scan to the nonlinear refractive index.

Table 9.1 Concentration-independent nonlinear parameters of water-soluble CdTe QDs.

	σ/GM^a	$\beta_{\mathrm{norm}}{}^b/$ cm GW^{-1}	$\gamma_{\mathrm{norm}}{}^b/$ cm^2 GW^{-1}	Im $\chi_{\mathrm{QD}}^{(3)}$/esu	Re $\chi_{\mathrm{QD}}^{(3)}$/esu	FOM^c
3.1 nm	2.1×10^2	4.4	-1.3×10^{-3}	1.2×10^{-10}	-5.6×10^{-9}	43
4.1 nm	5.5×10^3	32	-2.0×10^{-3}	8.7×10^{-10}	-8.6×10^{-9}	9.5
4.9 nm	1.7×10^4	61	-2.8×10^{-3}	1.6×10^{-9}	-1.2×10^{-8}	6.8

^{a}GM = 10^{-50} cm^4 s photon^{-1} particle^{-1}.
$^b\beta_{\mathrm{norm}}$ and γ_{norm} are normalized β and γ values by volume fraction and local field effect, respectively. $\beta_{\mathrm{norm}} = \beta/(\phi^4 f_v)$ and $\gamma_{\mathrm{norm}} = \gamma/(\phi^4 f_v)$, where f_v is the volume fraction of QDs, ϕ is the local-field correction given by $\phi = 3\varepsilon_{\mathrm{water}}/(2\varepsilon_{\mathrm{water}} + \varepsilon_{\mathrm{QD}})$, $\varepsilon_{\mathrm{water}}$ is the dielectric constant of water; $\varepsilon_{\mathrm{QD}}$ is the dielectric constant of CdTe.
^{c}FOM: figures of merit (Re $\chi_{\mathrm{QD}}^{(3)}$/Im $\chi_{\mathrm{QD}}^{(3)}$).

from β and used as an absolute parameter to determine the two-photon absorption properties of a single CdTe QD.

From these data, the following important points should be stressed:

1. β_{norm} or σ increases with increasing size of CdTe QDs.
2. The σ value of CdTe QDs is of the order of 10^4 GM (GM = 10^{-50} cm^4 s photon^{-1} particle^{-1}), which is 10~1000 times larger than that of normal organic compounds.
3. The figures of merit (FOM) of CdTe QDs defined by Re $\chi_{\mathrm{QD}}^{(3)}$/Im $\chi_{\mathrm{QD}}^{(3)}$ are larger than the bulk CdTe (FOM$_{\mathrm{bulk}}$ = 1.4) [18], and the value increases with decreasing size.

FOM is an important parameter to determine the performance of optical switching. If the nonlinear optical absorption, Im $\chi_{\mathrm{QD}}^{(3)}$, is too large, the change in refractive index will fall off rapidly as the optical beam propagates, which corresponds to the lower value of FOM. The above result suggests that the smaller sized CdTe QDs are suitable for optical switching applications. In four-wave mixing experiments, temporal response is very fast, similar to the excitation pulse width of a laser, suggesting the signal is mainly due to electronic polarization [17]. A two-photon absorption cross section σ as high as 10^4 GM has been observed in water soluble CdSe/ZnS core–shell QDs prepared by the colloidal method, although the size dependence of the nonlinear parameter is not so clear [19]. The large σ of CdTe QDs is suitable for three-dimensional imaging applications by using a near-IR laser pulse. The size dependence of $\chi^{(3)}$ in CdTe QDs is similar to that of CdS nanocrystals [13]. The size dependence on nonlinear parameters may originate from both the band gap E_g and the energy density of possible transitions [17].

9.3
Optical Trapping of CdTe QDs Probed by Nonlinear Optical Properties

Laser trapping is a technique to manipulate small sized materials, which was developed by Ashkin in 1970 [20, 21]. In this experiment, a laser beam is tightly focused by an objective lens with high numerical aperture (NA), and a dielectric particle near the focusing point experiences a force due to the momentum change of the incident optical field. The optical manipulation and photon pressure science have been developed by Masuhara and coworkers, who demonstrated that gold nanoparticles as small as 40 nm and carbazolyl-containing copolymers of 11~20 nm diameter in aqueous solution can be optically trapped with a cw laser of 100~300 mW [22–30]. The calculation of trapping laser power as described later is consistent with the above results. However, in the case of QDs with a few-nm diameter, an extremely high power (~20 W) cw laser is needed for optical trapping because of the dramatic decrease in particle size. Such high power is difficult to use in conventional optical trapping with a cw YAG laser beam, because the optics of a microscope are easily damaged at high laser power. Here, we introduce the pulse-laser optical trapping of nanometer-sized CdTe QDs at a relatively low intensity of ~100 mW with a high repetition-rate laser [31].

Two limiting cases should be considered for the optical trapping. For particle size $>\lambda/20$, Lorenz–Mie theory based on plane wave scattering is used to describe the optical trapping of an object placed in an arbitrary field distribution. For the particle size $\ll \lambda/20$, such a small particle behaves as an induced elementary dipole and the gradient and scattering forces act on it. In this approximation, the scattering and gradient force components are readily separated. Chaumet and Nieto-Vesperinas obtained the expression for the total time average force on a sphere, $F^i = (1/2)\text{Re}[\alpha E_{0j}\partial^i (E_0^j)^*]$, which includes $F_{\text{grad}} \propto a^3 \nabla I_0$ and $F_{\text{scatt}} \propto a^6 I_0$, where α is the polarizability of the dielectric particles, E_0 is the incident optical field [32, 33]. High incident beam intensity is needed to give a balance between the gradient force F_{grad} towards the origin direction and the scattering force F_{scatt} and drag force F_{drag} from a medium pushing the particle away from the focusing point. For stable trapping in all three dimensions, the axial gradient components of the force pulling the particle towards the focal region must exceed the scattering component of the force pushing it away from that region [32, 33]. For particle size $\ll \lambda/20$, for example, 10 nm, the scattering force can be ignored as compared with the gradient force since it is proportional to a^6 and is two orders weaker than the gradient force. As described by Ashkin and Chu [34], a sufficient trapping condition is that the time to pull particles into the trap potential should be much less than the time for the particles to diffuse out of the trap by Brownian motion. Based on Stokes law for a small sphere, the viscosity-induced drag force is expressed by, $F_{\text{drag}} = 6\pi\eta v a$, where η is the viscosity of D_2O, v is the velocity of the QDs driven by Brownian motion. When $\vec{F}_{\text{grad}} = \vec{F}_{\text{scatt}} + \vec{F}_{\text{drag}}$ is satisfied, QDs can be trapped in the focusing volume with $\pi w_0^2/\lambda$ length. With decreasing size, a much higher laser intensity is required to overcome the drag force, due to the rapid decrease in the gradient force from the a^3 term.

In the Lorenz–Mie regime, it was believed that there is no difference between cw laser and pulse laser trapping, and average power rather than peak power is the key parameter for optical trapping [35]. This comparison may not be adopted for small particles, such as semiconductor QDs with size $\ll \lambda/20$. When the high-repetition pulse laser is used to induce the gradient force, the highest dipole interaction between particles and the electromagnetic field of incident light appears at the peak position of the pulse between successive pulses. As driven by inertia, particles at the edge of the potential well will experience motion toward deep trapping because of the small diffusion distance (a few nm) of nm-sized QDs within the 80 MHz repetition rate. As a result, Rayleigh particles can be optically trapped with a high repetition rate ultrashort pulse laser beam. Figure 9.3 illustrates the photon pressure potential

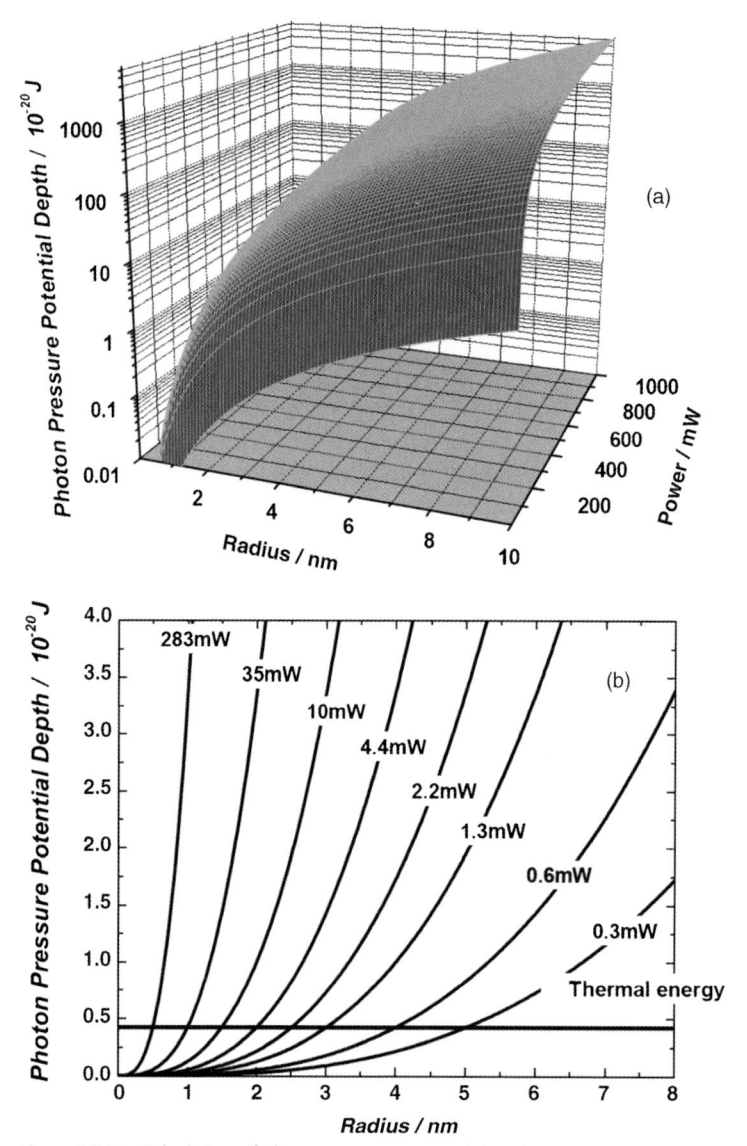

Figure 9.3 (a): Calculation of photon pressure potential against the radius and trapping power of CdTe QDs based on the peak power of YLF laser with 3.5 ps FWHM and 80 MHz repetition rate. (b) Parts of cross sections of (a), the corresponding average power is given in the figure.

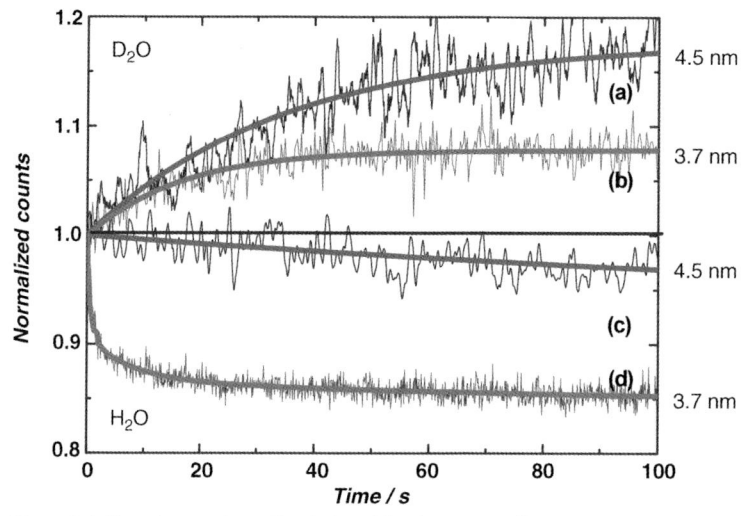

Figure 9.4 Two-photon absorption-induced luminescence of CdTe QDs in D_2O and H_2O obtained from the luminescence spectrum as a function of time. The diameter and solvent are (a) 4.5 nm in D_2O, (b) 3.7 nm in D_2O, (c) 4.5 nm in H_2O, (d) 3.7 nm in H_2O. Incident average laser power is 140 mW.

versus the radius of CdTe QDs based on the calculation of peak power transient trapping by a picosecond YLF laser (1047 nm, 3.5 ps FWHM, 80 MHz). If the photon pressure potential is above the thermal potential of the solution, the particles are supposed to be optically trapped. As clearly shown in the figure, 2-nm radius CdTe QDs can be optically trapped with an incident average power larger than 20 mW, suggesting that a high-repetition pulse laser can trap nanosized particles much more easily than a cw-mode laser with the same average power.

Figure 9.4 shows the time-dependent luminescence intensity of CdTe QDs in D_2O and in H_2O induced by the focused irradiation of a ps YLF laser at 1047 nm [31]. Since CdTe QDs have no absorption in the near-IR wavelength region and the luminescence intensity is proportional to the second power of the laser intensity, the luminescence is induced by a two-photon absorption process of CdTe QDs. As mentioned in our Z-scan experiments, the two-photon absorption cross section, σ, is of the order of 10^3–10^4 GM, which is much larger than a typical organic nonlinear material. In D_2O, two-photon luminescence of CdTe QDs increases with time, while the luminescence intensity of CdTe QDs in H_2O decreases with time. This result suggests that in D_2O the number of CdTe QDs increases with time within a focusing volume, indicating that CdTe QDs are optically trapped by a picosecond YLF laser. In H_2O, the number of CdTe QDs decreases with time within a focusing volume, due to the local heating of H_2O by the excitation of the vibrational overtone at 1047 nm resulting in an increase in thermal Brownian motion that exceeds the optical trapping force. As expected, large sized CdTe QDs are more easily trapped than small sized CdTe QDs, which is confirmed by the initial slope of the two-photon luminescence intensity against time in Figure 9.4. In addition, the intensity decrease in 4.5 nm CdTe QDs in H_2O is almost negligible

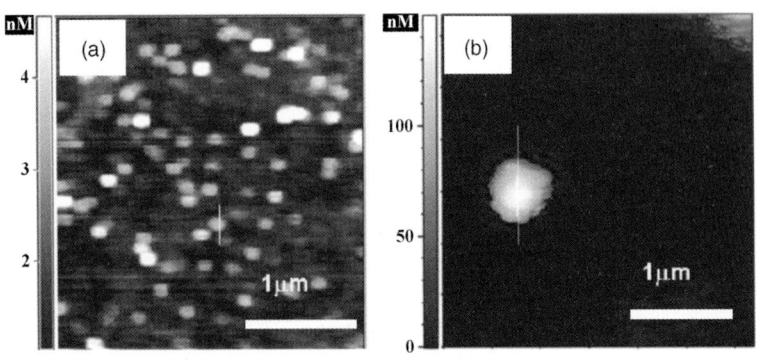

Figure 9.5 AFM images of CdTe QDs (3.3 nm) on a hydrophilic glass substrate after drying the D_2O solution without (a) and with (b) a trapping laser beam. Incident average laser power is 120 mW.

although faster decay is observed for 3.7 nm QDs, suggesting that the gradient force on 4.5 nm CdTe QDs is comparable to the thermal Brownian motion.

The optically trapped CdTe QDs can be fixed on a substrate. The sample was naturally dried with and without trapping laser beam. The volume of trapped particles may be similar to or less than that of a focusing beam. Figure 9.5 illustrates the surface topography of a naturally dried sample with and without a picosecond YLF laser. As clearly shown, an assembled structure of CdTe QDs less than 1 μm diameter is detected only when the trapping beam is applied. Furthermore, the height of the assembled structure is ~150 nm, much larger than CdTe QDs dried naturally without a trapping laser beam. By using a high repetition-rate near-IR pulse laser, as just mentioned above, ~nm-sized CdTe QDs can be optically trapped and fixed on a substrate, in which the two-photon absorption-induced luminescence of CdTe QDs as a nonlinear optical property is used to monitor the optical trapping process. These results will give a basic technique for quantum information science and technology to control and arrange semiconductor QDs freely [36–38].

9.4
Single Particle Spectroscopy of CdTe QDs

Single chromophore detection, that is, the study of single nano-objects such as molecules, QDs, metal colloids, and so on is a modern tool for investigating the physicochemical properties of a single object in interaction with surrounding environments. Several prominent features of single particle spectroscopy are the observation of luminescence intermittency or blinking [39–41], that is a random switching between an emitting and a non-emitting state, and spectral diffusion [42], that is a random shift of luminescence spectra. These observations cannot be detected in ensemble measurements and are characteristic for a single particle. Fluorescence correlation spectroscopy has also been used for analysis of single particle dynamics in solution [43]. Luminescence blinking of single QDs was first observed and analyzed by Nirmal *et al.* using single molecule spectroscopy and microscopy [44, 45]. In their investigations the luminescence intensity time-trajectories of single QDs were observed for an extended period of time, in contrast to single molecules which often suffer fast photobleaching. This high photostability of QDs offers better statistical accuracy with respect to the underlying kinetic phenomena, even on a single particle. In contrast to exponential processes like intersystem crossing-related photon bunching in single molecules, a single CdSe QD follows an inverse power law behavior over many decades in probability density and in time.

The control of the blinking behavior of single QDs is not so easy, which may become obstructive for imaging and single-photon source applications. Several studies have been carried out to understand the blinking mechanism and to suppress the blinking by changing environmental conditions or ligands [46, 47]. A time trace of luminescence intermittency (blinking) is shown in Figure 9.6 for water-soluble CdTe QDs embedded in polyvinyl alcohol (PVA) and trehalose [48]. From these two trajectories one can preliminarily predict that in trehalose media a longer "*on*" time is observed as compared to that in a PVA matrix. The plausible reason behind this is that we prepared the QDs in aqueous solution and hence the QDs preserve their luminescence properties better in an aqueous environment. Trehalose is a common non-reducing disaccharide of glucose found in a variety of organisms such as fungi, plants and animals, where it has a protective role in the case of water stress or dehydration and freezing. It is well known that trehalose helps to reduce the evaporation of water. These results suggest that trehalose is a better environmental matrix as compared to PVA, although both compounds have a hydrophilic OH group.

The blinking phenomenon of QDs and its unusual power-law distribution in the histogram of on–off emission events are still not clearly understood though the origin of luminescence intermittency of QDs has been studied in detail previously. It has been suggested that an electron tunneling from the excited QD to a trap state or Auger ionization may be responsible for the blinking of QDs [49, 50]. After electron transfer to trap states or ionization, the charged QD still absorbs, but is dark because of charge-induced nonradiative relaxation of the exciton energy. QDs become bright again when the trapped electron hops back. Thus the on state depends on the number of surface states or tunneling efficiency, and the off state depends on the stability of charged QDs and trapped electrons. Various parameters such as temperature, QD size, coated-shell thickness, excitation intensity, and so on are considered to affect the on state of QDs [39, 51].

We also found that the excitation intensity has a clear effect on the blinking behavior. Blinking dynamics and its dependence on excitation intensity have been examined by recording luminescence time trajectories and determining statistically the duration of the bright and dark periods for a large number of QDs at different excitation intensities from 0.28 to 7.1 kW cm^{-2}. As clearly shown in Figure 9.7, with increase in excitation intensity the duration of the bright interval values of CdTe QDs in PVA reduced dramatically. This is qualitatively expected for a competing photophysical branching process, the Auger effect. At higher excitation intensity multiple carriers are formed in a single CdTe QD, in which the possible ionization of CdTe QDs will increase due to

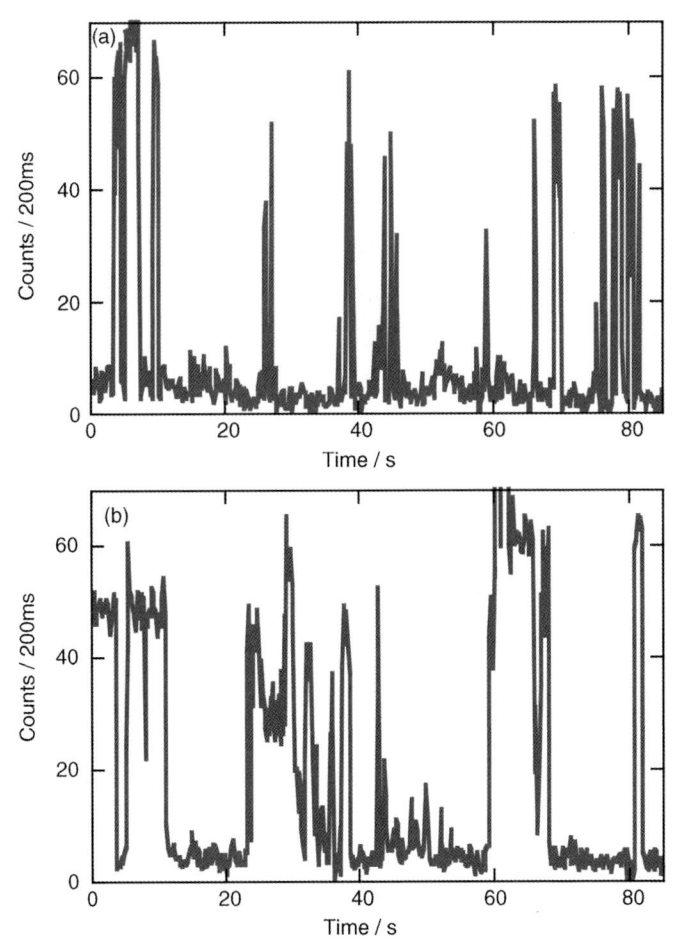

Figure 9.6 Typical luminescence intensity versus time trajectories of a single CdTe QD (4.6 nm) embedded in a PVA (a) and a trehalose (b) matrix dispersed on a cover glass surface at room temperature. The excitation intensity was 1.7 kW cm^{-2} and the integration time was 200 ms bin^{-1}.

carrier–carrier interactions. This nonlinear Auger process turns the particle luminescence off. In contrast, the statistics for "*off*" times is independent of the excitation intensity [48], so that the transition from a dark state to a bright state is not a light driven process and is probably a charge recombination process.

From a theoretical point of view, the blinking kinetics of these CdTe QDs can be quantified by analysis of the on- and off- time probability densities, $P(t_{on})$ and $P(t_{off})$, respectively. Figure 9.8 displays the luminescence intermittency statistics for CdTe QDs in trehalose environment. The distribution of off times involved in the blinking is almost linear on this scale, indicating that the lengths of off times events are distributed according to an inverse power-law of the type, $P(t_{off}) = P_0 t^{-\alpha}$, where P_0 is

9.4 Single Particle Spectroscopy of CdTe QDs

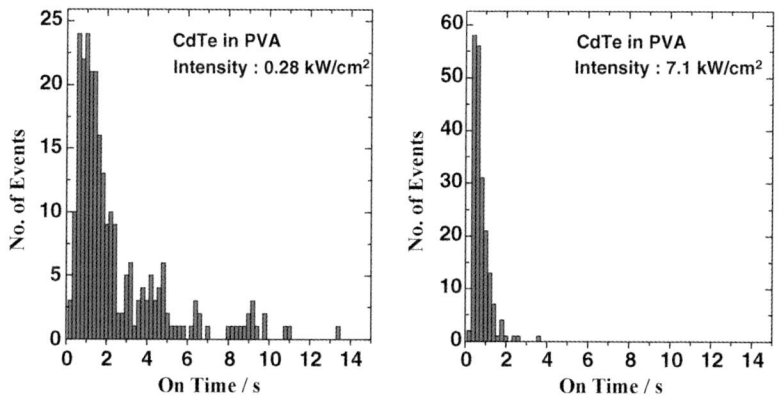

Figure 9.7 On-time histogram of single CdTe QDs in PVA matrix at 0.28 kW cm^{-2} and 7.1 kW cm^{-2}. All histograms were built from time traces recorded under identical experimental conditions except the excitation intensity.

the scaling coefficient and α is the power law exponent characterized by the statistics of each type of event. On the other hand, the on times were fitted with a power law with an exponentially decaying tail with the relaxation time τ_{on}, thereby limiting very long on times, $P(t_{on}) = P_0 t^{-\alpha} exp(-t/\tau_{on})$ [39, 52]. Fitting the data of $P(t_{off})$ to a power law distribution yields $\alpha_{off} = 1.55$ in PVA and $\alpha_{off} = 1.55$ in trehalose, and the data of $P(t_{on})$ to the power law distribution with exponential tail yields $\alpha_{on} = 1.40$ in PVA and $\alpha_{on} = 1.24$ in trehalose for all QDs. This result suggests that trehalose helps to reduce the formation of charged CdTe QDs. Trehalose, as compared with PVA, binds to a large number of water molecules that can be adsorbed on the surface of the QDs and passivate surface traps, which results in an increase in luminescence intensity. For the off time, on the other hand, Issac et al. have observed that an increase in the dielectric constant of the matrix surrounding the QDs leads to a decrease in the value of α_{off} [40]. The matrix in this case was thought to solvate the ejected charge carriers and stabilize the charge separated state. The value of the power law exponent of the off time (α_{off}) statistics scaled linearly with the reaction field factor [$f(\varepsilon)$], which

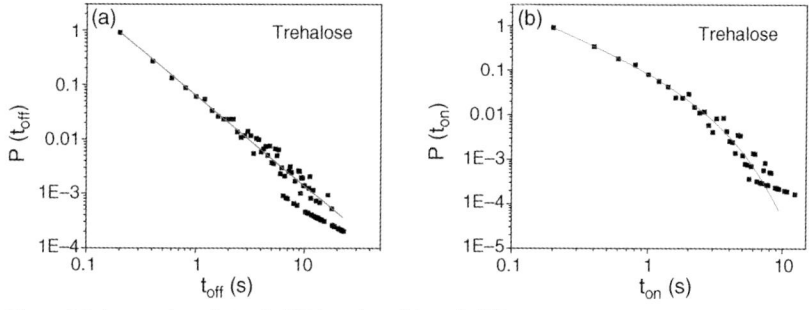

Figure 9.8 Log vs. log plots of off (a) and on (b) probability distribution compiled from a set of CdTe QDs in a trehalose matrix at room temperature.

correlates the stabilization energy with the dielectric properties of the matrix, $f(\varepsilon) = 1 - 1/\varepsilon$, where ε is the static dielectric constant of the matrix. The dielectric constant of trehalose and PVA ($\varepsilon = 14$) are high enough and hence the α_{off} values are considered to be similar to each other [48].

9.5
Summary

Third-order nonlinear optical properties of CdTe QDs were examined by Z-scan and FWM experiments in the nonresonant wavelength region. We found that the two-photon absorption cross section, σ, is as high as $\sim 10^4$ GM, although this value decreases with decreasing size. In addition, the nonlinear response is comparable to the pulse width of a fs laser and the figures of merit (FOM = $\text{Re}\,\chi_{\text{QD}}^{(3)}/\text{Im}\,\chi_{\text{QD}}^{(3)}$) is 40 times enhanced for small-sized CdTe QDs as compared with the bulk, suggesting that CdTe QDs are suitable for all optical switching applications. The optical trapping of nanometer-sized CdTe QDs in D_2O and fixation of QDs on a substrate were also demonstrated by high repetition-rate ps near-IR laser, in which the nonlinear optical properties of QDs were used to monitor the optical trapping process. By using single-particle spectroscopy, we have shown that the emission intermittency is dependent on the aqueous environment surrounding the CdTe QDs, and the lifetimes of the bright on states are prolonged in a trehalose matrix compared to a PVA one. Since the power law statistics for off times are excitation intensity independent, the process that couples a dark state to a bright state is probably a charge recombination process and not the light-driven process. The influence of the environment on the blinking behavior confirms that the charge rearrangements of QDs are relevant for the blinking statistics.

Recent developments for the synthesis of zero-, one-, and two-dimensional semiconductor nanomaterials such as nanowires, nanotubes, and nanosheets, and various microspectroscopy techniques provide systematic understanding of the photophysics and photochemistry of these nanomaterials. Linear and nonlinear optical properties, and Auger and inverse Auger effects and so on, will be analyzed as a function of size and shape of these semiconductor nanomaterials. Blinking dynamics of single nanomaterials are believed to be related to electron or charge transfer/shift processes, so that the real-time observation of electron or charge movements in semiconductor nanomaterials may be analyzed by time-resolved microspectroscopy techniques. Careful tuning of the environments also makes the semiconductor nanomaterials even more suitable for various applications or as model systems for different research.

Acknowledgments

The authors gratefully thank Dr K. Kamada, Dr V. Biju, and Dr M. Ishikawa. This work was partly supported by Grant-in-Aid for Scientific Research on Priority Areas of Molecular Nano Dynamics, from the Ministry of Education, Culture, Sports, Science and Technology (MEXT) of Japan.

References

1. Alivisatos, A. P., Harris, A. L., Levinos, N. J., Steigerwald, M. L. and Brus, L. E. (1988) Electronic states of semiconductor clusters – homogeneous and inhomogeneous broadening of the optical-spectrum. *J. Chem. Phys.*, **89**, 4001–4011.
2. Heath, J. R. (1995) The chemistry of size and order on a nanometer scale. *Science*, **270**, 1315–1316.
3. Alivisatos, A. P. (1996) Perspectives on the physical chemistry of semiconductor nanocrystals. *J. Phys. Chem.*, **100**, 13226–13239.
4. Murray, C. B., Norris, D. J. and Bawendi, M. G. (1993) Synthesis and characterization of nearly monodisperse CdE (E = sulfur, selenium, tellurium) semiconductor nanocrystallites. *J. Am. Chem. Soc.*, **115**, 8706–8715.
5. Qu, L. and Peng, X. (2002) Control of photoluminescence properties of CdSe nanocrystals in growth. *J. Am. Chem. Soc.*, **124**, 2049–2055.
6. Nirmal, M., Dabbousi, B. O., Bawendi, M. G., Macklin, J. J., Trautman, J. K., Harris, T. D. and Brus, L. E. (1996) Fluorescence intermittency in single cadmium selenide nanocrystals. *Nature*, **383**, 802–804.
7. Trindade, T. (2002) Synthetic studies on II/VI semiconductor quantum dots. *Curr. Opin. Solid State Mater. Sci.*, **6**, 347–353.
8. Bruchez, M. Jr, Moronne, M., Gin, P., Weiss, S. and Alivisatos, A. P. (1998) Semiconductor nanocrystals as fluorescent biological labels. *Science*, **281**, 2013–2016.
9. Chan, W. C. W. and Nie, S. (1998) Quantum dot bioconjugates for ultrasensitive nonisotopic detection. *Science*, **281**, 2016–2018.
10. Parak, W. J., Manna, L., Simmel, F. C., Gerion, D. and Alivisatos, P. (2004) *Nanoparticles – From Theory to Application* (ed. G. Schmid), Wiley-VCH, Weinheim, Chapter 2; A. Eychmüller, *ibid.*, Chapter 3.
11. Cho, K. (2004) *Optical Response of Nanostructures*, Springer.
12. Wang, Y. and Herron, N. (1992) Size-dependent nonresonant third-order nonlinear susceptibilities of CdS clusters from 7 to 120 Å. *Int. J. Nonlinear Opt. Phys.*, **1**, 683–698.
13. Prakash, G. V., Cazzanell, M., Gaburro, Z., Pavesi, L., Iacona, F., Franzò, G. and Priolo, F. (2002) Nonlinear optical properties of silicon nanocrystals grown by plasma-enhanced chemical vapor deposition. *J. Appl. Phys.*, **91**, 4607–4610.
14. Sheik-Bahae, M., Said, A. A. and Van Stryland, E. W. (1989) High-sensitivity, single-beam n_2 measurements. *Opt. Lett.*, **14**, 955–957.
15. Sheik-Bahae, M., Said, A. A., Wei, T. H., Hagan, D. J. and Van Stryland, E. W. (1990) Sensitive measurement of optical nonlinearities using a single beam. *IEEE J Quantum Electron.*, **26**, 760–769.
16. Kamada, K., Matsunaga, K., Yoshino, A. and Ohta, K. (2003) Two-photon-absorption-induced accumulated thermal effect on femtosecond Z-scan experiments studied with time-resolved thermal-lens spectrometry and its simulation. *J. Opt. Soc. Am. B*, **20**, 529–537.
17. Pan, L., Tamai, N., Kamada, K. and Deki, S. (2007) Nonlinear optical properties of thiol-capped CdTe quantum dots in nonresonant region. *Appl. Phys. Lett.*, **91**, 051902-1–051902-3.
18. Sun, C. K., Liang, J. C., Wang, J. C., Kao, F. J., Keller, S., Mack, M. P., Mishra, U. and DenBaars, S. P. (2000) Two-photon absorption study of GaN. *Appl. Phys. Lett.*, **76**, 439–441.
19. Larson, D. R., Zipfel, W. R., Williams, R. M., Clark, S. W., Bruchez, M. P., Wise, F. W. and Webb, W. W. (2003) Water-soluble quantum dots for multiphoton fluorescence imaging *in vivo*. *Science*, **300**, 1434–1436.
20. Ashkin, A. (1970) Acceleration and trapping of particles by radiation pressure. *Phys. Rev. Lett.*, **24**, 156–159. (1978) Trapping of atoms by resonance

radiation pressure. *Phys. Rev. Lett.*, **40**, 729–732.

21 Ashkin, A. (2000) History of optical trapping and manipulation of small-neutral particle, atoms, and molecules. *IEEE J. Select. Topics Quantum Electron.*, **6**, 841–856.

22 Hotta, J., Sasaki, K., Masuhara, H. and Morishima, Y. (1998) Laser-controlled assembling of repulsive unimolecular micelles in aqueous solution. *J. Phys. Chem. B*, **102**, 7687–7690.

23 Smith, T. A., Hotta, J., Sasaki, K., Masuhara, H. and Itoh, Y. (1999) Photon pressure-induced association of nanometer-sized polymer chains in solution. *J. Phys. Chem. B*, **103**, 1660–1663.

24 Won, J., Inaba, T., Masuhara, H., Fujiwara, H., Sasaki, K., Miyawaki, S. and Sato, S. (1999) Photothermal fixation of laser-trapped polymer microparticles on polymer substrates. *Appl. Phys. Lett.*, **75**, 1506–1508.

25 Ito, S., Yoshikawa, H. and Masuhara, H. (2001) Optical patterning and photochemical fixation of polymer nanoparticles on glass substrates. *Appl. Phys. Lett.*, **78**, 2566–2568; Ito, S., Yoshikawa, H. and Masuhara, H. (2002) Laser manipulation and fixation of single gold nanoparticles in solution at room temperature. *Appl. Phys. Lett.*, **80**, 482–484.

26 Wada, K., Sasaki, K. and Masuhara, H. (2002) Electric charge measurement on a single microparticle using thermodynamic analysis of electrostatic forces. *Appl. Phys. Lett.*, **81**, 1768–1770.

27 Hosokawa, Y., Matsumura, S., Masuhara, H., Ikeda, K., Shimo-oka, A. and Mori, H. (2004) Laser trapping and patterning of protein microcrystals: Toward highly integrated protein microarrays. *J. Appl. Phys.*, **96**, 2945–2948.

28 Yoshikawa, H., Matsui, T. and Masuhara, H. (2004) Reversible assembly of gold nanoparticles confined in an optical microcage. *Phys. Rev. E*, **70**, 061406-1–061406-6.

29 Hosokawa, C., Yoshikawa, H. and Masuhara, H. (2004) Optical assembling dynamics of individual polymer nanospheres investigated by single-particle fluorescence detection. *Phys. Rev. E*, **70**, 061410-1–061410-7; (2005) Cluster formation of nanoparticles in an optical trap studied by fluorescence correlation spectroscopy. *Phys. Rev. E*, **72**, 021408-1–021408-7.

30 Matsuhara, H., De Schryver, F. C., Kitamura, N. and Tamai, N. (eds) (1994) *Microchemistry: Spectroscopy and Chemistry in Small Domains*, North-Holland Delta Series, Amsterdam.

31 Pan, L., Ishikawa, A. and Tamai, N. (2007) Detection of optical trapping of CdTe quantum dots by two-photon-induced luminescence. *Phys. Rev. B*, **75**, 161305R-1–161305R-4.

32 Chaumet, P. C. and Nieto-Vesperinas, M. (2000) Time-averaged total force on a dipolar sphere in an electromagnetic field. *Opt. Lett.*, **25**, 1065–1067.

33 Harada, Y. and Asakura, T. (1996) Radiation forces on a dielectric sphere in the Rayleigh scattering regime. *Opt. Commun.*, **124**, 529–541.

34 Ashikin, A., Dziedzic, J. M., Bjorkholm, J. E. and Chu, S. (1986) Observation of a single-beam gradient force optical trap for dielectric particles. *Opt. Lett.*, **11**, 288–290.

35 Agate, B., Brown, C., Sibbett, W. and Dholakia, K. (2004) Femtosecond optical tweezers for in-situ control of two-photon fluorescence. *Opt. Express*, **12**, 3011–3017.

36 MacDonald, M. P., Paterson, L., Volke-Sepulveda, K., Arlt, J., Sibbett, W. and Dholakia, K. (2002) Creation and manipulation of three-dimensional optically trapped structures. *Science*, **296**, 1101–1103.

37 Chen, J. C., Chang, A. M. and Melloch, M. R. (2004) Transition between quantum states in a parallel-coupled double quantum dot. *Phys. Rev. Lett.*, **92**, 176801-1–176801-4.

38 Krenner, H. J., Stufler, S., Sabathil, M., Clark, E. C., Ester, P., Bichler, M.,

Abstreiter, G., Finley, J. J. and Zrenner, A. (2005) Recent advances in exciton-based quantum information processing in quantum dot nanostructures. *New J. Phys.*, **7**, 184-1–184-27.

39 Shimizu, K. T., Neuhauser, R. G., Leatherdale, C. A., Empedocles, S. A., Woo, W. K. and Bawendi, M. G. (2001) Blinking statistics in single semiconductor nanocrystal quantum dots. *Phys. Rev. B*, **63**, 205316-1–205316-5.

40 Issac, A., Borczyskowski, C. V. and Cichos, F. (2005) Correlation between photoluminescence intermittency of CdSe quantum dots and self-trapped states in dielectric media. *Phys. Rev. B*, **71**, 161302(R)-1–161302(R)-4.

41 Biju, V., Makita, Y., Nagase, T., Yamaoka, Y., Yokoyama, H., Baba, Y. and Ishikawa, M. (2005) Subsecond luminescence intensity fluctuations of single CdSe quantum dots. *J. Phys. Chem. B*, **109**, 14350–14355.

42 Empedocles, S. A. and Bawendi, M. G. (1999) Influence of spectral diffusion on the line shapes of single CdSe nanocrystallite quantum dots. *J. Phys. Chem. B*, **103**, 1826–1830.

43 Ito, S., Toitani, N., Pan, L., Tamai, N. and Miyasaka, H. (2007) Fluorescence correlation spectroscopic study on water-soluble cadmium telluride nanocrystals: fast blinking dynamics in the µs–ms region. *J. Phys.: Condens. Matter*, **19**, 486208-1–486208-10.

44 Nirmal, M., Dabbousi, B., Bawendi, M. G., Macklin, J. J., Trautman, J. K., Harris, T. D. and Brus, L. E. (1996) Fluorescence intermittency in single cadmium selenide nanocrystals. *Nature*, **383**, 802–803.

45 Nirmal, M. and Brus, L. E. (1999) Luminescence photophysics in semiconductor nanocrystals. *Acc. Chem. Res.*, **32**, 407–414.

46 Odoi, M. Y., Hammer, N. I., Sill, K., Emrick, T. and Barnes, M. D. (2006) Observation of enhanced energy transfer in individual quantum dot–oligophenylene vinylene nanostructures. *J. Am. Chem. Soc.*, **128**, 3506–3507.

47 Hohng, S. and Ha, T. (2004) Near-complete suppression of quantum dot blinking in ambient conditions. *J. Am. Chem. Soc.*, **126**, 1324–1325.

48 Mandal, A., Nakayama, J., Tamai, N., Biju, V. and Isikawa, M. (2007) Optical and dynamic properties of water-soluble highly luminescent CdTe quantum dots. *J. Phys. Chem. B*, **111**, 12765–12771; Mandal, A. and Tamai, N. (2008) Influence of acid on luminescence properties of thioglycolic acid-capped CdTe quantum dots. *J. Phys. Chem. C*, **112**, 8244–8250.

49 Shimizu, K. T., Woo, W. K., Fisher, B. R., Eisler, H. J. and Bawendi, M. G. (2002) Surface-enhanced emission from single semiconductor nanocrystals. *Phys. Rev. Lett.*, **89**, 117401-1–117401-4.

50 Verberk, R., Oijen, A. V. and Orrit, M. (2002) Simple model for the power-law blinking of single semiconductor nanocrystals. *Phys. Rev. B*, **66**, 233202-1–233202-4.

51 Banin, U., Bruchez, M., Alivisatos, A. P., Ha, T., Weiss, S. and Chemla, D. S. (1999) Evidence for a thermal contribution to emission intermittency in single CdSe/CdS core/shell nanocrystals. *J. Chem. Phys.*, **110**, 1195–1201.

52 Tang, J. and Marcus, R. A. (2005) Diffusion-controlled electron transfer processes and power-law statistics of fluorescence intermittency of nanoparticles. *Phys. Rev. Lett.*, **95**, 107401-1–107401-4; Tang, J. and Marcus, R. A. (2005) Mechanisms of fluorescence blinking in semiconductor nanocrystal quantum dots. *J. Chem. Phys.*, **123**, 054704-1–054704-12.

Part Two
Nanostructure Characteristics and Dynamics

10
Morphosynthesis in Polymeric Systems Using Photochemical Reactions

Hideyuki Nakanishi, Tomohisa Norisuye, and Qui Tran-Cong-Miyata

10.1
Introduction

In general, chemical reactions give rise to changes in chemical structures. For most cases, structures with spatial symmetry, such as lamellae or hexagons, cannot be solely produced by chemical reactions due to their short-range nature. In order to generate and, particularly, control structures with correlations over long ranges, it is necessary to couple chemical reactions, a *microscopic* phenomenon, to *macroscopic* phenomena such as flow induced by osmotic pressure or to the thermodynamics of the system. This coupling would result in long-range cooperative phenomena such as phase transition or phase separation. In this chapter, to express the unique roles of chemical reactions in the generation and control of polymer morphology via phase separation, we adopt the term *morphosynthesis* [1]. The scope of this chapter will be focused on inducing and controlling the morphology of multi-component polymers by taking advantage of photochemical reactions.

First, the significant roles of photochemical reactions in coupling with the thermodynamics of polymer mixtures are described. Subsequently, several examples are provided to demonstrate the advantages of this coupling over thermally activated reactions. We also show that light or photochemical reactions can be utilized as an efficient tool to drive the system from one location to another on the phase diagram of the mixture, that is, walking the mixture on its phase diagram. As a consequence, a number of unique morphologies were generated and subsequently controlled by using the intensity and wavelengths of the excited light. Finally, the mechanism for the formation of these structures is discussed in conjunction with the universality of systems with competing interactions [2].

10.2
Morphosynthesis of Polymeric Systems by Using Light

10.2.1
Significance of Photochemical Reactions

Among various photochemical reactions, photodimerization of *trans*-cinnamic acid and anthracene [3, 4] was chosen in this work for the morphosynthesis of multicomponent polymers. The scheme for these two specific reactions is illustrated in Figure 10.1. By labeling at a specific site on a polymer chain, anthracene derivatives were used as a cross-linker for UV-sensitive polymer systems, whereas the *trans*-cinnamic acid derivative can be photosensitized and used to crosslink polymers under visible light. The latter would be useful for computer-assisted irradiation (CAI) experiments, as described later. Furthermore, the photodimers of anthracene generated by irradiation with 365 nm UV light can undergo photodissociation to recover anthracene upon irradiation with shorter UV wavelengths (below 300 nm). Since the viscosity of the mixture containing anthracene-labeled polymers continues to increase during the photodimerization process, the photodissociation of anthracene photodimers can be used to regulate the viscosity of the reacting medium by

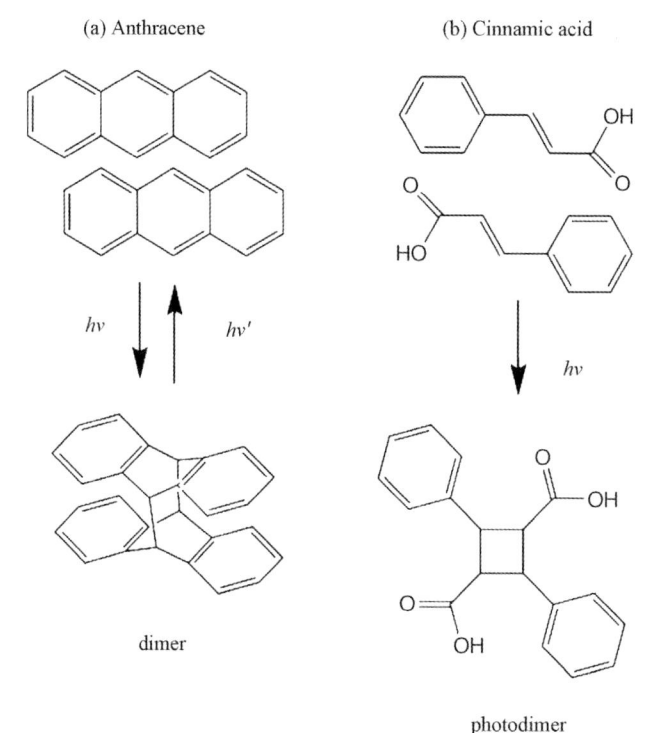

Figure 10.1 Photodimerization of anthracene (a) and *trans*-cinnamic acid (b).

de-crosslinking the forming networks. As described below, coupling this reversibility of anthracene to phase separation could provide a polymer blend exhibiting reversible phase separation behavior drivable by two different wavelengths.

10.2.2
Polymer Mixtures Used in this Study

Samples used in this work are the binary polymer mixtures with the characteristics illustrated in Table 10.1. Here, PSA and PSAF stand, respectively, for polystyrene labeled with anthracene and polystyrene doubly labeled with anthracene and fluorescein used as a fluorescent marker. On the other hand, PSC and PVME stands respectively for polystyrene labeled with *trans*-cinnamic acid and poly(vinyl methyl ether). The factor α in Table 10.1 indicates the label content of anthracene in the polystyrene chain in the unit of number of labels per one chain. For PSC, the label content is 1 cinnamic acid per 28 styrene monomers.

10.2.3
Polymers with Spatially Graded Morphologies Designed from Photo-Induced Interpenetrating Polymer Networks (IPNs)

Spatially graded structures observed in Nature, such as bamboo, have been of great interest to materials scientists for many years because of their strength against hostile environments [5]. This idea was first realized in several research laboratories for metallic materials, particularly ceramics and was later applied to polymer materials by a number of different experimental techniques. So far, the method of producing polymers with graded structures is essentially based on the gradient of the guest monomer generated in the host polymer matrix followed by their polymerization [6]. Alternatively, the method of generating spatial gradients of a guest polymer in a host polymer matrix by sorption was utilized for the synthesis of polymers with spatially gradient structures [7]. These methods have a common drawback caused by the difference in chemical affinity between the guest and the host at the experimental temperature that could induce phase separation and consequently alter the concentration gradient initially built in the host polymer. To resolve these problems, we have used light, instead of heat, to induce polymerization. The concentration gradient of the reactants was therefore generated by

Table 10.1 Characteristics of polymers used in this study. α indicates the label content.

Polymer	M_w	M_w/M_n	α (anthracenes/chain)
PSA	2.3×10^5	1.5	43
PSAF	3.4×10^5	2.6	44
PSC	3.0×10^5	2.8	—
PVME	1.0×10^5	2.5	—

taking advantage of the Lambert–Beer law applied to the UV absorption of the sample. From the viewpoint of phase separation, the characteristic length scale of the morphology emerging at the initial stage of phase separation is inversely proportional to the quench depth [8]. This quench depth is, in turn, governed by the change in the phase diagram upon reaction. Under irradiation with strong light, the reaction yield increases quickly with irradiation time, pushing the reacting mixture deeply into the two-phase region. As a result, the quench depth ΔT often increases with increasing light intensity, thus changing the characteristic length scales of the morphology. Figure 10.2 shows the co-continuous structures with a gradient of characteristic length scales developing along the direction of incident light [9]. By changing the light intensity, the gradient of the characteristic length scales can also be modified. The dependence of the structural gradient on irradiation intensity, illustrated in Figure 10.3, indicates that the spatial gradient of the characteristic length scales in the mixture can be modified by changing the light intensity, except for the region in the vicinity of the boundary between the glass

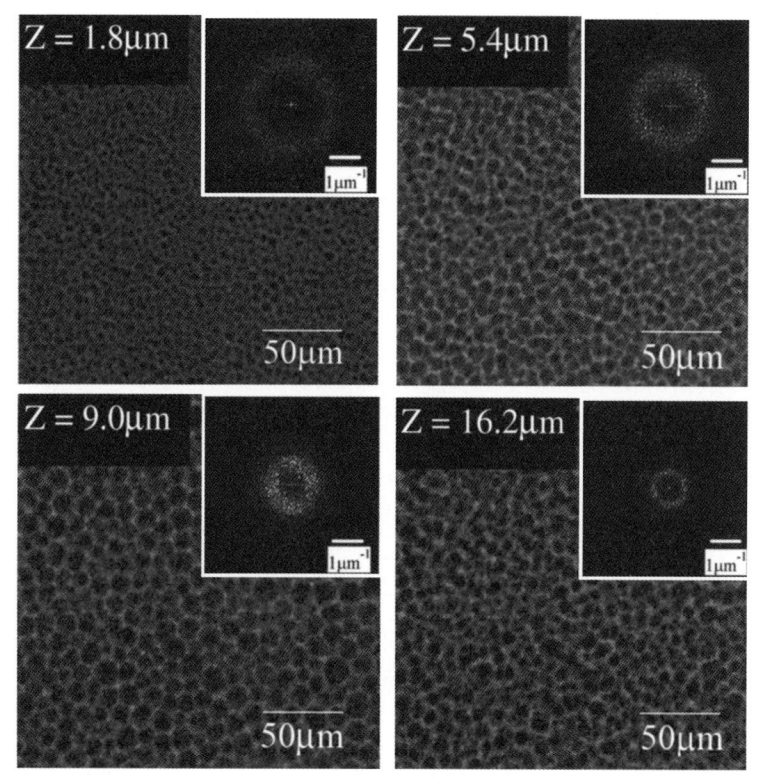

Figure 10.2 Graded co-continuous morphology obtained at different depths along the propagation direction of light in a PSAF/MMA (10/90) blend irradiated with 365 nm UV light at room temperature. The number on the upper left in each figure indicates the Z-coordinates of the sample.

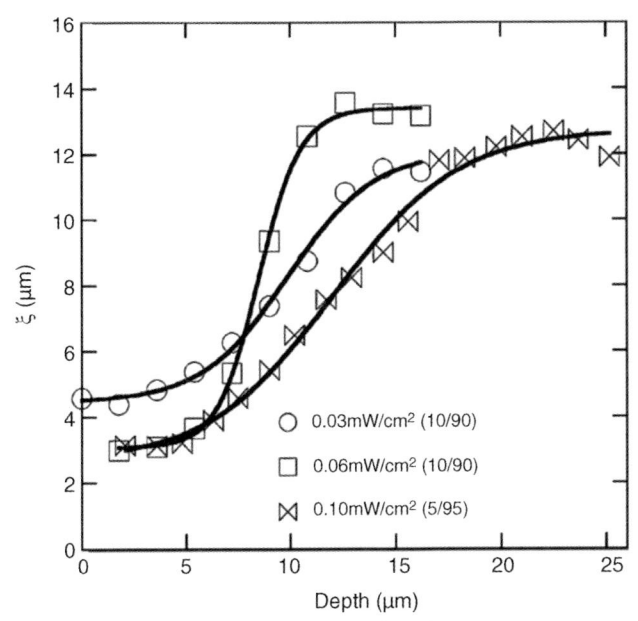

Figure 10.3 Dependence of the gradient of the characteristic length scales on the irradiation intensity observed for PSAF/MMA mixtures with different compositions at room temperature. The 2D power spectra corresponding to the morphologies are indicated in the inset.

substrate and the polymer mixtures where the effects of wetting are pronounced due to the selective adsorption of the PMMA component. It was found that the gradient of the characteristic length scales increases with increasing irradiation intensity. An example of a spatially graded structure is illustrated in Figure 10.4 which was obtained by stacking morphologies observed by a laser-scanning confocal microscope (LSCM) at different depths along the propagating direction of the UV light. These results suggest that by taking advantage of the Lambert–Beer law, polymers with spatially graded morphology can be generated and controlled by irradiation with UV light.

10.2.4
Designing Polymers with an Arbitrary Distribution of Characteristic Length Scales by the Computer-Assisted Irradiation (CAI) Method

In order to establish a general method of producing morphology with an arbitrary distribution of characteristic length scales, instead of utilizing the intensity gradient of light inside the sample, we developed a method of using a computer to design a light pattern with arbitrary characteristic length scales and time scales, the so-called computer-assisted irradiation (CAI) method [10]. Irradiation using a light pattern generated by a personal computer was initially performed to control chemical

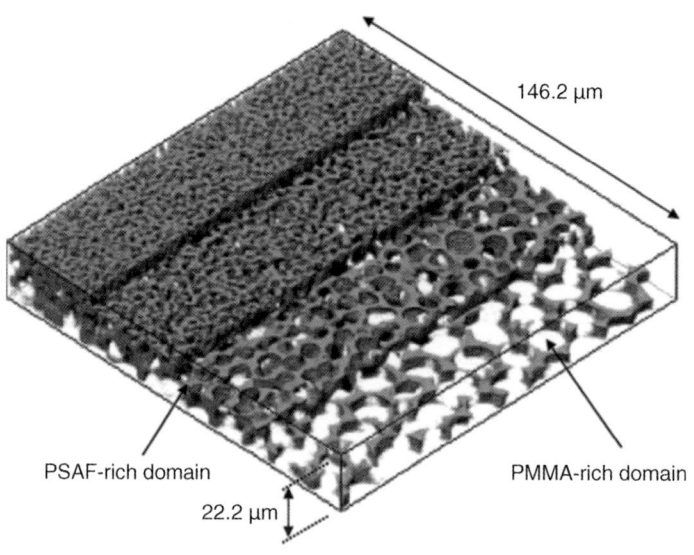

Figure 10.4 Three-dimensional morphology of the spatially graded morphology obtained for a PS/MMA (10/90) mixture: Dark region: PSAF-rich phase; transparent region: PMMA-rich phase [9].

reactions far-from-equilibrium such as the Belousov–Zhabotinsky (BZ) reaction in solution [11, 12]. Here, we have developed a method for controlling the phase separation of polymer mixtures in the bulk state. The block diagram of the instrument is given in Figure 10.5. A light pattern with a characteristic length and a characteristic time was generated on a laptop computer and then transferred to a digital projector. Subsequently, these patterns were projected onto a photosensitive polymer placed under a microscope through a series of lenses, as sketched in Figure 10.5. The time-evolution of the images of morphology under irradiation was observed *in situ* under a bright field optical microscope and stored in a second computer for further analysis. Since the digital projector only transmits visible light, polymer mixtures with the capability of being cross-linked upon irradiation with visible light were particularly designed for this CAI method. Since it is well known that photodimerization of cinnamic acid can be induced by irradiation with visible light via photosensitization, we took advantage of this mechanism and synthesized a polystyrene derivative labeled with *trans*-cinnamic acid (PSC). The chemical structure of the polymer is illustrated in Figure 10.6. The resulting polymer was mixed with poly (vinyl methyl ether)PVME) to form a blend which possesses a lower critical solution temperature (LCST), that is, undergoes phase separation with increasing temperature. The phase diagram obtained by light scattering detected at a fixed angle for this PSC/PVME blend is illustrated in Figure 10.7. The details of the experiments are given elsewhere [10]. Upon irradiation with 405 nm light, the photosensitizer 5-nitroacenaphthene is excited and transfers its excited energy to the triplet state of *trans*-cinnamic acid, triggering the dimerization of the cinnamic acid group. This

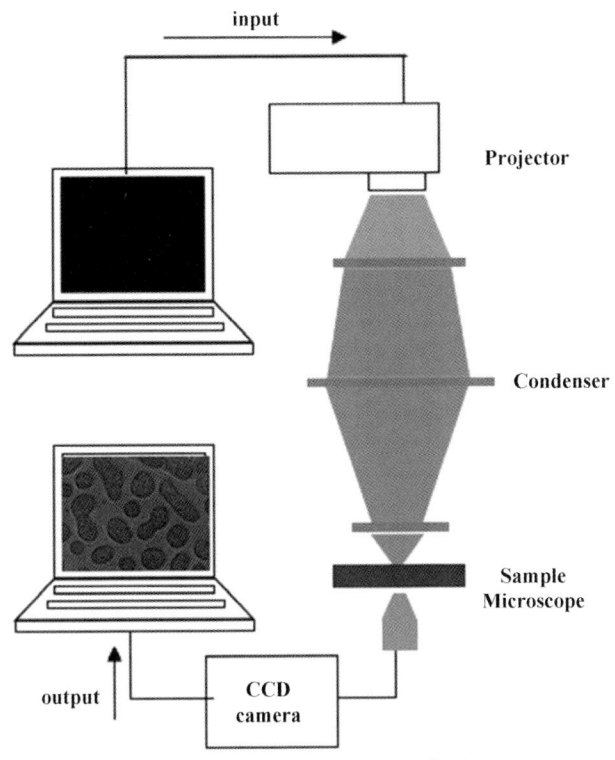

Figure 10.5 The block diagram of the apparatus for the computer-assisted- irradiation method.

photosensitized reaction of cinnamic acid leads to the formation of PSC networks in the blend. As the fraction of PSC networks exceeds a certain threshold, the PSC/PVME blend enters the two-phase region and undergoes phase separation. Under this circumstance, there exists a competition between phase separation and the

Figure 10.6 Chemical structure of the polystyrene component labeled with *trans*-cinnamic acid (PSC).

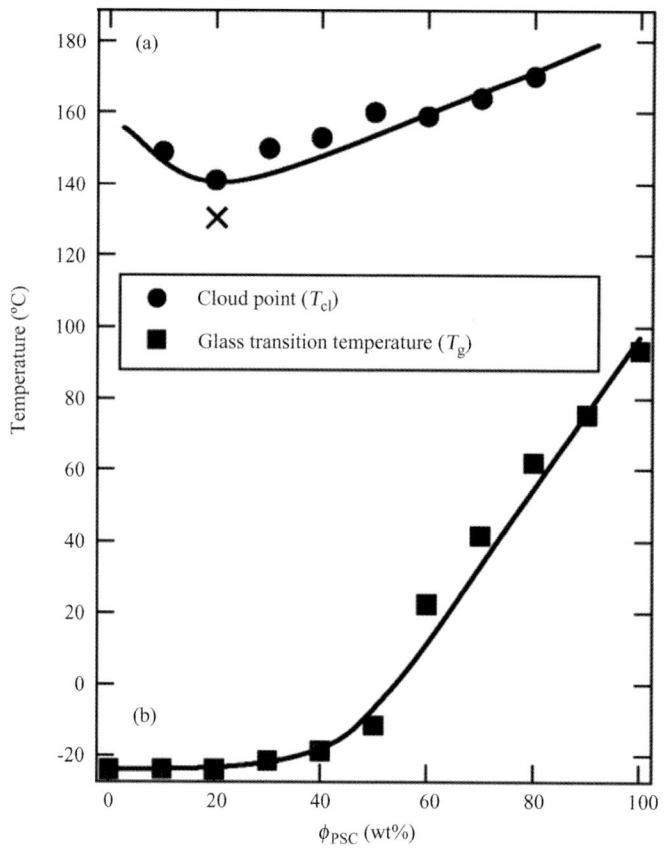

Figure 10.7 The phase diagram (a) and the glass transition temperatures (b) of a PSC/PVME mixture obtained, respectively, by light scattering and differential scanning calorimetry (DSC). Irradiation experiments were performed in the miscible region at 127 °C indicated by (X) in the figure of *trans*-cinnamic acid-labeled polystyrene/poly(vinyl methyl ether) blends.

network formation process of the PSC component. The characteristic length scales of the morphology are determined by this competition. Illustrated in Figure 10.8 are the morphologies of a PSC/PVME (20/80) blend irradiated by a 405 nm light stripe with a width of 150 μm. The intensity of the stripe at the center is 1.0 mW cm^{-2} and its background is 0.01 mW cm^{-2}. It was found that the blend underwent phase separation after 40 min of irradiation, exhibiting spinodal structures in the region irradiated with the light stripe, whereas no structure was observed at either side of the stripe with much lower intensity. This result clearly indicates that phase separation of the mixture can be locally induced at an arbitrary position in the blend by using this CAI method. Further investigation is currently underway to examine

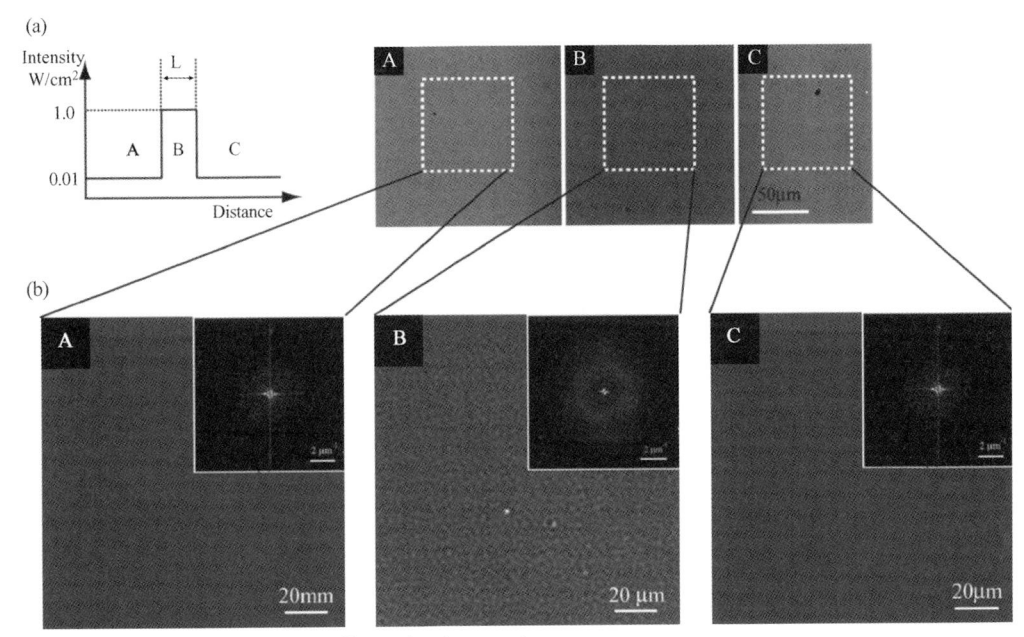

Figure 10.8 (a) The light intensity profile used in this experiment and the morphology obtained at low magnification obtained for a PSC/PVME (20/80) blend irradiated at 127 °C for 240 min. (b) Morphologies of the same sample observed under high magnification at different positions corresponding to the light intensity profile. $L = 150\,\mu m$.

phase separation phenomena under spatio-temporally non-uniform conditions imposed by this experimental technique.

10.2.5
Reversible Phase Separation Driven by Photodimerization of Anthracene: A Novel Method for Processing and Recycling Polymer Blends

Among a number of reversible photochemical reactions, photodimerization of anthracene would be one of the most convenient reactions to drive the phase separation process of polymer blends because of its easy accessibility in the range of absorption wavelengths and, particularly, without catalyst. Figure 10.9 shows the reversibility of the photodimerization of anthracene moieties labeled on polystyrene in a blend with poly(vinyl methyl ether) monitored under irradiation with two wavelengths 365 and 297 nm at room temperature [13]. Obviously, the reaction rate increases with increasing the light intensity upon irradiation with 365 nm. On the other hand, the recovery process of anthracene via photodissociation of its photodimers induced by 297 nm UV light becomes faster with increasing light intensity. These results clearly indicate that the cross-linking process of the PSAF component

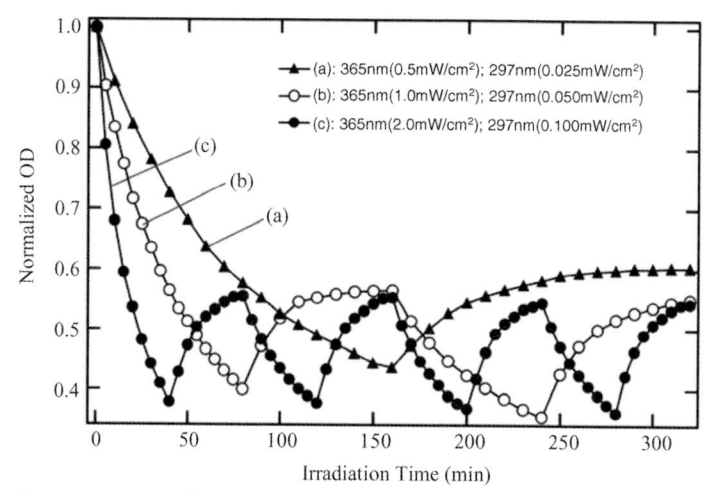

Figure 10.9 Reversible photodimerization of anthracene induced by light of two wavelengths 365 and 295 nm in a PSAF/PVME (20/80) mixture observed at 25 °C.

in the blend can be reversibly induced by using two wavelengths 365 and 297 nm. Consequently, the phase separation of PSA/PVME blends can also be reversibly driven by these two wavelengths. As expected, phase separation of the blend was promoted by irradiation with 365 nm light, as shown in Figure 10.10a where the scattering profile with the characteristic peaks appears as a result of the spinodal decomposition process induced by photodimerization of anthracenes. The peak moves toward the side of *large* wavenumbers, indicating that the phase-separated structures have shrunk as the phase separation proceeds. This unusual behavior of the scattering profiles was reported for the phase separation of epoxy systems under curing, but different explanations were provided for this peculiar behavior [14]. To elucidate the mechanism of this unusual scattering behavior, the local deformation of the same sample was monitored under the same irradiation conditions by using Mach–Zehnder interferometry (MZI) [15]. By comparison of the scattering results with the deformation data observed from the MZI for the same blend under the same irradiation conditions, it was found that there exists a strong correlation between the two sets of data. Furthermore, as seen in Figure 10.10b, the scattering intensity decreases gradually with time upon dissociation of the cross-link network by irradiation with 297 nm light, suggesting that the homogenization process of the phase-separated blends was induced by the photodissociation of anthracene photodimers. The phase separation behavior depicted in Figure 10.10a and b is quite different from the scattering data obtained previously in the reverse quenched experiments [16], revealing the effects of crosslink-induced shrinkage on the critical behavior of polymer blends. The experimental results described above suggest not only a potential method for photo-recycling of polymer blends, but could also provide a novel method of processing multiphase polymer materials using two UV wavelengths.

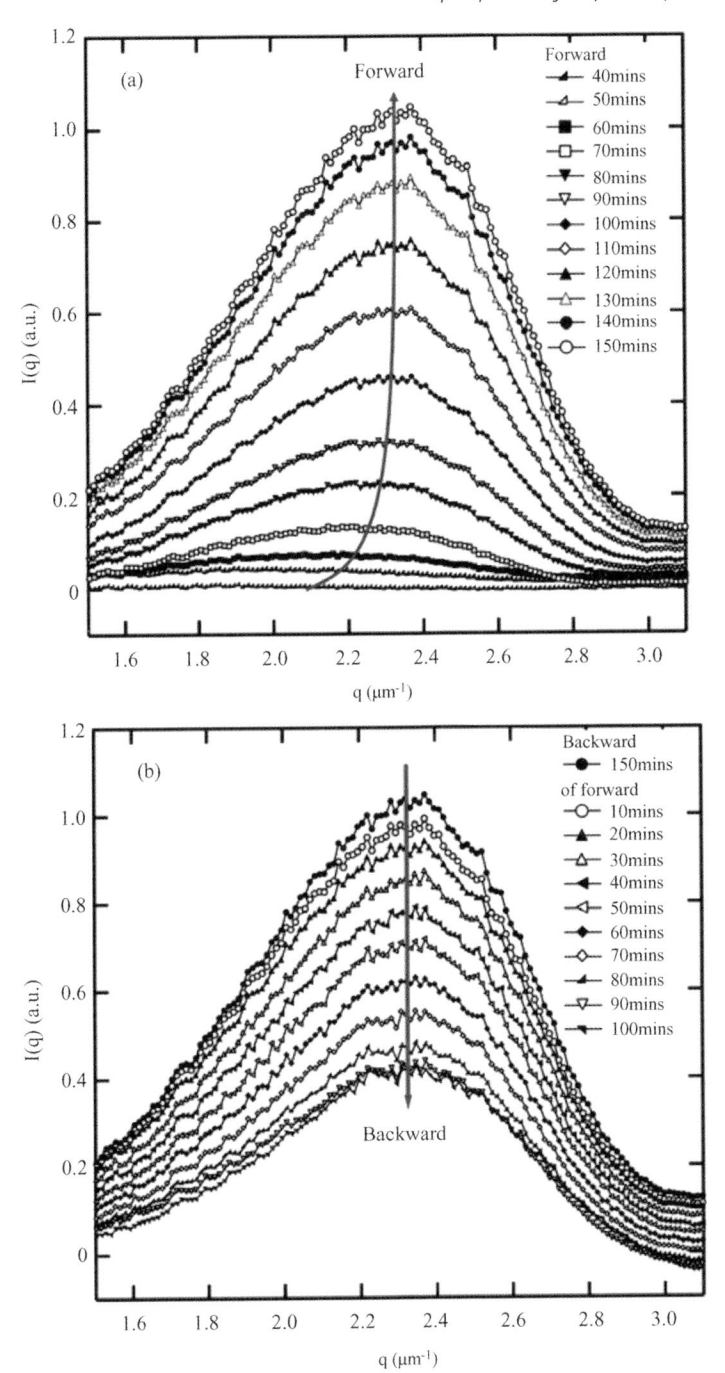

Figure 10.10 Scattering profiles of a PSA/PVME (20/80) blend driven by irradiation with light of wavelength (a) 365 nm and (b) 297 nm.

10.3
Concluding Remarks

We have shown that by coupling chemical reactions to cooperative phenomena such as phase separation, chemical reactions become a long-range effect via which a wide variety of morphologies can be generated. Particularly, the long-range effects of this coupling are further enhanced by the viscoelasticity of polymeric networks. Since photochemical reactions can be started and stopped independently from the thermodynamics of the polymer systems, light can be used as a useful tool to control the morphology of polymeric systems. On the other hand, light-induced phase separation is also an interesting physical phenomenon because of its relation to the mode-selection processes in systems far from equilibrium. By systematically examining this reaction-induced phase separation, the mode selection process in polymeric systems can be understood and the information obtained would provide a novel way of designing morphology of polymer materials. Finally, the morphologies with a variety of ordered structures, obtained by the methods described here, could be utilized as templates for dispersion of nanofillers such as dendrimers, fullerenes or carbon nanotubes for practical applications in materials science.

Abbreviations

CAI	computer-assisted irradiation method
EGDMA	ethyleneglycol dimethacrylate
IPNs	interpenetrating polymer networks
LCST	lower critical solution temperature
PSA	anthracene-labeled polystyrene
PSAF	polystyrene doubly labeled with anthracene and fluorescein
PSC	polystyrene labeled with *trans*-cinnamic acid
PVME	poly(vinyl methyl ether)

Acknowledgments

This work was supported by Grant-in-Aid for Scientific Research on Priority Area "Molecular Nanodynamics" from the Ministry of Education, Culture, Sports, Science and Technology, Japan. We thank Professor Hiroshi Masuhara (Nara Institute of Science and Technology, Nara, Japan) and Professor Hiroshi Fukumura (Tohoku University, Sendai, Japan) for their encouragement throughout the course of this study. The efforts of former graduate students Y. Adachi, J. Fukuda, S. Ishino, N. Namikawa and X.-A. Trinh in our research group are greatly appreciated.

References

1 Antonietti, M. and Ozin, G. A. (2004) Promises and problems of mesoscale materials chemistry or why meso. *Chem-Eur. J*, **10**, 28–41.

2 For example, see, Vedmedenko, E. Y. (2007) *Competing Interactions and Pattern Formation in Nanoworld*, Wiley-VCH, Weinheim.

3 See, for example, Cowan, D. C. and Drisko, R. L. (1976) *Elements of Photochemistry*, Plenum Press, New York.

4 See, for example, McCullough, J. J. (1987) Photoadditions of aromatic compounds. *Chem. Rev.*, **87**, 811–860.

5 Rabin, B. H. and Shiota, I. (1995) Functionally gradient materials. *MRS Bull.*, **20**, 14–18.

6 Akovali, G., Biliyar, K. and Shen, M. (1976) Gradient polymers by diffusion polymerization. *J. Appl. Polym. Sci.*, **20**, 2419–2427.

7 Agari, Y., Shimada, M., Ueda, A. and Nagai, S. (1996) Preparation, characterization and properties of gradient polymer blends: discussion of poly(vinyl chloride)/poly(methyl methacrylate) blend films containing a wide compositional gradient phase. *Macromol. Chem. Phys.*, **197**, 2017–2033.

8 For example, see: Cahn, J. W. (1961) On spinodal decomposition. *Acta Metal.*, **9**, 795–801.

9 Nakanishi, H., Namikawa, N., Norisuye, T. and Tran-Cong-Miyata, Q. (2006) Autocatalytic phase separation and graded co-continuous morphology generated by photocuring. *Soft Matter*, **2**, 149–156.

10 Ishino, S., Nakanishi, H., Norisuye, T., Awatsuji, Y. and Tran-Cong-Miyata, Q. (2006) Designing a polymer blend with phase separation tunable by visible light for computer-assisted irradiation experiments. *Macromol. Rapid. Commun.*, **27**, 758–762.

11 Petrov, V., Ouyang, Q. and Swinney, H. L. (1997) Resonant pattern formation in a chemical system. *Nature*, **388**, 655–657.

12 Kadar, S., Wang, J. and Showalter, K. (1998) Noise-supported travelling waves in sub-excitable media. *Nature*, **391**, 770–772.

13 Trinh, X. A., Fukuda, J., Adachi, Y., Nakanishi, H., Norisuye, T. and Tran-Cong-Miyata, Q. (2007) Effects of elastic deformation on phase separation of a polymer blend driven by a reversible photo-cross-linking reaction. *Macromolecules*, **40**, 5566–5574.

14 Kyu, T. and Lee, J.-H. (1996) Nucleation initiated spinodal decomposition in a polymerizing system. *Phys. Rev. Lett.*, **76**, 3746–3749.

15 Inoue, K., Komatsu, S., Trinh, X. A., Norisuye, T. and Tran-Cong-Miyata, Q. (2005) Local deformation in photo-crosslinked polymer blends monitored by Mach-Zehnder interferometry. *J. Polym. Sci. Part B: Polym. Phys.*, **43**, 2898–2913.

16 Akcasu, Z. A., Bahar, I., Erman, B., Feng, Y. and Han, C. C. (1992) Theoretical and experimental study of dissolution of inhomogeneities formed during spinodal decomposition in polymer mixtures. *J. Chem. Phys.*, **97**, 5782–5793.

11
Self-Organization of Materials Into Microscale Patterns by Using Dissipative Structures
Olaf Karthaus

11.1
Self-Organization and Self-Assembly

First, it has to be noted that there is no clear-cut definition of self-organization, nor is there a sharp boundary between self-organization and self-assembly [1]. Both, self-citation for organization and self-assembly, give rise to ordered one-, two-, or three-dimensional structures on various length scales from a disordered precursor state without outside guidance. Some authors want to restrict the phrase "self-assembly" to the formation of monolayers of reactive organic compounds, often functionalized alkanes, on solid substrates. The most famous examples of these types of layers are the self-assembled monolayers (SAMs) of alkylthiols on gold or silver [2], and monolayers of mixed alkoxy-alkyl silanes on glass or SiO_2 [3]. By restricting the phrase "self-assembly" to these few examples, all other types of structures formed by spontaneous aggregation of atoms and molecules have to be categorized as "self-organized" structures.

Another school of thought makes a distinction between self-assembly and self-organization on the basis of their building blocks, their dynamics and energetics. Here, self-assembled structures are made from preformed building blocks and the structures are static and near the thermodynamic equilibrium. Self-organized structures on the other hand are independent of the component shape, they are dynamic and far away from equilibrium. In that sense, self-assembly is a directed process in which the obtained structure is stable and predictable. Interactions between the building blocks, atoms or molecules, are short range and can be between the same kind of building block (homologous) or between different kinds of building blocks (heterologous). Self-organization, on the other hand, is only present in states that are far from equilibrium, and relies on the collective interaction of many parts in a dynamic fashion.

It should be noted that in this definition a system that ends up having a self-assembled structure may also start far from equilibrium and may be dynamic until the final structure has been reached.

The following structures are examples that are grouped according to this latter definition of self-assembly and self-organization:

Self-assembly:

- Snow crystals [4]: Their macroscopic structure is different from a bulk three-dimensional ice crystal, but they are formed by homologous pair–pair interaction between water molecules and are static and in thermodynamic equilibrium. It should be noted, however, that dendritic crystal growth is a common phenomenon for metals [5–7] and polymers. The crystals grow under non-equilibrium conditions, but the final crystal is static.

- Self-assembled monolayers (SAMs) [8]: The layers are formed by heterologous interaction between reactive groups, such as thiols, and noble metals, such as gold or silver. Since the molecules are selectively adsorbed on these metals, film growth stops after the first monolayer is completed. The molecular aggregation is enthalpy driven, and the final structure is in thermodynamic equilibrium.

- Blockcopolymer microphase separation [9]: Depending on the length of chemically different blocks of monomers in a block copolymer, ordered nanostructures can be obtained in bulk samples and thin films. The film morphology can differ significantly from the bulk morphology, but because the structure is determined by the pair–pair interaction of monomers and/or an interface, and it is a thermodynamically stable structure, it is classified as self-assembly.

- Micelles: the mostly spherical nanoscale aggregates formed by amphiphilic compounds above their critical micelle concentration in aqueous solution have a narrow size distribution and are dynamic, because there is a fast exchange of amphiphiles in solution and those incorporated in micelles. However, micelles are defined as self-assembled structures, since the structure is in thermodynamical equilibrium.

Self organization:

- Vesicles [10, 11]: these aggregates of insoluble natural or artificial amphiphiles in water can have various shapes (spherical, cylindrical). Depending on the preparation conditions, small unilamellar or large multilamellar vesicles can be produced. The structures meet the self-organization criterion, because they are, albeit on a long time scale, dynamic and not in thermodynamic equilibrium, which would in many cases be a macroscopically phase separated lamellar phase.

- Belouzov–Zhabotinsky reaction [12, 13]: This chemical reaction is a classical example of non-equilibrium thermodynamics, forming a nonlinear chemical oscillator [14]. Redox-active metal ions with more than one stable oxidation state (e.g., cerium, ruthenium) are reduced by an organic acid (e.g., malonic acid) and re-oxidized by bromate forming temporal or spatial patterns of metal ion concentration in either oxidation state. This is a self-organized structure, because the reaction is not dominated by equilibrium thermodynamic behavior. The reaction is far from equilibrium and remains so for a significant length of time. Finally, when the reducing organic acid and the oxidizing bromate concentrations are too low for further redox cycles, the reaction stops.

- Biological cell: This is an often cited prime example of a self-organized structure. A cell is made up of many different building blocks that interact with each other in various ways and in which compartments form ordered structures. A living cell requires energy (light or chemical energy) to maintain its function and structure. However, a cell is actually more than just a complex self-organized structure, in order to be functional it needs preexisting information, passed down from the previous generation through self-replication of hereditary material (DNA or RNA).

It should be noted that the glassy state of amorphous polymers, metals, or other inorganics is neither self-assembled, nor self-organized. Even though it is not an equilibrium structure, it lacks the characteristic order and regular structure that is inherent in self-assembly or self-organization systems.

Even though the concept of self-organization can be followed back to Descartes, who theorized that the ordinary laws of nature tend to produce organization [15], the phrase 'self-organization' did not appear until 1947 when it was used by the psychologist and engineer W. Ross Ashby [16]. Even after that, the concept lay dormant with little progress until the 1980s. Only more recently has a significant increase in scientific dissertations been noted, reaching nearly 600 for the years 1991–2000.

Self-organization seems to be counterintuitive, since the order that is generated challenges the paradigm of increasing disorder based on the second law of thermodynamics. In statistical thermodynamics, entropy is the number of possible microstates for a macroscopic state. Since, in an ordered state, the number of possible microstates is smaller than for a more disordered state, it follows that a self-organized system has a lower entropy. However, the two need not contradict each other: it is possible to reduce the entropy in a part of a system while it increases in another. A few of the system's macroscopic degrees of freedom can become more ordered at the expense of microscopic disorder. This is valid even for isolated, closed systems. Furthermore, in an open system, the entropy production can be transferred to the environment, so that here even the overall entropy in the entire system can be reduced.

11.2
Dissipative Structures

The flow of matter and energy through an open system allows the system to self-organize, and to transfer entropy to the environment. This is the basis of the theory of dissipative structures, developed by Ilya Prigogine. He noted that self-organization can only occur far away from thermodynamic equilibrium [17].

Dissipative, open systems that allow for the flux of energy and matter may exhibit non-linear and complex behavior. Following the above argumentation, complex systems are usually far from thermodynamic equilibrium but, despite the flux, there may be a stable pattern, which may arise from small perturbations that cause a larger, non-proportional effect. These patterns can be stabilized by positive (amplifying)

```
Open System
   │
   │  Flow of matter and energy
   ▼
Fluctuations
   │
   │  Positive feedback
   ▼
Enhancement
   │
   │  Collectivity
   ▼
Temporal and spatial patterns
```

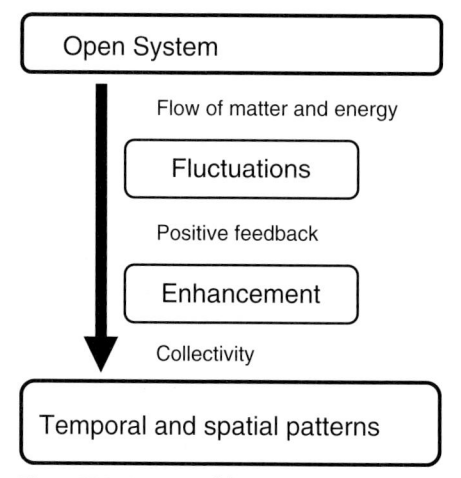

Figure 11.1 Concept of dissipative structures for the emergence of spatio-temporal patterns.

feedback as illustrated in Figure 11.1. Another quality of complex systems is that they often exhibit hysteresis or periodic behavior and may be nested (hierarchical), meaning that the parts of a complex system are complex themselves. Dissipative structures can occur from the nanometer to the megameter scale. At the smaller end of the scale, it is argued that dissipative structures in liquid water are responsible for homeopathic effects [18]. Large-scale dissipative structures are found, for example, in the hydrodynamic instabilities of the ozone layer [19] and the "eye" in the atmosphere of the planet Jupiter. The complexity of life requires complex order principles, and it is only logical that dissipative structures have been proposed, both for the origin of life [20], as well as for biological rhythms [21] and noise [22]. It goes without saying that pattern formation occurs not only during physical processes such as drying [23] and phase separation [24], but also during chemical reactions [25], a specific example being the passivation of metals [26].

In the following, two types of dissipative structures are explained in more detail, because they can be used to produce nano- and micropatterns of organic materials.

- Bénard convection cells [27, 28]: a liquid with an inverse temperature gradient (hot below and cool on top) may exhibit thermal convection. Less dense parts of the liquid well upward whereas denser parts show down-welling. The convection cells may arrange in hexagonal order in which the center of each cell wells downwards and the rim wells upwards. The cells stem from the concerted movement of many molecules and cease when the temperature gradient is below a threshold at which the thermal equilibrium can be reached solely by thermal conduction and not convection.

- Tears-of-Wine: The surface motion of mixed liquids was described in ancient times [29], and scientifically in a paper by James Thompson, the brother of Lord Kelvin [30] in 1855. Recently Neogi [31], and Cazabat [32] reported on the formation of ordered structures in evaporating solutions of two liquids. The meniscus of wine in a glass is drawn upward on the glass surface and forms a thin film. Due to its lower boiling point, evaporation of alcohol leads to a water-rich liquid surface that has a higher surface tension. This leads to a surface tension gradient and a flow by which accumulated liquid forms "tears", very often at regular intervals along the edge of the liquid film. The tears-of-wine phenomenon can be generalized to describe the flow near the surface and the edge of liquid films driven by surface tension gradients that lead to ordered structures.

These two examples show that regular patterns can evolve but, by definition, dissipative structures disappear once the thermodynamic equilibrium has been reached. When one wants to use dissipative structures for patterning of materials, the dissipative structure has to be fixed. Then, even though the thermodynamic instability that led to and supported the pattern has ceased, the structure would remain. Here, polymers play an important role. Since many polymers are amorphous, there is the possibility to "freeze" temporal patterns. Furthermore, polymer solutions are non-linear with respect to viscosity and thus strong effects are expected to be seen in evaporating polymer solutions. Since a macromolecule is a nanoscale object, conformational entropy will also play a role in nanoscale ordered structures of polymers.

11.3
Dynamics and Pattern Formation in Evaporating Polymer Solutions

The two dissipative structures in the examples given above are observable with the naked eye. Tears of wine have a periodicity of a few mm. The size of Benard convection cells depends on the thickness of the liquid layer, because the convection cells have the same height and diameter. Since a macroscopic layer of liquid is often used to demonstrate this effect, the cells have a diameter of at least a few mm. By reducing the thickness of the layer, convection cells are expected to become smaller. On the other hand, an effect of a thinner layer on the tears of wine would not be that obvious. Instead of heating a solution from below to induce convection, the cooling of the surface should be as effective in creating a temperature gradient. When a polymer is dissolved in a volatile solvent, such surface cooling occurs spontaneously by allowing the solvent to evaporate. The heat of evaporation is enough to drive the system out of equilibrium. For example, by placing a few µl of a dilute chloroform solution of a fluorescent-labeled polymer on a substrate we could observe the formation of Benard-like convection cells on a sub-millimeter scale [33] as can be seen in Figure 11.2. The polymer is a polyion complex [34] of poly(styrene sulfonate) having bisoctadecyldimethylammonium as counterion, containing 1 mol% of Rhodamine B, an ionic red-fluorescent dye. By exciting with green light in an epifluorescence microscope, areas of the solution that have higher polymer concentration are brighter. Because the polymer solution droplet has a spherical-cap shape, the solution height is not constant along its radius. Thus it is impossible to develop a two-dimensional regular convection-cell pattern. Instead, there is just one ring of laterally connected Benard cells. At thinner areas, closer to the edge of the polymer solution droplet, the fluorescence is weaker, because the amount of polymer in the light-pass is less.

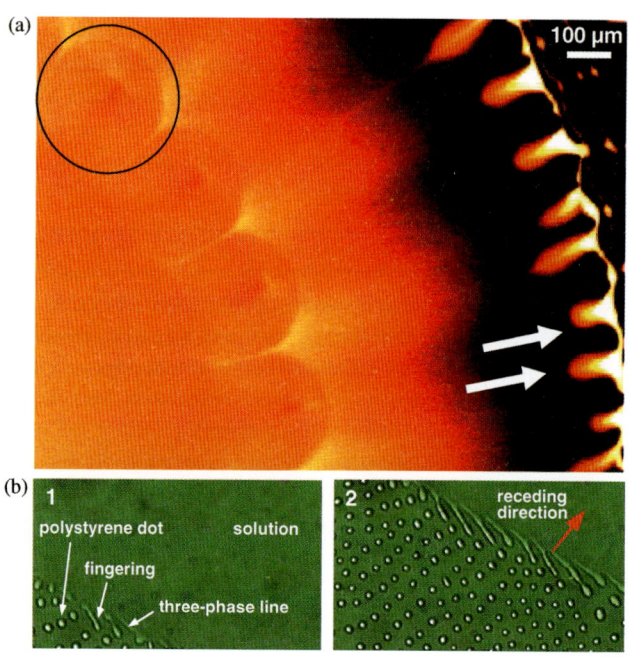

Figure 11.2 (a) Microscope image of Benard convection cells (indicated by the circle in the upper left corner) and "tears of wine" (indicated by the white arrows) in an evaporating fluorescence-labeled polymer solution (adapted from Ref. 33). (b) Two microscope snapshots of an evaporating polystyrene solution on silicon wafer. The time between the two frames is approximately 100 ms. The polymer droplets have a diameter of approximately 3 µm.

In addition, a periodic intensity profile along the edge of the droplet can be seen. Areas with strong fluorescence are separated by areas that do not fluoresce. This indicates a periodic concentration profile, very similar to the tears-of-wine effect in which surface tension gradients lead to periodic build-up of solution droplets. Similar to the tears of wine, this concentration profile is dynamic. Monitoring the time course revealed that the higher concentrated areas merge, and new areas of higher concentration evolve between existing concentrated areas [35]. These dynamic processes, convection cells and polymer-rich "finger" structures, are due to solvent evaporation. Consequently, the volume of the solution droplet becomes constantly smaller, and thus the diameter of the solution droplets begins to decrease shortly after the dynamic structures are formed. Hence, the dynamic structures, because they are linked to the edge of the solution droplet, move across the substrate. In this process, polymer may be deposited on the substrate, and droplet or line patterns are commonly observed. Line patterns are formed when the polymer is continuously deposited at the highly concentrated finger positions [36]. Droplets are formed when the emerging line exhibits another instability, the so-called Plateau–Rayleigh instability, in which the line decomposes into droplets, very much like a dripping faucet [37]. These line and droplet patterns are similar to those obtained by heating a thin polymer film that has been spun cast on a substrate. Here, too, a fingering

11.3 Dynamics and Pattern Formation in Evaporating Polymer Solutions

instability is observed while circular holes develop in the film [38–41]. Such a process occurs because the surface and interfacial energies of polymer and substrate favor droplet formation, that is, dewetting occurs when the polymer is above its glass transition temperature and thus mobile. Droplet formation is triggered by randomly generated holes in the continuous polymer films, thus no regular array of droplets can be formed, even thought the droplets may have a narrow size distribution and have an average nearest neighbor distance [42]. By contrast, the dewetted polymer solution may produce two-dimensionally ordered droplet arrays, because the droplet formation occurs only at the edge of the evaporating solution. Typically, the droplets have a diameter of 200 nm to 10 μm, and they have been used to immobilize organic dyes [43] and inorganic nanoparticles [44] in an orderly fashion on a substrate. [45] Dye aggregates may also accumulate along the rim of the microscopic polymer droplet, leading to hierarchically ordered structures (dye molecule–nanoscale dye aggregate–circular arrangement within droplet–two-dimensional array of droplets), as can be seen in Figure 11.3. In the case of organic cyanine dyes a dependence of the fluorescence spectra on the droplet size has been found [46], which indicates that the nanoscale aggregation of dye molecules can be regulated by micrometer size confinement.

Recently it has been reported that even colloidal particle suspensions themselves, without added polymers, can form dissipative structures. Periodic stripes of colloidal particles (monodisperse particles of diameter 30 nm and 100 nm, respectively) and polystyrene particles (monodisperse; diameters from 0.5 to 3 μm) can be formed from dilute aqueous suspensions. The stripes are parallel to the receding direction of the edge of the suspension droplet and thus indicate that a fingering instability

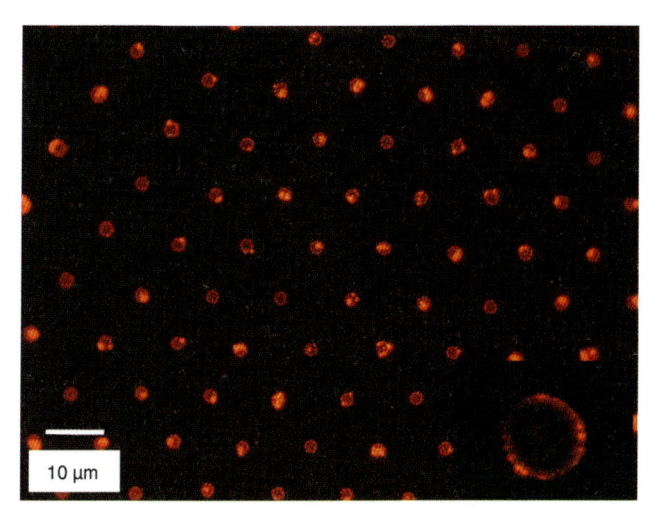

Figure 11.3 Hierarchical pattern of cyanine-dye J-aggregates. Dye molecules form strongly fluorescent nanoscale J-aggregates; J-aggregates arrange at the rim of the micrometer-sized polymer droplet; droplets arrange in a regular mm-sized two-dimensional array (adapted from Ref. 39).

is the driving force for patterning [47, 48]. The thickness of the stripes is one monolayer of particles and their width is from three to ten particles, depending on the preparation conditions.

As explained above, the formation of regular droplet arrays stems from the regular fingering instability at the liquid edge of the solution droplet and the smooth receding of the solution. Thus, the surface roughness of the substrate plays an important role. Regular two-dimensional arrays can easily be produced on atomically flat mica and semiconductor-grade polished silicon wafers with a surface roughness of less than 1 nm. Rougher surfaces, like float glass or indium tin oxide (ITO) covered substrates rarely yield regular patterns since surface inhomogeneities (either topological or chemical) disturb the regular fingering patterns. Stick–slip motion of the edge of the solution causes the three-phase-line to recede with non-uniform speed.

One possibility to prepare regular dewetted patterns on rough substrates is by pattern transfer. A regular pattern is formed on mica, on which then another substrate is placed. The pattern is released from the mica and fixed on the other substrate at the original droplet positions [49].

In order to allow for more control during the evaporation, the solution can be placed in a motor-controlled sliding [50] or rolling apparatus [51] which uses capillary force to confine the solution between two glass surfaces. The motor determines the speed at which the solution edge is drawn over the substrate and is one of the main parameters to control patterning. Droplet, stripe and ladder patterns have been observed.

Recently, we explored the effect of molecular weight on the pattern and employed post-dewetting processes to alter the shape of the dewetted polymer droplets. Since the viscosity of a polymer solution is nonlinear with respect to concentration and also strongly dependent on polymer weight, we expected a drastic effect. Figure 11.4

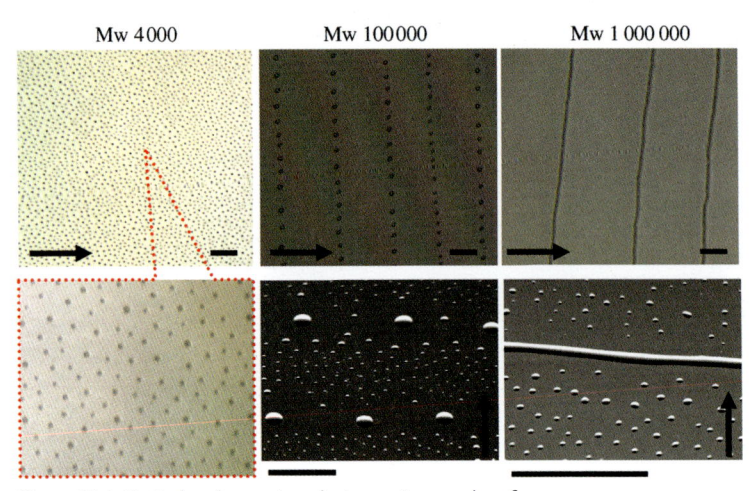

Figure 11.4 Optical and scanning electron micrographs of patterns obtained by using polystyrene with different molecular weight. The black arrows indicate the rolling direction. The black scale bar is 10 μm.

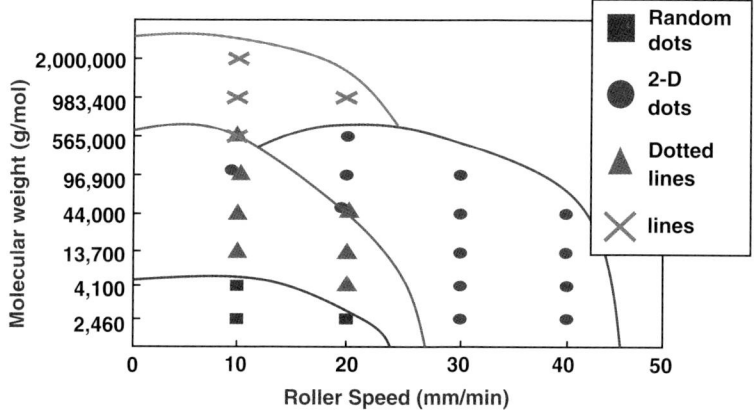

Figure 11.5 Dependence of the polymer dewetting pattern on molecular weight and dewetting speed.

shows examples of the different patterns that were obtained by using polystyrene with different molecular weights. The regular fingering instability that leads to regular 2-D patterns is only achieved with medium molecular weights. Below those, irregular patterns prevail. At higher molecular weights, the fingering instability is suppressed and gelation occurs at the solution edge. If the molecular weight is below 1 000 000, a Plateau–Rayleigh instability leads to the formation of lines of droplets, while at molecular weights above 1 000 000, the lines perpendicular to the receding direction remain. The periodic gelation of polymers at the edge of an evaporating solution has been modeled [52] and is in good agreement with the experimental data. Figure 11.5 shows the approximate regions where the patterns emerge.

Dewetting of polymers leads to droplets that have a shallow contact angle of 5–10° [39, 53]. When considering applications, it would be beneficial to be able to control the contact angle, thus allowing for various droplet shapes. By applying a mixed solution containing a good solvent as well as a miscible non-solvent for a polymer, it is possible to swell the polymer without dissolving it. For polystyrene a good solvent is tetrahydrofuran (THF) and a non-solvent is water. By adjusting the volume ratio of both liquids, we succeeded in swelling the dewetted polymer microdroplets. Electron microscope imaging shows high aspect-ratio spherical droplets after the THF–water treatment. The swelling does not overcome the polymer–substrate interaction, so that the position of the droplets is not changed. These microdroplets have been demonstrated to act as a microlens arrays [54] as shown in Figure 11.6. The shape change is reversible by raising the temperature above the glass transition temperature of the polymer [55].

Wetting and dewetting depend strongly on the surface energy of the substrate and the interfacial energy of the polymer and the substrate. Thus dewetting on a patterned substrate with areas of different surface energies gives rise to selective deposition

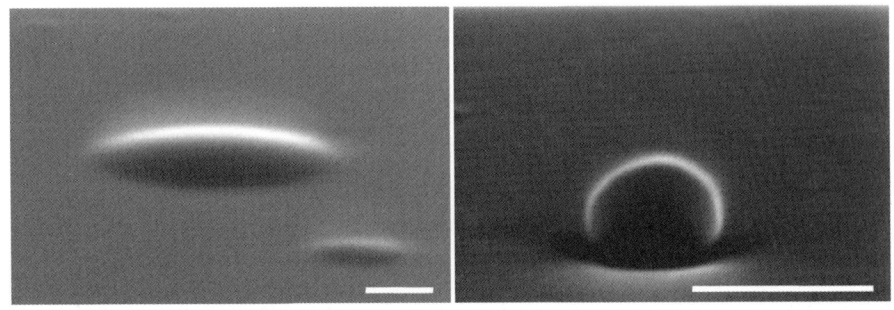

Figure 11.6 Scanning electron micrographs of a dewetted polymer droplet before and after solvent treatment. Viewing angle is 45° and the scale bar is 1 μm.

of the polymer on one surface functionalization only. Selective deposition of conducting polymers has been demonstrated for dewetted structures on Au patterns on Si wafer [56]. Similarly, dewetting can be used to form micron-sized gaps that can be used for the self-organized assembly of electrodes for organic field effect transistors [57].

Besides polymers, other amorphous materials such as dendrimers or even small molecules with a molar mass as low as $300 \,g\, mol^{-1}$ can also be dewetted. The reason for this is that dissipative structures in solution are independent of the molecular structure of the solute, as long as the solute does not crystallize during pattern formation. Hence, dendrimers have been reported to form submicron-sized droplets on mica [58], and submicron-sized droplets of arylamines can be used as hole conductors in organic light emitting diodes [59]. Since the amorphous state of small molecules is thermodynamically unstable, dewetted droplets of organic molecules that are initially amorphous tend to crystallize. Depending on the compound and the annealing conditions, several crystal morphologies can be obtained. Single crystalline, polycrystalline or fibrous crystals have been reported [60].

11.4
Applications of Dewetted Structures in Organic Photonics and Electronics

Controlled micro-structuring of surfaces is important for a wide variety of applications. Using organics in electronics devices has multiple advantages. Organics are lightweight and easy to apply via solution processes or vacuum evaporation. Many organics are flexible and transparent. Since their electrical, mechanical and magnetic properties can be tailored, the interaction with magnetic, electric and electromagnetic fields offers the possibility to control electric current, light emission or absorption. The first recombination of electric charges that led to light emission in anthracene crystals was reported in the 1960s [61–63]. Because trapping sites in crystals require high quality single crystals for efficient luminescence, research quickly moved to amorphous materials as charge carriers and in the 1980s vacuum-deposited thin films of low molar mass compounds were shown to exhibit strong electroluminescence [64]. Patterning is crucial for display applications, and shadow-mask evaporation, ink-jet printing, or screen printing is widely used. Besides these top-down methods, bottom-up self-assembly has alsobeen reported as an effective patterning tool. A phase separated mixture of two polymers, a matrix polymer and an electroluminescent guest polymer, shows electroluminescence from the micron-sized phase-separated guest polymer [65]. However, since the phase separation occurs randomly in the thin film, only some guest domains are in contact with the ITO electrode and emit light. Furthermore, the spinodal decomposition of the host and guest leads to a broad size distribution of the luminescent domains. By dewetting a solution of a hole-transport material, alq3, we were able to produce a device with submicron-sized emitting dots, a size that cannot be reached by the conventional top-down patterning methods described above. Low molar mass organics tend to crystallize and crystals have a series of advantages over amorphous materials in electronic applications, once the difficulties with trap sites due to crystal heterogeneities can be overcome. Thus we attempted to create an array of microcrystals of the hole conductor by annealing the dewetted samples. Bright luminescence was observed, even though the aggregation state of the molecules is still unclear [66].

Another, related organic device that is receiving increasing attention recently is the organic field effect transistor (OFET). Dewetting can be used for fabricating short-channel polymer field effect transistors with conducting polymer employed as electrodes. Source and drain contacts can be patterned by splitting a conducting polymer water solution over a hydrophobic self-assembled monolayer [67], or by self-aligning ink-jet printed gold contacts [68].

Besides forming the source-drain gap by dewetting, patterning of the organic semiconductor itself can also be achieved by self-organization. Pentacene is a promising material, and OFETs have been prepared by a conventional vacuum evaporation process [69]. Grain boundaries inevitably present in the resulting polycrystalline film prepared by this method restrict the carrier mobility. Spin coating from hot trichlorobenzene solution led to a preferred radial crystal orientation, but grain boundaries were still present [70]. By using the roller apparatus to cast a hot trichlorobenzene solution, we were able to produce unidirectionally oriented crystal fibers with a length of several hundred micrometers [71] as shown in Figure 11.7. As in the case of polymers, a concentration profile develops along the solution edge during solvent evaporation and at areas with highest concentration, crystal seeds are formed. Since the edge of the solution is dragged over the substrate, the crystal seed can only grow in one direction – perpendicular to the solution edge. Thus ordered arrays of unidirectionally aligned crystals are formed, when the receding speed of the edge matches the crystal growth rate. Even though the source-drain gap was only partially covered by pentacene, charge mobilities of $0.1\,\mathrm{cm}^2\,\mathrm{V}^{-1}\,\mathrm{s}^{-1}$ have been obtained.

It can be expected that, in the future, other organic electronic devices and circuits, such as sensors [72], radio-frequency identification tags (RFIDs) [73], and ring oscillators [74] may be fabricated using dissipative structures.

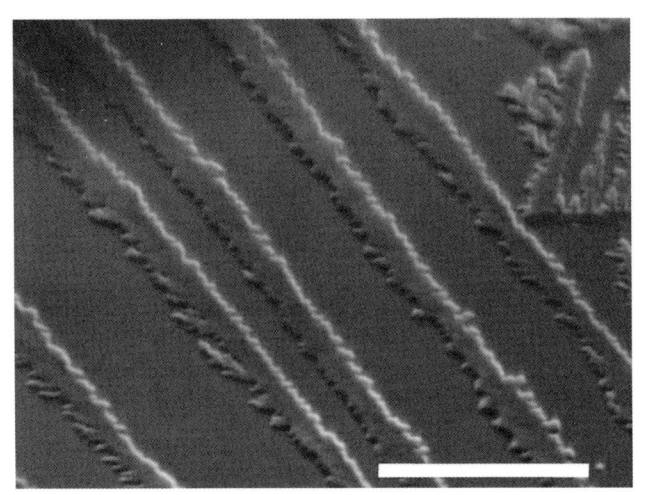

Figure 11.7 Scanning electron micrograph of pentacene fibers prepared by dewetting of a hot trichlorobenzene solution using the roller apparatus. The fibers are aligned along the rolling direction (reprinted with permission from Ref. 71). The scale bar is 5 μm.

11.5 Summary

Dissipative structures may form when a system is in a non-equilibrium state that allows the flow of energy and/or matter. The self-organized structures formed cover a wide range of length scales and even though they are temporal by definition, it is possible to fix them on solid substrates to produce nano- to micrometer-sized regular patterns. By using functional organic materials, functional patterns can be prepared. This opens up the possibility of using bottom-up self-organization in connection with top-down patterning processes to produce functional devices for use in photonics and electronics.

References

1 Whitesides, G. M. and Grzyboski, B. (2002) Self-assembly at all scales. *Science*, **295**, 2418–2421.
2 Whitesides, G. M., Mathias, J. P. and Seto, C. T. (1991) Molecular self-assembly and nanochemistry: a chemical strategy for the synthesis of nanostructures. *Science*, **254**, 1312–1319.
3 Cohen, S. R., Naaman, R. and Sagiv, J. (1986) Thermally induced disorder in organized organic monolayers on solid substrates. *J. Phys. Chem.*, **90**, 3054–3056.
4 Nakaya, U. (1954) *Snow Crystals: Natural and Artificial*, Harvard University Press.
5 Billia, B. and Trivedi, R. (1993) *Handbook of Crystal Growth*, vol. **1** (ed. D. T. J. Hurle), Elsevier, Amsterdam, Chapter 14.
6 Wunderlich, B. (1973–1980) *Macromolecular Physics*, vols. 1–3, Academic, New York.

7 Trivedi, R. and Kurz, W. (1994) Dendritic growth. *Int. Mater. Rev.*, **39**, 49–74.

8 Love, J. C., Estroff, L. A., Kriebel, J. K., Nuzzo, R. G. and Whitesides, G. M. (2005) Self-assembled monolayers of thiolates on metals as a form of nanotechnology. *Chem. Rev.*, **105**, 1103–1169.

9 Hadjichristidis, N., Pispas, S. and Floudas, G. (2002) *Block Copolymers: Synthetic Strategies, Physical Properties and Applications*, Wiley-Interscience, New York.

10 Discher, D. E. and Eisenberg, A. (2002) Polymer Vesicles. *Science*, **297**, 967–973.

11 Antonietti, M. and Foerster, S. (2003) Vesicles and liposomes: a self-assembly principle beyond lipids. *Adv. Mater.*, **15**, 1323–1533.

12 Belousov, B. P. (1959) A periodic reaction and its mechanism. *Compilation Abs. Radiation Med.*, **147**, 145.

13 Zhabotinsky, A. M. (1964) Periodic processes of malonic acid oxidation in a liquid phase. *Biofizika*, **9**, 306–311 (in Russian).

14 Mikhailov, A. S. and Showalter, K. (2006) Control of waves, patterns and turbulence in chemical systems. *Phys. Rep.*, **425**, 79–194.

15 Vartanian, A. (1976) *Diderot and Descartes*, Greenwood Press.

16 Ross Ashby, W. (1947) Principles of the self-organizing dynamic system. *J. Gen. Psych.*, **37**, 125–128.

17 Nicolis, G. and Prigogine, I. (1977) *Self-Organization in Nonequilibrium Systems*, Wiley Interscience, New York.

18 Elia, V., Napoli, E. and Germano, R. (2007) The 'Memory of Water': an almost deciphered enigma. Dissipative structures in extremely dilute aqueous solutions. *Homeopathy*, **96**, 163–169.

19 Myagkov, N. N. (2006) On hydrodynamic instability of the ozone layer. *Phys. Lett. A*, **359**, 681–684.

20 Trevors, J. T. and Abel, D. L. (2006) Self-organization vs. self-ordering events in life-origin models. *Phys. Life Rev.*, **3**, 211–228.

21 Goldbeter, A. (2007) Biological rhythms as temporal dissipative structures. *Adv. Chem. Phys.*, **135**, 253–295.

22 Blomberg, C. (2006) Fluctuations for good and bad: The role of noise in living systems. *Phys. Life Rev.*, **3**, 133–161.

23 Okubo, T., Yokota, N. and Tsuchida, A. (2007) Drying dissipative patterns of dyes in ethyl alcohol on a cover glass. *Colloid. Polym. Sci.*, **285**, 1257–1265.

24 Wessling, B. (1996) Cellular automata simulation of dissipative structure formation in heterogeneous polymer systems, formation of networks of a dispersed phase by flocculation. *J. Phys. II*, **6**, 395–404.

25 Tabony, J. (2006) Historical and conceptual background of self-organization by reactive processes. *Biol. Cell*, **98**, 589–602.

26 Malyshev, V. N. (1996) Coating formation by anodic-cathodic microarc oxidation. *Protect. Met.*, **32**, 607–611.

27 Benard, H. (1900) Les tourbillons cellulaires dans une nappe liquide. *Rev. Gen. Sci*, **12**, 1261–1271.

28 Mutabari I., Wesfreid J. E. and Guyon E.(eds) (2004) *Dynamics of Spatio-temporal Cellular Structures: Henri Benard Centenary Review: Springer Tracts in Modern Physics*, Springer, New York.

29 The Bible, King James Version, Book of Proverbs, Chapter 23, Verse 31.

30 Thomson, J. J. (1855) On certain curious motions observable at the surfaces of wine and other alcoholic liquors. *Philos. Mag.*, **10**, 330–333.

31 Neogi, P. (1985) Tears-of-wine and related phenomena. *J. Colloid. Interface Sci.*, **105**, 94–101.

32 Vuilleumier, R., Ego, V., Neltner, L. and Cazabat, A. M. (1995) Tears of wine: The stationary state. *Langmuir*, **11**, 4117–4121.

33 Maruyama, N., Koito, T., Sawadaishi, T., Karthaus, O., Ijiro, K., Nishi, N., Tokura, S., Nishimura, S. and Shimomura, M. (1998) Mesoscopic pattern formation of nanostructured polymer assemblies. *Supramol. Sci.*, **5**, 331–336.

34 Higashi, N., Kunitake, T. and Kajiyama, T. (1987) Efficient oxygen enrichment by a Langmuir-Blodgett film of the polyion complex of a double-chain fluorocarbon amphiphile. *Polym. J.*, **19**, 289–291.

35 Karthaus, O., Grasjo, L., Maruyama, N. and Shimomura, M. (1999) Formation of ordered mesoscopic polymer arrays by dewetting. *Chaos*, **9**, 308–314.

36 Maruyama, N., Koito, T., Nishida, J., Cieren, X., Ijiro, K., Karthaus, O. and Shimomura, M. (1998) Mesoscopic patterns of molecular aggregates on solid substrates. *Thin Solid Films*, **327–329**, 854–856.

37 Rayleigh, J. W. S. (1878) On the instability of jets. *Proc. London Math. Soc.*, **10**, 4–13.

38 Sharma, A. (1993) Relationship of thin film stability and morphology to macroscopic parameters of wetting in the apolar and polar systems. *Langmuir*, **9**, 861–869.

39 Reiter, G. (1993) Unstable thin polymer films: Rupture and dewetting processes. *Langmuir*, **9**, 1344–1351.

40 Faldi, A., Composto, R. J. and Winey, K. I. (1995) Unstable polymer bilayers. 1. Morphology of dewetting. *Langmuir*, **11**, 4855–4861.

41 Stange, T. G., Evans, D. F. and Hendrickson, W. A. (1997) Nucleation and growth of defects leading to dewetting of thin polymer films. *Langmuir*, **13**, 4459–4465.

42 Miyamoto, Y., Yamao, H. and Sekimoto, K. (2004) *ACS Symposium Series No. 869* (eds J. A. Pojman and Q. Tran-Cong-Miyata), American Chemical Society, Washington, DC.

43 Karthaus, O., Kaga, K., Sato, J., Kurimura, S., Okamoto, K. and Imai, T. (2004) *ACS Symposium Series No. 869* (eds J. A. Pojman and Q. Tran-Cong-Miyata), American Chemical Society, Washington, DC.

44 Yamaguchi, T., Suematsu, N. J. and Mahara, H. (2004) *ACS Symposium Series No. 869* (eds J. A. Pojman and Q. Tran-Cong-Miyata), American Chemical Society, Washington, DC.

45 Suematsu, N. J., Ogawa, Y., Yamamoto, Y. and Yamaguchi, T. (2007) Dewetting self-assembly of nanoparticles into hexagonal array of nanorings. *J. Colloid. Interface Sci.*, **310**, 684–652.

46 Karthaus, O., Okamoto, K., Chiba, R. and Kaga, K. (2002) Size effect of cyanine dye J-aggregates in micrometer-sized polymer "Domes". *Int. J. Nanosci.*, **1**, 461–464.

47 Sawadaishi, T. and Shimomura, M. (2005) Two-dimensional patterns of ultra-fine particles prepared by self-organization. *Colloid. Surf. A*, **257–258**, 71–74.

48 Sawadaishi, T. and Shimomura, M. (2006) Control of structures of two-dimensional patterns of nanoparticles by dissipative process. *Mol. Cryst., Liq. Cryst.*, **464**, 227–231.

49 Suematsu, N. J., Nishimura, S. and Yamaguchi, T. (2008) Release and transfer of polystyrene dewetting pattern by hydration force. *Langmuir*, **24**, 2960–2962.

50 Yabu, H. and Shimomura, M. (2005) Preparation of self-organized mesoscale polymer patterns on a solid substrate: continuous pattern formation from a receding meniscus. *Adv. Func. Mater.*, **15**, 575–581.

51 Karthaus, O., Mikami, S. and Hashimoto, Y. (2006) Control of droplet size and spacing in microsize polymeric dewetting patterns. *J. Colloid. Interface Sci.*, **301**, 703–705.

52 Nonomura, M., Kobayashi, R., Nishiura, Y. and Shimomura, M. (2003) Periodic precipitation during droplet evaporation on a substrate. *J. Phys. Soc. Jpn.*, **72**, 2468–2471.

53 Karthaus, O., Ijiro, K. and Shimomura, M. (1996) Ordered self-assembly of nanosize polystyrene aggregates on mica. *Chem. Lett.*, 821–822.

54 Kiyono, Y. and Karthaus, O. (2006) Reversible shape change of polymer microdomes. *Jpn. J. Appl. Phys.*, **45**, 588–590.

55 Nussbaum, P., Völkel, R., Herzig, H. P., Eisner, M. and Haselbeck, S. (1997) Design, fabrication and testing of

microlens arrays for sensors and Microsystems. *Pure Appl. Opt.*, **6**, 617–636.

56 Karthaus, O., Terayama, T. and Hashimoto, Y. (2006) Selective dewetting of polymers on metal/silicon micropatterns. *e-J. Surf. Sci. Nanotechnol.*, **4**, 656–660.

57 Wang, J. Z., Zheng, Z. H., Li, H. W., Huck, W. T. S. and Sirringhaus, H. (2004) Dewetting of conducting polymer inkjet droplets on patterned surfaces. *Nature Mater.*, **3**, 171–176.

58 Hellmann, J., Hamano, M., Karthaus, O., Ijiro, K., Shimomura, M. and Irie, M. (1998) Aggregation of dendrimers with a photochromic dithienylethene core group on the mica surface – atomic force microscopic imaging. *Jpn. J. Appl. Phys.*, **37**, L816–L819.

59 Karthaus, O., Adachi, C., Kurimura, S. and Oyamada, T. (2004) Electroluminescence from self-organized microdomes. *Appl. Phys. Lett.*, **84**, 4696–4698.

60 Karthaus, O., Imai, T., Sato, J., Kurimura, S. and Nakamura, R. (2005) Control of crystal morphology in dewetted films of thienyl dyes. *Appl. Phys. A*, **80**, 903–906.

61 Pope, M., Kallmann, H. P. and Magnante, P. (1963) Electroluminescence in organic crystals. *J. Chem. Phys.*, **38**, 2042–2043.

62 Mehl, W. and Funk, B. (1967) Dark injection of electrons and holes and radiative recombination in anthracene with metallic contacts. *Phys. Lett. A*, **25**, 364–365.

63 Helfrich, W. and Schneider, W. G. (1965) Recombination radiation in anthracene crystals. *Phys. Rev. Lett.*, **14**, 229–231.

64 Tang, W. and van Slyke, S. A. (1987) Organic electroluminescent diodes. *Appl. Phys. Lett.*, **51**, 913–915.

65 Adachi, C., Hibino, S., Koyama, T. and Taniguchi, Y. (1997) Submicrometer-sized organic light emitting diodes with a triphenylamine-containing polycarbonate as a guest molecule in a polymer blend. *Jpn. J. Appl. Phys.*, **36**, L827–L830.

66 Karthaus, O., Adachi, C., Arakaki, S., Endo, A. and Wada, T. (2006) Preparation of micropatterned organic light emitting diodes by self-organization. *Mol. Cryst., Liq. Cryst.*, **444**, 87–94.

67 Wang, J. Z., Zheng, Z. H., Li, H. W., Huck, W. T. S. and Sirringshaus, H. (2004) Polymer field effect transistors fabricated by dewetting. *Synth. Met.*, **146**, 287–290.

68 Noh, Y.-Y., Cheng, X., Sirringhaus, H., Sohn, J. I., Welland, M. E. and Kang, D. J. (2007) Ink-jet printed ZnO nanowire field effect transistors. *Appl. Phys. Lett.*, **91**, 043109.

69 Jurchescu, O. D., Baas, J. and Palstra, T. T. M. (2004) Effect of impurities on the mobility of single crystal pentacene. *Appl. Phys. Lett.*, **84**, 3061–3063.

70 Minakata, T. and Natsume, Y. (2005) Direct formation of pentacene thin films by solution process. *Syn. Met.*, **153**, 1–4.

71 Kai, K. and Karthaus, O. (2007) Growth of unidirectionally oriented pentacene nanofibers by a roller method'. *e-J. Surf. Sci. Nanotechnol.*, **5**, 103–105.

72 Crone, B. K., Dodabalapur, A., Sarpeshkar, R., Gelperin, A., Katz, H. E. and Bao, Z. (2001) Organic oscillator and adaptive amplifier circuits for chemical vapor sensing. *J. Appl. Phys.*, **91**, 10140.

73 Baude, P. F., Ender, D. A., Haase, M. A., Kelley, T. W., Muyres, D. V. and Theiss, S. D. (2003) Pentacene-based radio-frequency identification circuitry. *Appl. Phys. Lett.*, **82**, 3964–3966.

74 Klauk, H., Halik, M., Zschieschang, U., Eder, F., Schmid, G. and Dehm, Ch. (2003) Pentacene organic transistors and ring oscillators on glass and on flexible polymeric substrates. *Appl. Phys. Lett.*, **82**, 4175–4177.

12
Formation of Nanosize Morphology of Dye-Doped Copolymer Films and Evaluation of Organic Dye Nanocrystals Using a Laser

Akira Itaya, Shinjiro Machida, and Sadahiro Masuo

12.1
Introduction

The formation of nanopatterned functional surfaces is a recent topic in nanotechnology. As is widely known, diblock copolymers, which consist of two different types of polymer chains connected by a chemical bond, have a wide variety of microphase separation structures, such as spheres, cylinders, and lamellae, on the nanoscale, and are expected to be new functional materials with nanostructures. Further modification of the nanostructures is also useful for obtaining new functional materials. In addition, utilization of nanoparticles of an organic dye is also a topic of interest in nanotechnology.

In most studies using block copolymers, the block copolymer films are used as a scaffold for patterning inorganic nanoparticles such as metals and are removed from the substrate at the end of a process. On the other hand, there are fewer studies on the functionalization of a block copolymer film by the introduction of a functional chromophore into a part of the nanoscale microphase separation structure. However, the introduction of a functional chromophore, that exhibits charge transport [1–8], a photovoltaic effect [5–8], nonlinear optical effects [4, 9, 10], emission [11], photochromism [12, 13], and so on, into a block copolymer film is very attractive in view of the recent development of organic electronic and photonic devices. Several methods are adopted for the site-selective introduction of a functional chromophore into a block copolymer film: (i) chemical introduction of a functional chromophore to one component or the junction of a block copolymer [1–11], (ii) immersion of a block copolymer film into the solution of a functional chromophore using a solvent that can dissolve only one component [14], and (iii) vapor transportation of a functional chromophore whose compatibility with the two components of a diblock copolymer is different [12, 13]. Although chemical introduction is the best method with regard to position selectivity, it lacks versatility, requires intensive processing, and is expensive. Vapor transportation is also costly since it needs vacuum and high temperature but the immersion method is simple and convenient.

Molecular Nano Dynamics, Volume I: Spectroscopic Methods and Nanostructures
Edited by H. Fukumura, M. Irie, Y. Iwasawa, H. Masuhara, and K. Uosaki
Copyright © 2009 WILEY-VCH Verlag GmbH & Co. KGaA, Weinheim
ISBN: 978-3-527-32017-2

Laser ablation of polymer films has been extensively investigated, both for application to their surface modification and thin-film deposition and for elucidation of the mechanism [15]. Dopant-induced laser ablation of polymer films has also been investigated [16]. In this technique ablation is induced by excitation not of the target polymer film itself but of a small amount of the photosensitizer doped in the polymer film. When dye molecules are doped site-selectively into the nanoscale microdomain structures of diblock copolymer films, dopant-induced laser ablation is expected to create a change in the morphology of nanoscale structures on the polymer surface.

First, we describe the position-selective arrangement of polystyrene microspheres with a diameter of 20 or 50 nm onto a diblock copolymer film of polystyrene-*block*-poly(4-vinylpyridine) (PS-*b*-P4VP) with a nanosize sea–island microphase structure. The microspheres contain a fluorescent chromophore, and their surface is modified with carboxylic acid groups. Hydrogen bonding between the pyridyl groups of P4VP and the carboxylic acid groups of the microspheres is expected to be the driving force for position-selective adsorption [17].

Polystyrene-*block*-poly(4-vinylpyridine)
(PS-*b*-P4VP)

Tetrakis-5,10,15,20-(4-carboxyphenyl)porphyrin
(TCPP)

Aurintricarboxylic
(ATA)

N,N'-bis(2,6-dimethylphenyl)-3,4,9,10-perylenedicarboxyimide
(DMPBI)

Secondly, we describe the site-selective introduction of a functional molecule, tetrakis-5,10,15,20-(4-carboxyphenyl)porphyrin (TCPP), into the microphase separation structure of a diblock copolymer film of PS-*b*-P4VP. Since porphyrin derivatives show various functionalities such as sensitization, redox activity, and nonlinear optical effect, a polymer nanodot array containing a porphyrin at a high concentration would be applicable to a light-harvesting and charge transporting nanochannel.

Site-selective doping was performed by immersing the copolymer film into the dopant solutions with different solubilities for the two components. For the site-selective doping, we also utilized multipoint hydrogen bonding between the four carboxylic acid groups of TCPP and pyridyl groups of P4VP [18].

Further modification of the above nanostructures is useful for obtaining new functional materials. Thirdly, we apply the dopant-induced laser ablation technique to site-selectively doped thin diblock copolymer films with spheres (sea–island), cylinders (hole-network), and wormlike structures on the nanoscale [19, 20]. When the dye-doped component parts are ablated away by laser light, the films are modified selectively. Concerning the laser ablation of diblock copolymer films, Lengl et al. carried out the excimer laser ablation of diblock copolymer monolayer films, forming spherical micelles loaded with an Au salt to obtain metallic Au nanodots [21]. They used the laser ablation to remove the polymer matrix. In our experiment, however, the laser ablation is used to remove one component of block copolymers. Thereby, we can expect to obtain new functional materials with novel nanostructures.

Organic nanoparticles, including nanosized crystals and aggregates, have attracted growing attention in recent years because of their potential application to optoelectronics, pharmaceuticals, cosmetics, and so on [22, 23]. Finally, we describe how the emission from a single nanocrystal consisting of organic dye molecules, namely, N,N'-bis(2,6-dimethylphenyl)-3,4,9,10-perylenedicarboxyimide (DMPBI) shows photon antibunching [24]. That is, even molecular assemblies can be made to behave as single-photon sources by controlling the size on the nanometer scale. The photon antibunching means that the probability of detecting two simultaneous photons drops to zero; therefore, materials that exhibit photon antibunching are called single-photon emitters or single-photon sources. A single molecule is a typical single-photon source because the molecule cannot emit two photons simultaneously [25, 26]. Generally, single-photon emission can be observed from so-called "single quantum systems". However, it is considered that the multi-quantum systems, namely, multichromophoric systems, can also be single-photon sources if one exciton remains as a result of exciton–exciton annihilation among the generated excitons. This phenomenon is demonstrated.

12.2
Position-Selective Arrangement of Nanosize Polymer Microspheres Onto a PS-b-P4VP Diblock Copolymer Film with Nanoscale Sea–island Microphase Structure

PS-b-P4VP (M_n; PS:P4VP = 301 000 : 19 600) films were prepared by spin-casting of the toluene solution onto glass substrates. The adsorption of microspheres on PS-b-P4VP films was performed by immersing the film in a microsphere/methanol suspension and subsequently rinsing the film in methanol to remove the microspheres that were not adsorbed. On the basis of the brightness of the fluorescence microscope images of P4VP homopolymer films after immersing in a microsphere/water suspension and rinsing in water, we determined the most appropriate immersion time of the block copolymer films for adsorption and rinsing.

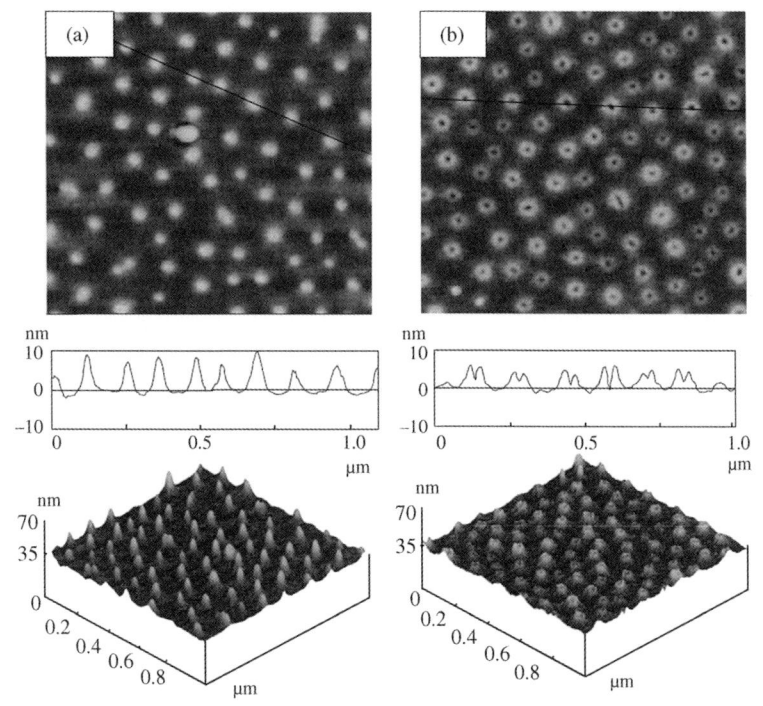

Figure 12.1 AFM images of a PS-*b*-P4VP (301 000 : 19 600) film (a) before and (b) after immersion in methanol for 75 min and the height profiles. S. Machida, H. Nakata, K. Yamada, A. Itaya: Position-selective arrangement of nanosized polymer microsphere on diblock copolymer film with sea–island microphase structure. *Jpn. J. Appl. Phys.* **2006**, *45*, 4270–4273. Copyright Wiley InterScience. Reproduced with permission.

First, we observed the AFM images of a PS-*b*-P4VP film before and after immersion in methanol for 75 min. For the AFM image before immersion (Figure 12.1a), a clear sea–island structure is observed. On the basis of the molecular weight ratio of the PS-*b*-P4VP, we can safely judge that the island part corresponds to the P4VP domain and the sea part to PS. The AFM image observed after immersion in methanol (Figure 12.1b) shows islands with a smaller height and a larger width than those before immersion. Moreover, a hole is created at the center of each island. This morphological change is attributed to the difference in solubility between P4VP and PS chains in methanol. A similar morphological change (formation of a hole in the P4VP domain) of a PS-*b*-P4VP film was reported to be also induced by exposure of the film to methanol vapor [27].

Figure 12.2 shows the AFM images and height profiles of PS-*b*-P4VP films immersed in the microsphere/methanol suspension (diameter: 20 nm for (a) and 50 nm for (b)) for 75 min and subsequently rinsed in pure methanol for 90 min. The height profile in Figure 12.2a shows a larger island-top curvature radius and a larger island height (15–30 nm) than those before immersion (Figure 12.1a). This morphological change is

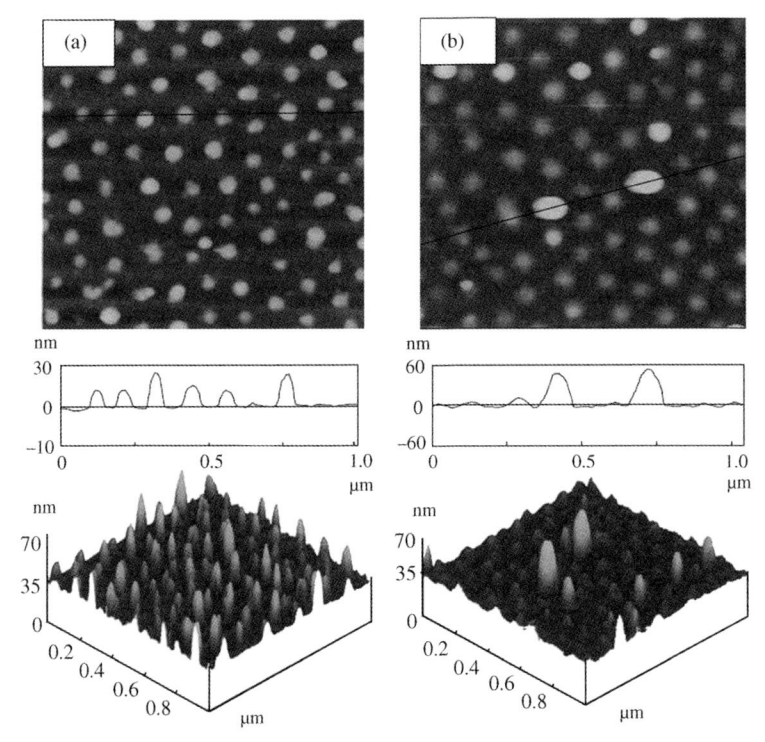

Figure 12.2 AFM images of a PS-*b*-P4VP (301 000 : 19 600) film after immersion in a microsphere/methanol ((a): 20 nm, (b): 50 nm) suspension for 75 min and rinsing in methanol for 90 min and the height profiles. S. Machida, H. Nakata, K. Yamada, A. Itaya: Position-selective arrangement of nanosized polymer microsphere on diblock copolymer film with sea–island microphase structure. *Jpn. J. Appl. Phys.* **2006**, *45*, 4270–4273. Copyright Wiley InterScience. Reproduced with permission.

attributed to the adsorption of a single microsphere onto each island of the block copolymer film. For most islands in the height profile, the increase in height is smaller than the diameter of the microsphere (20 nm). This is reasonable because P4VP chains swell in methanol so that the microspheres would sink into the islands. The bird's-eye view in Figure 12.2a, compared with those in Figure 12.1a and b, indicates that most of the island parts in the film adsorb 20 nm microspheres. On the other hand, in the case of the 50 nm microspheres, only a small number of the islands show an increase in height (Figure 12.2b). In the height profile, the height and width of the two islands are approximately equal to the microsphere diameter (50 nm). Hence, we conclude that, although some 50 nm microspheres are adsorbed onto the block copolymer film, most islands do not adsorb 50 nm microspheres.

Here, we have demonstrated that it is possible to arrange successfully polystyrene microspheres with a diameter of 20 nm on each island (P4VP domain) of a PS-*b*-P4VP block copolymer film using hydrogen bonds. A 50 nm-large microsphere was rarely adsorbed to the PS-*b*-P4VP film. Since the present technique does not require an expensive apparatus and the chemical modification of block copolymers, it can be applied to various types of organic functional nanoparticle.

12.3
Nanoscale Morphological Change of PS-b-P4VP Block Copolymer Films Induced by Site-Selective Doping of a Photoactive Chromophore

12.3.1
Nanoscale Surface Morphology of PS-b-P4VP Block Copolymer Films

It is widely known that the microphase separation structures of block copolymers depend on the volume fraction of the components [28] and film preparation conditions [29–33]. As aforementioned, when PS-b-P4VP (M_n; PS:P4VP = 301 000 : 19 600) films were prepared by the spin-coating of the toluene solution, a clear sea–island microphase separation structure was formed (Figure 12.1a). On the other hand, when PS-b-P4VP (162 400 : 87 400) and chloroform solvent were used with a polymer concentration of 2.5 wt%, regular nanoscale networklike phase separation structures with P4VP-cylinder parts as holes in the PS-matrix parts are obtained (Figure 12.3a). The film thickness was about 250 nm, and the P4VP-cylinder parts were about 2 nm lower in height than the PS-matrix parts (Figure 12.3g). Since the solvent is one of the key factors that affects the resultant microphase separation structures [29], 3-pentanone, which has a high boiling point, was used instead of chloroform, which has a lower boiling point. The film spin-cast from a PS-b-P4VP/3-pentanone solution of 0.52 wt%, with a thickness of about 25 nm, showed a surface morphology with network-like phase separation structures (Figure 12.4a and d), which is similar to that of the film prepared using chloroform solvent (Figure 12.3a). However, the structure is nonuniform, and both the average diameter and depth of the holes (P4VP-cylinder parts) are larger than those of the above-mentioned film. On the other hand, symmetric PS-b-P4VP (20 000 : 19 000) diblock copolymer films spin-cast from a 3-pentanone solution (0.2 wt%) show nanoscale worm-like phase separation structures with an average roughness height of about 3 nm, as shown in Figure 12.5a and d. In the AFM image, the bright (high) and dark (low) parts correspond to P4VP and PS components, respectively.

12.3.2
Nanoscale Surface Morphological Change of PS-b-P4VP Block Copolymer Films Induced by Site-Selective Doping of a Photoactive Chromophore

The selective doping of TCPP chromophore into the nanoscale domains of P4VP was carried out by immersing the thin films in a methanol solution of TCPP and subsequently rinsing them in pure methanol. As aforementioned, the driving force of the selective doping is the hydrogen bonds between the nitrogen atoms of the pyridyl groups and the carboxylic groups of TCPP.

Figure 12.6a and b shows AFM images of PS-b-P4VP (301 000 : 19 600) films before and after doping of TCPP, respectively. The regular sea–island structure with

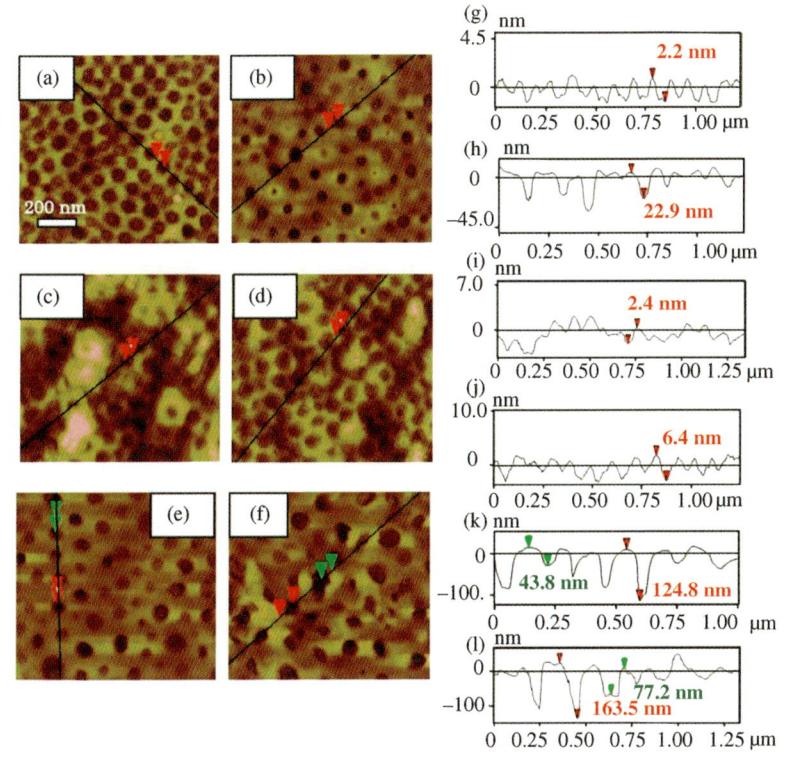

Figure 12.3 AFM images of thin PS-b-P4VP (162 400 : 87 400) films (chloroform solvent) with network-like structures of P4VP cylinders in PS matrices on glass substrates, height profiles of horizontal lines in these images. (a) Before and (b) after immersion in methanol, (c) after doping with TCPP, (d) TCPP-doped films irradiated using one shot with fluence of about 170 mJ cm^{-2} in air, (e) TCPP-doped films irradiated using one shot with fluence of about 150 mJ cm^{-2} in methanol and (f) 100-shot irradiation with the same fluence. (g)–(l) Show the height profiles of the horizontal lines shown in the AFM images of (a)–(f), respectively. Z. Wang, S. Masuo, S. Machida, A. Itaya: Site-selective doping of dyes into polystyrene-block-poly(4-vinyl pyridine) diblock copolymer films and selective laser ablation of the dye-doped films. *Jpn. J. Appl. Phys.* **2007**, *46*, 7569–7576. Copyright the Japan Society of Applied Physics. Reproduced with permission.

nanoscale size is observed clearly for both of the films before and after doping. The average heights of the island parts are estimated to be about 8 and 12 nm before and after doping, respectively. On the basis of these AFM images, an AFM image of PS-b-P4VP films after immersing in methanol (Figure 12.1b), and the results of the doping of TCPP into PS and P4VP homopolymer films, we can judge that the selective doping of TCPP into P4VP-island parts is successful [18].

TCPP was doped selectively into the P4VP-cylinder domains of the PS-b-P4VP (162 400 : 87 400) films prepared from the chloroform solution. Figure 12.3c and i show an AFM image of a film doped with TCPP and the height profile of the line indicated in the image, respectively. On the surface of the TCPP-doped films, there

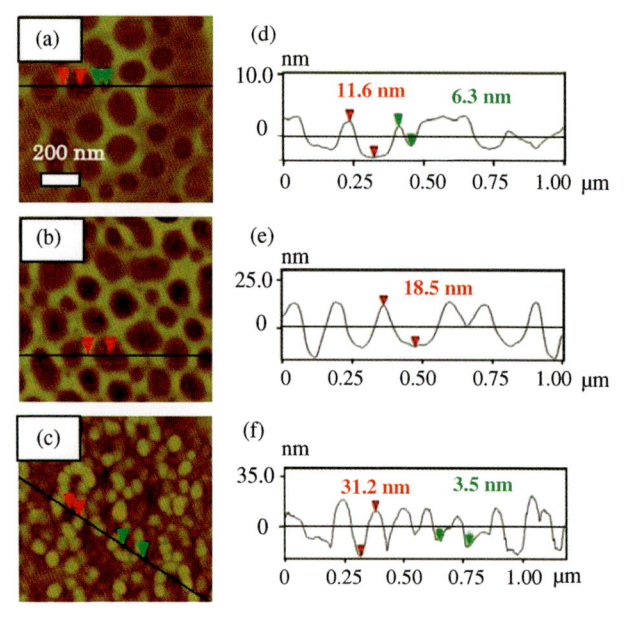

Figure 12.4 AFM images of thin PS-b-P4VP (162 400 : 87 400) films (3-pentanone solvent) with phase separation structures of P4VP cylinders in PS matrices on glass substrates, and height profiles of horizontal lines in these images. (a), (d) Before and (b), (e) after immersion in methanol; (c), (f) after being doped with TCPP; (d)–(f) are the height profiles of the horizontal lines shown in the AFM images of (a)–(c), respectively. Z. Wang, S. Masuo, S. Machida, A. Itaya: Site-selective doping of dyes into polystyrene-block-poly(4-vinyl pyridine) diblock copolymer films and selective laser ablation of the dye-doped films. Jpn. J. Appl. Phys. **2007**, 46, 7569–7576. Copyright the Japan Society of Applied Physics. Reproduced with permission.

are some conglomeration parts (P4VP + TCPP), which are attributed to the TCPP doped in the P4VP cylinders. When the undoped films are immersed in TCPP/methanol solution, the hole volume in the matrices does not change, because methanol is a poor solvent for PS matrices, that is, there is no redundant space to accept the swollen P4VP domains doped with TCPP. Thus, the conglomeration parts remain on the film surface, even after being dried under vacuum. The coloration of the doped film was observed clearly, suggesting that a large number of TCPP molecules were doped into the thick films. In the case of the PS-b-P4VP (162 400 : 87 400) films prepared from the 3-pentanone solution (Figure 12.4c and f), the protruding parts consisting of many small spheres are formed around the holes; thus, the size of the holes is decreased. The small spheres are likely to be composed of P4VP chains cross-linked by many TCPP molecules.

As aforementioned, symmetric PS-b-P4VP (20 000 : 19 000) diblock copolymer films show the wormlike phase separation structures with an average roughness height of about 3 nm (Figure 12.5a and d). The doping of TCPP into the P4VP domain of the high part induces an increase in the average roughness height of about 8 nm (Figure 12.5b and e).

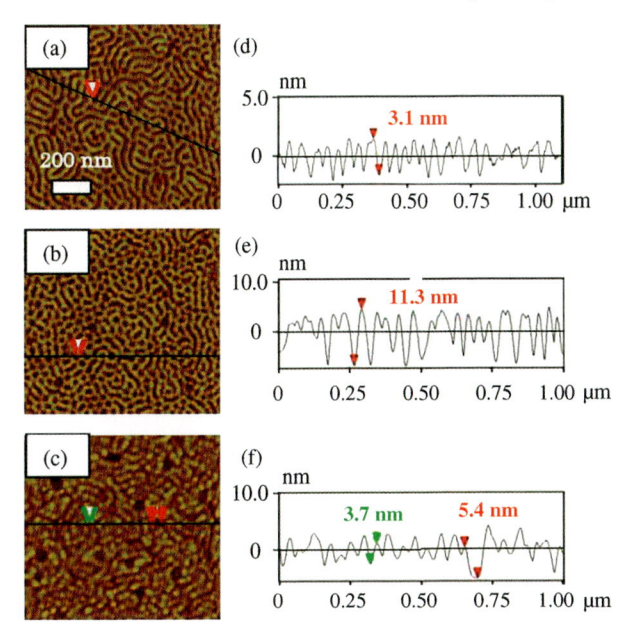

Figure 12.5 AFM images of thin PS-b-P4VP (20 000 : 19 000) films (3-pentanone solvent) with worm-like structures on mica substrates and height profiles of horizontal lines in these images. (a), (d) Before and (b), (e) after being doped with TCPP. (c), (f) TCPP-doped films irradiated with fluence of about 150 mJ cm^{-2} in air. (d)–(f) show the height profiles of the horizontal lines shown in the AFM images of (a)–(c), respectively. S. Machida, H. Nakata, K. Yamada, A. Itaya: Position-selective arrangement of nanosized polymer microsphere on diblock copolymer film with sea–island microphase structure. Jpn. J. Appl. Phys. **2006**, 45, 4270–4273. Copyright Wiley InterScience. Reproduced with permission.

When PS-b-P4VP diblock copolymers were spin-cast using different solvents, the surface morphology of the films depended on the solvent used and the number-average molecular weight. Utilizing the multiple hydrogen bonds between the nitrogen atoms of pyridyl groups and the carboxylic groups, a selective doping into the films with different morphologies of nanoscale phase separation structures was carried out by immersing the films in a methanol solution of TCPP chromophore, resulting in a further nanoscale surface morphological change of the films. One can see schematic illustrations of the surface modification process of PS-b-P4VP films during doping in the TCPP/methanol solution in Refs. [18] and [20].

12.4
Site-Selective Modification of the Nanoscale Surface Morphology of Dye-Doped Copolymer Films Using Dopant-Induced Laser Ablation

As aforementioned, laser ablation of polymer films themselves and dopant-induced laser ablation of polymer films have been extensively investigated. The photochemical or photothermal mechanism has been discussed. The feature of the dopant-

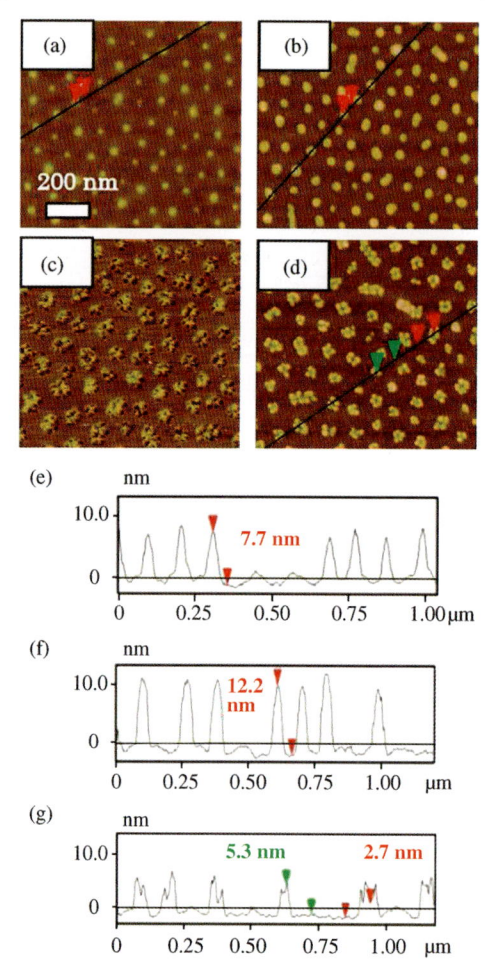

Figure 12.6 AFM images of thin PS-b-P4VP (301 000 : 19 600) films with sea–island structures (sea parts of PS components and island parts of P4VP components) on glass substrates, and height profiles of horizontal lines in these images. (a) Before and (b) after being doped with TCPP. Doped films irradiated using one shot with fluences of about (c) 110 and (d) 150 mJ cm^{-2} in methanol. Height profiles of (e)–(g) correspond to the horizontal lines shown in the AFM images of (a), (b), and (d), respectively. Z. Wang, S. Masuo, S. Machida, A. Itaya: Application of dopant-induced laser ablation to site-selective modification of sea–island structures of polystyrene-block-poly(4-vinylpyridine) films. *Jpn. J. Appl. Phys.* **2005**, *44*, L402–L404. Copyright the Japan Society of Applied Physics. Reproduced with permission.

induced laser ablation is that ablation is induced by excitation not of the target polymer film itself but of a small amount of the photosensitizer doped in the polymer film. By choosing a molecule with a large π-electronic conjugated system as a dopant, the ablation of polymer films is induced easily with longer wavelength lasers and lower fluences than those employed in laser ablation with excitation of the polymer film itself. For photostable dopants, the "cyclic-multiphoton absorption mechanism" has been proposed for the dopant-induced laser ablation [34]. This mechanism is applied to molecules whose transient states have substantial absorption coefficients at the excitation wavelength. In the mechanism, relaxation from the excited states of the transient states of a dopant to the transient states involves internal conversion in the dopant and subsequent intermolecular vibrational energy transfer from the dopant to the surrounding polymer matrix. Thus, the polymer matrix is heated up repeatedly during the multiphoton absorption. The irradiation of a laser with high fluences results in rapid thermal decomposition of the polymer matrix, that is, ablation.

As aforementioned, diblock copolymer films have a wide variety of nanosized microphase separation structures such as spheres, cylinders, and lamellae. As described in the above subsection, photofunctional chromophores were able to be doped site-selectively into the nanoscale microdomain structures of the diblock copolymer films, resulting in nanoscale surface morphological change of the doped films. The further modification of the nanostructures is useful for obtaining new functional materials. Hence, in order to create further surface morphological change of the nanoscale microdomain structures, dopant-induced laser ablation is applied to the site-selectively doped diblock polymer films.

First, the laser ablation behavior of PS and P4VP homopolymer films and TCPP-doped P4VP homopolymer films was investigated. PS and P4VP homopolymer films show no absorption at a laser wavelength of 532 nm (second harmonic output of a Nd^{3+}:YAG laser with full width at half maximum of about 8 ns); however, the ablation phenomena of these films were observed for laser irradiation with this wavelength. Hence, the ablation is likely to be attributed to multiphoton absorption of these films. The ablation thresholds were determined to be about 190 and 320 mJ cm^{-2} for PS and P4VP films, respectively. The ablation thresholds of the TCPP-doped P4VP films were about 220 and 150 mJ cm^{-2} for dopant concentrations of 1.0 and 3.5 wt%, respectively, which corresponds to the previous reports that indicate that the ablation threshold strongly depends on the absorbance, that is, the concentration of the dopant [16, 35–38]. Since these values are smaller than that of the neat P4VP films, the dopant-induced laser ablation of TCPP-doped P4VP films is successfully induced upon excitation at this wavelength. According to the above ablation thresholds, the concentration of TCPP is considered to be one of the important parameters for ablating the P4VP parts effectively. However, it is difficult to measure the concentration of TCPP doped in the P4VP parts of PS-b-P4VP films. Thus, it is difficult to estimate the ablation thresholds of TCPP-doped P4VP parts. Hence, we irradiated TCPP-doped PS-b-P4VP films with laser fluences lower than the threshold fluence of PS parts to prevent damage to the PS parts but to induce the selective ablation of the TCPP-doped P4VP parts.

Next, laser ablation of the TCPP-doped PS-b-P4VP (301 000 : 19 600) films with a sea–island structure was carried out in air. However, ablation of the P4VP-island parts did not occur after one-shot laser irradiation with a fluence of 170 mJ cm^{-2}, but both the P4VP-island and PS-sea parts were ablated by laser irradiation with fluences higher than 190 mJ cm^{-2}. Hence, the ablation threshold of the P4VP-island parts doped with TCPP may be even higher than 170 mJ cm^{-2} in air.

As mentioned, methanol is a good solvent for P4VP and a poor solvent for PS. When the TCPP-doped thin films are immersed in methanol, the TCPP-doped P4VP-island parts swell. However, TCPP molecules doped in the P4VP-island parts do not dissolve in methanol, because the TCPP molecules are connected with P4VP chains via multiple hydrogen bonds. This is supported by the fact that the surface morphology of the TCPP-doped thin films was unaffected by immersing the film in methanol. Thus, the laser ablation experiment of the films was carried out in methanol. Figure 12.6c and d show AFM images of the films irradiated using one laser shot with fluences of 110 and 150 mJ cm^{-2}, respectively. For the irradiation in methanol, a selective ablation of the TCPP-doped P4VP-island parts is obtained. However, for ablation with a fluence of 110 mJ cm^{-2}, a small number of fragments are redeposited. The irradiation with a fluence of 150 mJ cm^{-2} induces the ablation of the topside of the island parts, and the shape of the island parts becomes that of a flower. The average heights of the edge and center in the flowerlike-shaped island parts were about 5 and 3 nm from the surface of the surrounding sea parts, respectively (Figure 12.6g). These small etch depths suggest that the distribution of TCPP in the P4VP-island parts is not homogeneous; the concentration of TCPP in the topside is larger than that in the underside.

For ArF excimer laser (193 nm) ablation of hydrated collagen gel films in the swollen state, the boiling of the water is reported to be responsible for the ablation [39]. In the present case, the photon energy is low (532 nm), and laser fluences are not particularly high. In addition, the swollen island parts are covered with methanol. Hence, laser irradiation is unlikely to induce the boiling of methanol in the swollen parts. That is, the ablation of the swollen P4VP-island parts is not attributable to the boiling of methanol. A portion of the irradiated energy is considered to be absorbed by TCPP, and dopant-induced laser ablation occurs in the swollen state, but not in the dry state. The dopant-induced laser ablation can likely be attributed to the photothermal effect due to the cyclic multiphotonic absorption mechanism. The reason why the ablation threshold for the swollen state is lower than that for the dry state is still unclear. In this experiment, however, it is suggested that the swollen state causes the selective ablation of TCPP-doped P4VP-island parts to become easier. When the doped films were irradiated with a fluence of 150 mJ cm^{-2} in a methanol vapor environment, where the TCPP-doped P4VP-island parts were in a swollen state as they were in methanol, selective ablation was not induced as it was in the dry condition. This can likely be attributed to the low degree of the swollen state of the TCPP-doped P4VP parts in solvent vapor compared with the degree in the solvent.

In the present experiment, the concentration of TCPP is one of the most important parameters for obtaining a sufficient laser energy for ablating the island parts. In order to remove further P4VP-island parts, we tried to dope TCPP into the sample films again after the first laser irradiation and to irradiate the sample film. However, AFM images showed that the average height of the P4VP-island parts did not increase after the second doping and that the remaining parts of the island could no longer be ablated. These results suggest that the first laser irradiation induces not only the ablation of P4VP-island parts but also a cross-linking reaction among P4VP chains. The presence of the cross-linking reaction was suggested by the fact that the solubility

12.4 Site-Selective Modification of the Nanoscale Surface Morphology | 215

rate of the surface of the TCPP-doped P4VP homopolymer films in methanol was smaller for the part irradiated in acetone solvent than for the nonirradiated one.

As for PS-*b*-P4VP (162 400 : 87 400) films with a different surface morphology (Figure 12.3), the conglomeration parts on the film surface are ablated away selectively to a certain extent by one-shot laser irradiation with a fluence of about 170 mJ cm^{-2} in air (Figure 12.3d and j). On the other hand, one-shot laser irradiation with a fluence of about 150 mJ cm^{-2} in methanol results in a further selective ablation of P4VP-cylinder parts, although the degree of ablation is not homogeneous for each cylinder (Figure 12.3e and k). The latter result suggests that the concentration of TCPP doped into each cylinder is not homogeneous and/or that the phase separation structures in the interior of the present thick films are not uniform throughout the films [29]. The same film was also irradiated using 100 shots of the laser pulse with the same fluence in methanol. Figure 12.3f and l show an AFM image of this irradiated film and the height profile of the line indicated in the image, respectively. The depths of the ablated cylinder parts indicate that the TCPP-doped P4VP-cylinder parts are selectively ablated away to a larger extent by 100-shot laser irradiation. However, the laser irradiation of 100 shots results in a morphological change of the PS-matrix parts. That is, the PS parts were also ablated in part.

As aforementioned, site-selective laser ablation was also successful using the diblock copolymer films with a nanosize regular network-like phase separation structure with P4VP cylinders in PS matrices. In the present case, because of the large film thickness, a large amount of TCPP was doped into the P4VP-cylinder parts compared with the amount doped into the above-mentioned films with the sea–island structure. This larger concentration of TCPP also induced selective ablation in air, although the degree of ablation was low. A further selective ablation of the doped matrices was also achieved by irradiation in methanol.

For symmetric PS-*b*-P4VP (20 000 : 19 000) diblock copolymer films with the wormlike phase separation structures, the TCPP-doped films were irradiated using one laser shot with a fluence of 150 mJ cm^{-2} in air. The ablation phenomenon is observed for this irradiation fluence (Figure 12.5c and f), but it is difficult to conclude that this is a selective ablation of the doped-P4VP parts. We cannot deny the possibility that the decomposition of the P4VP parts affects the PS parts because of the existence of large interfaces between the two symmetric blocks in wormlike structures. Thus, for the site-selective ablation of diblock copolymer films, the surface morphology of the phase separation structures is one of the most important parameters.

Next, in order to investigate the effect of the dye on selective ablation, another dye, namely, aurintricarboxylic (ATA), was doped selectively into the PS-*b*-P4VP (162 400 : 87 400) diblock copolymer films spin-coated from a 3-pentanone solution using the same immersion method. As shown in Figure 12.7, the surface morphology of the films was changed markedly by the doping with ATA (Figure 12.7a), and the morphology of the ATA-doped films is very different from that of the TCPP-doped films (Figure 12.4c). The ATA-doped copolymer films were irradiated using one laser shot with a fluence of 150 mJ cm^{-2} in air. An AFM image of the irradiated films and the height profile of the line indicated in the image are shown in

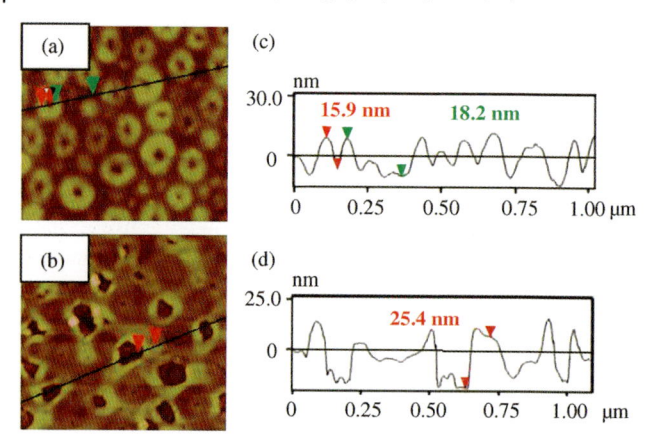

Figure 12.7 AFM images of thin PS-b-P4VP (162 400 : 87 400) films (3-pentanone solvent) on glass substrates, height profiles of horizontal lines in these images. (a), (c) After being doped with ATA, and (b), (d) ATA-doped films irradiated using one shot with fluence of about 150 mJ cm^{-2} in air. (c) and (d) Show the height profiles of the horizontal lines shown in the AFM images of (a) and (b), respectively. Z. Wang, S. Masuo, S. Machida, A. Itaya: Site-selective doping of dyes into polystyrene-*block*-poly(4-vinyl pyridine) diblock copolymer films and selective laser ablation of the dye-doped films. *Jpn. J. Appl. Phys.* **2007**, *46*, 7569–7576. Copyright the Japan Society of Applied Physics. Reproduced with permission.

Figure 12.7b and d, respectively. Selective laser ablation of the ATA-doped P4VP parts is observed, even though some parts are not ablated completely. According to Figure 12.7d, some matrices are ablated markedly, particularly where the substrate surface comes out. In the present case, we also achieved selective ablation under dry conditions, which is attributed to a sufficient number of ATA molecules being present in the P4VP components.

In conclusion, when the diblock copolymer films, whose P4VP parts were selectively doped with dyes, were irradiated using appropriate laser fluences lower than the threshold fluences of the PS and P4VP parts, selective ablation of the P4VP parts in the PS-b-P4VP films was obtained. The concentration of dyes in the P4VP parts was an important parameter for ablating P4VP parts effectively. For the effect of irradiation environments, further selective ablation of the dye-doped parts was obtained in methanol with respect to the ablation both in solvent vapor and under dry conditions, suggesting that the dye-doped parts have a lower ablation threshold in the swollen state and that the ablation thresholds depend on the degree of the swollen state. For the effect of the dye, it was suggested that the difference in the cross-linked structures between the TCPP-doped and the ATA-doped P4VP parts is an important factor both for the morphology of the nanoscale phase-separation structures of the dye-doped films and for the photothermal-energy release relating to ablation. For the effect of the number-average molecular weight of the diblock copolymers, we considered the following two factors: (i) The amount of dye per diblock copolymer; diblock copolymers with a large number-average molecular weight of P4VP blocks can contain a large number of dye molecules. (ii) The morphology of the phase separation structures of the neat films. In sea–island and network-like structures, selective ablation was achieved easily when the doped parts were the island parts or cylinder parts, respectively. However, in worm-like structures, both PS and P4VP parts were easily ablated away together, because of the larger interfaces between the two blocks. To our knowledge, there has been no report concerning the dopant-induced selective ablation of block copolymer films. No completely selective ablation was obtained in the present study, but we were able to confirm the possibility of obtaining a completely selective ablation of block copolymer films. We consider that a completely selective ablation of block copolymer films can be carried out when the selective doping of dopant is improved such that an increase in the dopant concentration in the target domains occurs.

12.5
Photon Antibunching Behavior of Organic Dye Nanocrystals on a Transparent Polymer Film

As mentioned, photon antibunching is when the probability of detecting two simultaneous photons drops to zero. Hence, the materials that exhibit photon antibunching are called single-photon emitters or single-photon sources. Since molecules cannot emit two photons simultaneously, a single molecule is the typical single-photon source [25, 26]. However, it is considered that multichromophoric systems can also be single-photon sources if one exciton remains as a result of exciton migration and subsequent exciton–exciton annihilation. It was demonstrated, on the basis of nanosecond laser flash photolysis of poly(N-vinylcarbazole) in solution, that only one excited-state remains in the single polymer chain, even when more than one excited-state is generated by an intense excitation pulse [40]. Densely generated excited-states undergo mutual interaction such as singlet–singlet annihilation during their migration along the polymer chain, and this sequential annihilation results in one excited-state in the single chain. This idea was extended to multichromophore-substituted dendrimer molecules which have definitively designed interchromophore distances and the number of chromophores, and single-photon emission from single dendrimers was observed by single molecule spectroscopy [41, 42]. The results suggest that multichromophoric systems also behave as single-photon sources so long as efficient exciton migration and exciton–exciton annihilation are possible. In this section, we report that the emission from a single nanocrystal consisting of N,N'-bis(2,6-dimethylphenyl)-3,4,9,10-perylenedicarboxyimide (DMPBI) molecules shows photon antibunching.

The nanoparticles of DMPBI prepared by the reprecipitation method [43, 44] are square or rectangular in shape (Figure 12.8a), suggesting strongly that they are in a crystalline state. This was also confirmed by the X-ray diffraction measurement. The size distribution of the nanocrystals (Figure 12.8b) indicates that the size of the long-axis ranges from 35 to 85 nm, and that the average size is 50 nm. Figure 12.8c shows the absorption and fluorescence spectra of the DMPBI nanocrystal-dispersed aqueous solution and the DMPBI chloroform solution. The spectra of the nanocrystals are quite different from those of the DMPBI molecules in solution. These differences

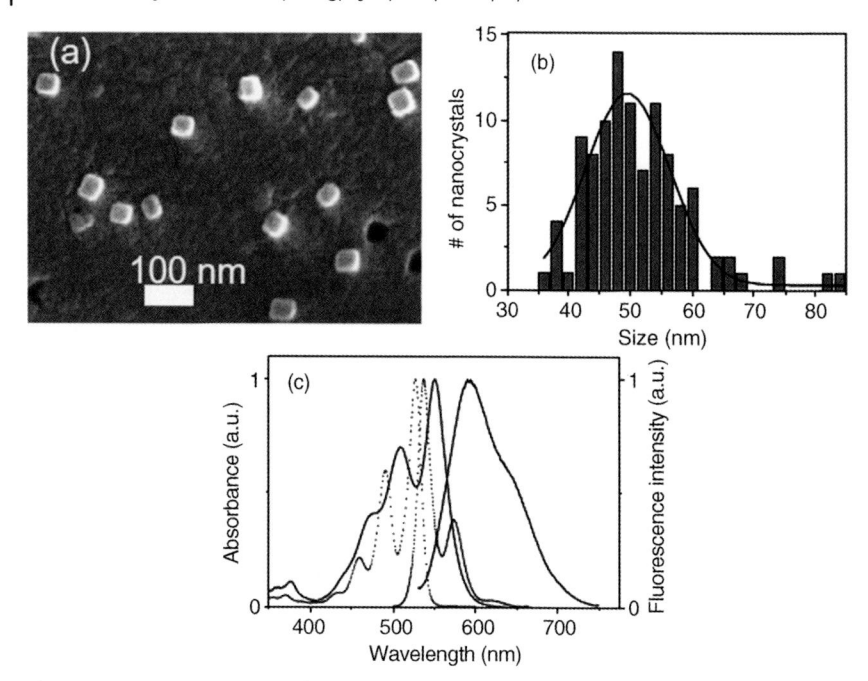

Figure 12.8 (a) FESEM image of prepared nanocrystals. (b) Size distribution of the nanocrystals evaluated from the FESEM image (a). This distribution was built up by scaling the long-axis of 100 individual nanocrystals. (c) Absorption and fluorescence spectra of the DMPBI nanocrystal-dispersed aqueous solution (solid line), and of the DMPBI chloroform solution (dotted line). S. Masuo, A. Masuhara, T. Akashi, M. Muranushi, S. Machida, H. Kasai, H. Nakanishi, H. Oikawa, A. Itaya: Photon antibunching in the emission from a single organic dye nanocrystal. *Jpn. J. Appl. Phys.* **2007**, *46*, L268–L270. Copyright the Japan Society of Applied Physics. Reproduced with permission.

indicate that a strong intermolecular interaction exists in the nanocrystals, and the fluorescence spectrum of the nanocrystals is attributed to a self-trapped exciton state (excimer-like emission) [45].

The photon antibunching measurement of single nanocrystals was carried out using a classical Hanbury-Brown and Twiss type photon correlation set-up [46] combined with pulsed laser excitation (488 nm, 8.03 MHz, 100 fs fwhm). Single nanocrystals were selected from a scanned image and positioned in the laser focus of a confocal microscope (objective: $100 \times$, numerical aperture $= 1.3$) in order to collect the emission from the single nanocrystal. The emitted photons over all wavelengths were collected by the same objective, divided by a 50/50 nonpolarizing beam splitter, and detected by two avalanche photodiodes.

An aqueous solution with an appropriate concentration of the nanocrystals was spin-coated onto PMMA-coated clean coverglasses. The samples were measured under nitrogen atmosphere at room temperature.

First, the excitation-power dependence of the emission count rate of single DMPBI nanocrystals was examined (Figure 12.9). The emission count rate of the single

Figure 12.9 The excitation-power dependence of the emission count rate of single DMPBI nanocrystals (dots), and a saturation curve calculated from a two-level model (solid line). One count rate value to one laser power was calculated as an average of 30 nanocrystals. S. Masuo, A. Masuhara, T. Akashi, M. Muranushi, S. Machida, H. Kasai, H. Nakanishi, H. Oikawa, A. Itaya: Photon antibunching in the emission from a single organic dye nanocrystal. *Jpn. J. Appl. Phys.* **2007**, *46*, L268–L270. Copyright the Japan Society of Applied Physics. Reproduced with permission.

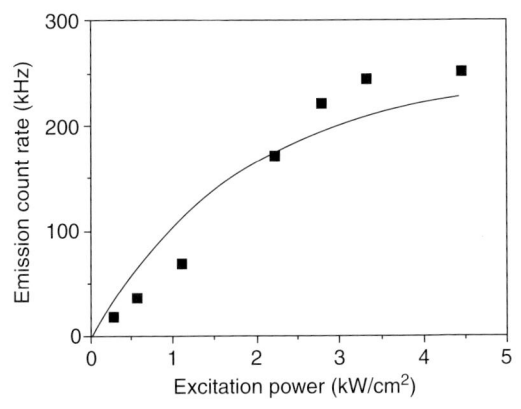

nanocrystals saturates for laser powers over $4\,\text{kW}\,\text{cm}^{-2}$, and is not well fitted by the saturation law of a two-level system, which is the same as the case of the multichromophoric dendrimers [41]. If the single nanocrystals emit one photon every excitation pulse, the maximum count rate should reach 400 kHz by taking into account the excitation laser repetition rate of 8.03 MHz and the detection efficiency of about 5%. However, the measured count rate reaches a maximum value of 250 kHz; therefore, the emission efficiency of the nanocrystals is evaluated to be 0.63. For the photon correlation measurements, excitation laser power of $2\text{–}4\,\text{kW}\,\text{cm}^{-2}$ was used, and thereby some excitons were generated in a single nanocrystal by a single excitation pulse.

Typical results of simultaneous measurements for a single nanocrystal (time traces of emission intensity and lifetime, and photon correlation) are shown in Figure 12.10. The time traces of both the emission intensity and lifetime show a constant value (Figure 12.10a and b), and the lifetime is about 3.7 ns, indicating that the nanocrystal behaves like a single molecule. The central peak at 0 ns in the photon correlation histogram (Figure 12.10c) corresponds to the photon pairs induced by the same laser pulse. In all other cases, the interphoton times are distributed around a multiple of the repetition rate of the laser pulses (8.03 MHz), that is, one peak every ~125 ns. In Figure 12.10c, there is apparently no central peak at 0 ns, only the background is observed. This means that the emission from the nanocrystal shows photon antibunching, that is, this nanocrystal behaves as a single-photon source. It was reported that the ratio (N_C/N_L) of the number of photon pairs contributing to the central peak, N_C, to the average number of counts in the lateral peaks, N_L, can be used to estimate the number of photons induced by the same laser pulse. N_C/N_L ratios of 0.0, 0.5, 0.67, and 0.75 are expected for 1, 2, 3, and 4 photons per one laser pulse, respectively [47].

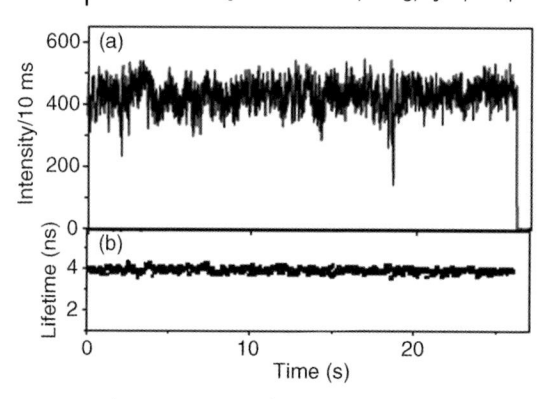

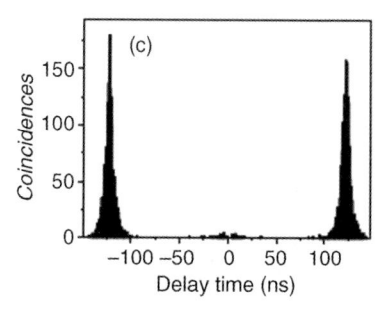

Figure 12.10 Typical time traces of (a) emission intensity and (b) lifetime, measured from a single DMPBI nanocrystal. (c) Photon correlation histogram obtained from the time trace of the emission intensity (a). The lifetimes were obtained by fitting a single exponential function to the decay curves constructed for every 2000 photons. S. Masuo, A. Masuhara, T. Akashi, M. Muranushi, S. Machida, H. Kasai, H. Nakanishi, H. Oikawa, A. Itaya: Photon antibunching in the emission from a single organic dye nanocrystal. *Jpn. J. Appl. Phys.* **2007**, *46*, L268–L270. Copyright the Japan Society of Applied Physics. Reproduced with permission.

However, a nonzero background in the experiment always produces a small contribution from the detection events that fall in the central N_C peak. For example, a signal-to-background ratio of 7 leads to an N_C/N_L ratio of about 0.2. In the present case, an N_C/N_L ratio of 0.06 was determined. Hence, the N_C/N_L value also indicates that the present nanocrystal behaves as a single-photon source.

To investigate the antibunching behavior of other nanocrystals, the photon correlations of 213 different nanocrystals were measured, and the histogram of their N_C/N_L ratios is shown in Figure 12.11. Most of the ratios are distributed around a fairly low value of about 0.1. This result indicates that most of the single nanocrystals of size 35–85 nm behave as single-photon sources. An intense single excitation pulse of the present excitation condition generates simultaneously several excitons in the single nanocrystals. In order to behave as a single-photon source even when more than one exciton is generated, efficient exciton–exciton annihilation has to occur in the nanocrystal, while inter-exciton distances for the efficient annihilation have to be below a few nanometers. Thus, efficient exciton migration also has to occur in the nanocrystals in order to show photon antibunching. As shown in Figure 12.8c, spectral overlap between the absorption and fluorescence spectra of DMPBI nanocrystals is observed; consequently, excitons can migrate in the present nanocrystal. Hence, it can be regarded that the photogenerated excitons can access each other by migration and efficient exciton–exciton annihilation can occur, resulting in the photon antibunching. However, the maximum size of the nanocrystal which shows the photon antibunching has not yet been revealed. In general, the migration length of an exciton in a molecular assembly depends strongly on both the spectroscopic properties of the chromophore and the molecular alignment in the assembly. In addition, defects in the assembly also play an important role in the exciton dynamics. In another experiment, we confirmed that several-micrometer-sized DMPBI crystals

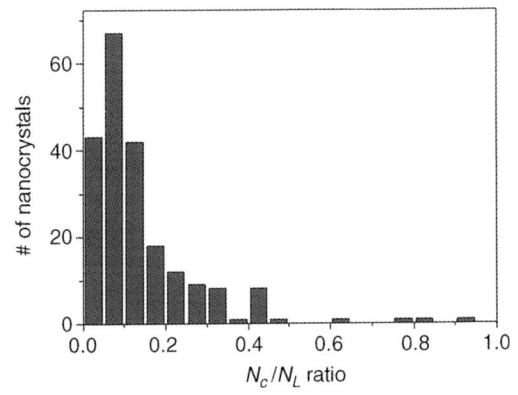

Figure 12.11 Histogram of N_C/N_L ratios obtained from the photon correlation measurement of 213 nanocrystals. S. Masuo, A. Masuhara, T. Akashi, M. Muranushi, S. Machida, H. Kasai, H. Nakanishi, H. Oikawa, A. Itaya: Photon antibunching in the emission from a single organic dye nanocrystal. *Jpn. J. Appl. Phys.* **2007**, *46*, L268–L270. Copyright the Japan Society of Applied Physics. Reproduced with permission.

show no antibunching; therefore, the antibunching behavior of the DMPBI crystals is limited to nanometer-sized crystals.

In this section, we have demonstrated that a single organic dye nanocrystal comprised of many chromophores shows photon antibunching when the size is sufficiently small. The present results indicate that molecular assemblies can also be considered as candidates for new single-photon sources.

Acknowledgments

The present work was partly supported by a Grant-in –Aid for Scientific Research in the Priority Area "Molecular Nano Dynamics" from the Ministry of Education, Culture, Sports, Science and Technology, Japan. The authors wish to express their sincere thanks to Professor H. Nakanishi, Professor H. Oikawa, Dr H. Kasai, Dr A. Masuhara, Dr Z. Wang, Dr K. Yamada, H. Nakata, T. Akashi, and M. Muranushi.

References

1 Tew, G. N., Pralle, M. U. and Stupp, S. I. (2000) Supramolecular materials with electroactive chemical functions. *Angew. Chem.*, **112**, 527–531.
2 Goren, M. and Lennox, R. B. (2001) Nanoscale polypyrrole patterns using block copolymer surface micelles as templates. *Nano Lett.*, **1**, 735–738.
3 Liu, J., Sheina, E., Kowalewski, T. and McCullough, R. D. (2002) Tuning the electrical conductivity and self-assembly of regioregular polythiophene by block

copolymerization: nanowire morphologies in new di- and triblock copolymers. *Angew. Chem. Int. Ed.*, **41**, 329–332.

4 Behl, M., Hattemer, E., Brehmer, M. and Zentel, R. (2002) Tailored semiconducting polymers: living radical polymerization and NLO-functionalization of triphenylamines. *Macromol. Chem. Phys.*, **203**, 503–510.

5 Peter, K. and Thelakkat, M. (2003) Synthesis and characterization of bifunctional polymers carrying tris(bipyridyl)ruthenium(II) and triphenylamine units. *Macromolecules*, **36**, 1779–1785.

6 Stalmbach, U., deBoer, B., Videlot, C., van Hutten, P. F. and Hadziioannou, G. (2000) Semiconducting diblock copolymers synthesized by means of controlled radical polymerization techniques. *J. Am. Chem. Soc.*, **122**, 5464–5472.

7 Linder, S. M. and Thelakkat, M. (2004) Nanostructures of n-type organic semiconductor in a p-type matrix via self-assembly of block copolymers. *Macromolecules*, **37**, 8832–8835.

8 Kietzke, T., Neher, D., Kumke, M., Montenegro, R., Landfester, K. and Scherf, U. (2004) A nanoparticle approach to control the phase separation in polyfluorene photovoltaic devices. *Macromolecules*, **37**, 4882–4890.

9 Pan, J., Chen, M., Warner, W., He, M., Dalton, L. and Hogen-Esch, T. E. (2000) Synthesis of block copolymers containing a main chain polymeric NLO segment. *Macromolecules*, **33**, 4673–4681.

10 Li, L., Zubarev, E. R., Acker, B. A. and Stupp, S. I. (2002) Chemical structure and nonlinear optical properties of polar self-assembling films. *Macromolecules*, **35**, 2560–2565.

11 Cong, Y., Fu, J., Cheng, Z., Li, J., Han, Y. and Lin, J. (2005) Self-organization and luminescent properties of nanostructured europium (III)-block copolymer complex thin films. *J. Polym. Sci. B*, **43**, 2181–2189.

12 Mizokuro, T., Mochizuki, H., Xiaoliang, M., Horiuchi, S., Tanaka, N., Tanigaki, N. and Hiraga, T. (2003) Addition of functional characteristics of organic photochromic dye to nano-structures by selective doping on a polymer surface. *Jpn. J. Appl. Phys.*, **42**, L983–L985.

13 Mizokuro, T., Mochizuki, H., Kobayashi, A., Horiuchi, S., Yamamoto, N., Tanigaki, N. and Hiraga, T. (2004) Selective doping of photochromic dye into nanostructures of diblock copolymer films by vaporization in a vacuum. *Chem. Mater.*, **16**, 3469–3475.

14 Machida, S., Nakata, H., Yamada, K. and Itaya, A. (2002) Position-selective adsorption of functional nanoparticles on block copolymer films. Prepr. IUPAC Polym. Conf., p. 443.

15 Lippert, T. and Dickinson, J. T. (2003) Chemical and spectroscopic aspects of polymer ablation: special features and novel directions. *Chem. Rev.*, **103**, 453–486.

16 Masuhara, H., Hiraoka, H. and Domen, K. (1987) Dopant-induced ablation of poly(methyl methacrylate) by a 308-nm excimer laser. *Macromolecules*, **20**, 450–452.

17 Machida, S., Nakata, H., Yamada, K. and Itaya, A. (2006) Position-selective arrangement of nanosized polymer microsphere on diblock copolymer film with sea–island microphase structure. *Jpn. J. Appl. Phys.*, **45**, 4270–4273.

18 Machida, S., Nakata, H., Yamada, K and Itaya, A. (2007) Morphological change of a diblock copolymer film induced by selective doping of a photoactive chromophore. *J. Polym. Sci. B, Polym. Phys. Ed.*, **45**, 368–375.

19 Wang, Z., Masuo, S., Machida, S. and Itaya, A. (2005) Application of dopant-induced laser ablation to site-selective modification of sea–island structures of polystyrene-*block*-poly(4-vinylpyridine) films. *Jpn. J. Appl. Phys.*, **44**, L402–L404.

20 Wang, Z., Masuo, S., Machida, S. and Itaya, A. (2007) Site-selective doping of dyes into polystyrene-*block*-Poly(4-vinyl pyridine) diblock copolymer films and selective laser

ablation of the dye-doped films. *Jpn. J. Appl. Phys.*, **46**, 7569–7576.

21 Lengl, G., Plettl, A., Ziemann, P., Spatz, J. P. and Möller, M. (2001) Excimer laser ablation of gold-loaded inverse polystyrene-block-poly (2-vinylpyridine) micelles. *Appl. Phys. A*, **72**, 679–685.

22 Oikawa, H., Masuhara, A., Kasai, H., Mitsui, T., Sekiguchi, T. and Nakanishi, H. (2004) Nanophotonics: integrating photochemistry, optics and nano/bio materials studies, in *Proceedings of the International Nanophotonics Symposium Handai*, vol. 1 (eds H. Masuhara and S. Kawata), Elsevier Science, Amsterdam, pp. 205–224.

23 Horn, D. and Rieger, J. (2001) Organic nanoparticles in the aqueous phase – theory, experiment, and use. *Angew. Chem. Int. Ed.*, **40**, 4330–4361.

24 Masuo, S., Masuhara, A., Akashi, T., Muranushi, M., Machida, S., Kasai, H., Nakanishi, H., Oikawa, H. and Itaya, A. (2007) Photon antibunching in the emission from a single organic dye nanocrystal. *Jpn. J. Appl. Phys.*, **46**, L268–L270.

25 Lounis, B. and Moerner, W. E. (2000) Single photons on demand from a single molecule at room temperature. *Nature*, **407**, 491–493.

26 Basché, T., Moerner, W. E., Orrit, M. and Talon, H. (1992) Photon antibunching in the fluorescence of a single dye molecule trapped in a solid. *Phys. Rev. Lett.*, **69**, 1516–1519.

27 Zhao, J., Jiang, S., Ji, X., Lijia, L.An. and Jiang, B. (2005) Study of the time evolution of the surface morphology of thin asymmetric diblock copolymer films under solvent vapor. *Polymer*, **46**, 6513–6521.

28 Hong, B. K. and Jo, W. H. (2000) Effects of molecular weight of SEBS triblock copolymer on the morphology, impact strength, and rheological property of syndiotactic polystyrene/ethylene–propylene rubber blends. *Polymer*, **41**, 2069–2079.

29 Gong, Y., Huang, H., Hu, Z., Chen, Y., Chen, D., Wang, Z. and He, T. (2006) Inverted to normal phase transition in solution-cast polystyrene-poly(methyl methacrylate) block copolymer thin films. *Macromolecules*, **39**, 3369–3376.

30 Funaki, Y., Kumano, K., Nakao, T., Jinnai, H., Yoshida, H., Kimishima, K., Tsutsumi, K., Hirokawa, Y. and Hashimoto, T. (1999) Influence of casting solvents on microphase-separated structures of poly(2-vinylpyridine)-*block*-polyisoprene. *Polymer*, **40**, 7147–7156.

31 Zhang, Q., Tsui, O. K. C., Du, B., Zhang, F., Tang, T. and He, T. (2000) Inverted to normal phase transition in solution-cast polystyrene-poly(methyl methacrylate) block copolymer thin films. *Macromolecules*, **33**, 9561–9567.

32 Huang, H., Zhang, F., Hu, Z., Du, B., He, T., Lee, F. K., Wang, Y. and Tsui, O. K. C. (2003) Study on the origin of inverted phase in drying solution-cast block copolymer films. *Macromolecules*, **36**, 4084–4092.

33 Gong, Y., Hu, Z., Chen, Y., Huang, H. and He, T. (2005) Ring-shaped morphology in solution-cast polystyrene-poly(methyl methacrylate) block copolymer thin films. *Langmuir*, **21**, 11870–11877.

34 Fukumura, H. and Masuhara, H. (1994) The mechanism of dopant-induced laser ablation. Possibility of cyclic multiphotonic absorption in excited states. *Chem. Phys. Lett.*, **221**, 373–378.

35 Bolle, M., Luther, K., Troe, J., Lhlemann, J. and Gerhardt, H. (1990) Photochemically assisted laser ablation of doped polymethyl-methacrylate. *Appl. Surf. Sci.*, **46**, 279–286.

36 Fukumura, H., Mibuka, N., Eura, S. and Masuhara, H. (1991) Porphyrin-sensitized laser swelling and ablation of polymer films. *Appl. Phys. A*, **53**, 255–259.

37 Fukumura, H., Mibuka, N., Eura, S., Masuhara, H. and Nishi, N. (1993) Mass spectrometric studies on laser ablation of polystyrene sensitized with anthracene. *J. Phys. Chem.*, **97**, 13761–13766.

38 Fukumura, H., Takahashi, E.-I. and Masuhara, H. (1995) Time-resolved spectroscopic and photographic studies on laser ablation of poly(methyl methacrylate) film doped with biphenyl. *J. Phys. Chem.*, **99**, 750–757.

39 Tsunoda, K., Sugiura, M., Sonoyama, M., Yajima, H., Ishii, T., Taniyama, J. and Itoh, H. (2001) Characterization of water contribution to excimer laser ablation of collagen. *J. Photochem. Photobiol. A*, **145**, 195–200.

40 Masuhara, H., Ohwada, S., Mataga, N., Itaya, A., Okamoto, K. and Kusabayashi, S. (1980) Laser photochemistry of poly(N-vinylcarbazole) in solution. *J. Phys. Chem.*, **84**, 2363–2368.

41 Masuo, S., Vosch, T., Cotlet, M., Tinnefeld, P., Habuchi, S., Bell, T. D. M., Oesterling, I., Beljonne, D., Champagne, B., Müllen, K., Sauer, M., Hofkens, J. and De Schryver, F. C. (2004) Multichromophoric dendrimers as single-photon sources: a single-molecule study. *J. Phys. Chem. B.*, **108**, 16686–16696.

42 Tinnefeld, P., Weston, K. D., Vosch, T., Cotlet, M., Weil, T., Hofkens, J., Müllen, K., De Schryver, F. C. and Sauer, M. (2002) Antibunching in the emission of a single tetrachromophoric dendritic system. *J. Am. Chem. Soc.*, **124**, 14310–14311.

43 Baba, K., Kasai, H., Okada, S., Oikawa, H. and Nakanishi, H. (2000) Novel fabrication process of organic microcrystals using microwave-irradiation. *Jpn. J. Appl. Phys.*, **39**, L1256–L1258.

44 Kasai, H., Nalwa, N. S., Oikawa, H., Okada, S., Matsuda, H., Minami, N., Kakuta, A., Ono, K., Mukoh, A. and Nakanishi, H. (1992) A novel preparation method of organic microcrystals. *Jpn. J. Appl. Phys.*, **31**, L1132–L1134.

45 Gesquiere, A. J., Uwada, T., Asahi, T., Masuhara, H. and Barbara, P. F. (2005) Single molecule spectroscopy of organic dye nanoparticles. *Nano Lett.*, **5**, 1321–1325.

46 Hanbury-Brown, R. and Twiss, R. (1956) Correlation between photons in two coherent beams of light. *Nature*, **177**, 27–29.

47 Weston, K. D., Dyck, M., Tinnefeld, P., Müller, C., Herten, D. P. and Sauer, M. (2002) Measuring the number of independent emitters in single-molecule fluorescence images and trajectories using coincident photons. *Anal. Chem.*, **74**, 5342–5347.

13
Molecular Segregation at Periodic Metal Nano-Architectures on a Solid Surface
Hideki Nabika and Kei Murakoshi

13.1
Molecular Manipulation in Nano-Space

Exploring new molecular diffusion modes in confined spaces has long been an attractive challenge for both science and technology. When free space is reduced to sizes below a few hundred nanometers, interaction between molecules and their confining walls cannot be ignored. Especially for a macromolecule such as DNA or proteins, geometrical constraint significantly alters the molecular diffusion dynamics. One trend in this field is to apply these unique diffusion phenomena to a molecular separation or sieving system in microfluidic devices [1–3]. These findings help to promote the development of future integrated bio-analytical devices. In order to manipulate biomaterials in these microfluidic systems, it is useful to use a lipid bilayer as a molecular manipulation and separation medium in order to avoid alternating the structure and properties of the biomaterials. Below we will first describe how a molecule diffuses in a two-dimensional lipid bilayer. Then, several attempts to apply the phenomenon for molecular manipulation and separation techniques in both solid-supported and self-spreading lipid bilayers are introduced.

13.1.1
Lipid Bilayer and its Fluidic Nature

The lipid molecule is the main constituent of biological cell membranes. In aqueous solutions amphiphilic lipid molecules form self-assembled structures such as bilayer vesicles, inverse hexagonal and multi-lamellar patterns, and so on. Among these lipid assemblies, construction of the lipid bilayer on a solid substrate has long attracted much attention due to the many possibilities it presents for scientific and practical applications [4]. Use of an artificial lipid bilayer often gives insight into important aspects of biological cell membranes [5–7]. The wealth of functionality of this artificial structure is the result of its own chemical and physical properties, for example, two-dimensional fluidity, bio-compatibility, elasticity, and rich chemical composition.

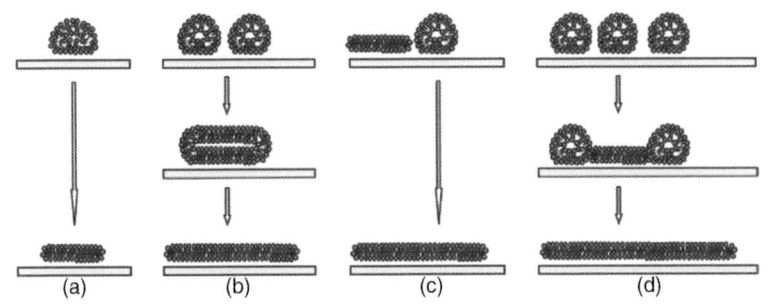

Figure 13.1 Schematic illustrations of vesicle fusion process. (a) An adsorbed vesicle ruptures and forms a bilayer. (b) Two adjacent vesicles fuse and eventually rupture. (c) A ruptured bilayer patch promote the rupture of the adjacent vesicle. (d) The cooperative action of the vesicle rupture. Adapted from Ref. [9] with permission.

The artificial lipid bilayer is often prepared via the vesicle-fusion method [8]. In the vesicle fusion process, immersing a solid substrate in a vesicle dispersion solution induces adsorption and rupture of the vesicles on the substrate, which yields a planar and continuous lipid bilayer structure (Figure 13.1) [9]. The Langmuir–Blodgett transfer process is also a useful method [10]. These artificial lipid bilayers can support various biomolecules [11–16]. However, we have to take care because some transmembrane proteins incorporated in these artificial lipid bilayers interact directly with the substrate surface due to a lack of sufficient space between the bilayer and the substrate. This alters the native properties of the proteins and prohibits free diffusion in the lipid bilayer [17]. To avoid this undesirable situation, polymer-supported bilayers [7, 18, 19] or tethered bilayers [20, 21] are used.

Due to the high compatibility of artificial lipid bilayers with biomolecules, they are often used as a bio-assay or bio-detection system [22]. For development of combinatorial analysis based on the lipid bilayer, a substrate on which has been deposited a well-defined micro- or nano-patterned lipid bilayer has been fabricated in various ways [22–30]. Well-defined patterning of the lipid bilayer can be used to fabricate substrates to be used as bio- or chemical-sensors and in drug screening. Normally, patterning based on the vesicle fusion process yields a single component pattern of the lipid bilayer [23, 26, 30]. Alternatively, multi-component bilayer patterns can be obtained via multistep vesicle fusion [24] or dip-pen nanolithography (DPN) (Figure 13.2) [29].

For an artificial lipid bilayer of any size scale, it is a general feature that the bilayer acts as a two-dimensional fluid due to the presence of the water cushion layer between the bilayer and the substrate. Due to this fluidic nature, molecules incorporated in the lipid bilayer show two-dimensional free diffusion. By applying any bias for controlling the diffusion dynamics, we can manipulate only the desired molecule within the artificial lipid bilayer, which leads to the development of a molecular separation system.

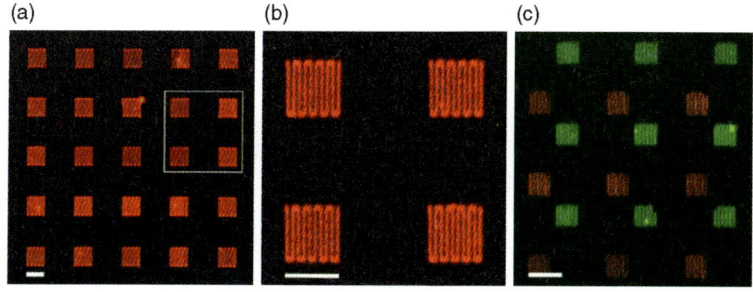

Figure 13.2 Fluorescence micrographs of DOPC multi-layer patterns fabricated by dip-pen nanolithography. (a) An array of 25 contiguous line features. Red color is from doped rhodamine-labeled lipid. (b) A higher magnification of the region highlighted by the white square in (a). (c) Two-component patterns containing two different dyes. Green color is from doped NBD-labeled lipid. Scale bars: 5 μm. Adapted from Ref. [29] with permission.

13.1.2
Controlling Molecular Diffusion in the Fluidic Lipid Bilayer

Direct observation of molecular diffusion is the most powerful approach to evaluate the bilayer fluidity and molecular diffusivity. Recent advances in optics and CCD devices enable us to detect and track the diffusive motion of a single molecule with an optical microscope. Usually, a fluorescent dye, gold nanoparticle, or fluorescent microsphere is used to label the target molecule in order to visualize it in the microscope [31–33]. By tracking the diffusive motion of the labeled-molecule in an artificial lipid bilayer, random Brownian motion was clearly observed (Figure 13.3) [31]. As already mentioned, the artificial lipid bilayer can be treated as a two-dimensional fluid. Thus, an analysis for a two-dimensional random walk can be applied. Each trajectory observed on the microscope is then numerically analyzed by a simple relationship between the displacement, r, and time interval, τ,

$$\langle r^2 \rangle = 4D\tau \tag{13.1}$$

where $\langle r^2 \rangle$ and D are the mean-square displacement (MSD) and the two-dimensional (lateral) diffusion constant, respectively. MSD plots shown in Figure 13.3 demonstrate

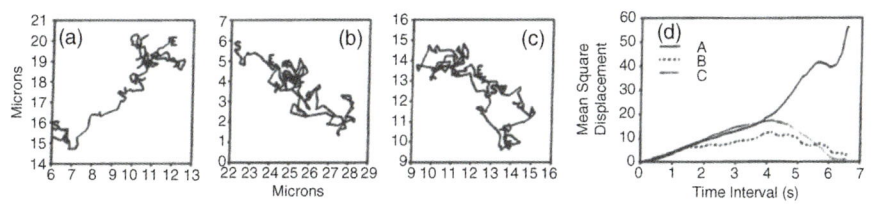

Figure 13.3 (a–c) Trajectories of the diffusion motion of a gold nanoparticle probe on a planar lipid membrane. (d) Mean-square displacement plots for the diffusion shown in (a–c). Adapted from Ref. [31] with permission.

a linear increase in a short time interval region, which is characteristic of random diffusion. It should be noted here that molecular diffusion is strongly suppressed when the lipid matrix is in the gel phase [34]. In this chapter, only the lipid bilayer in the liquid phase will be treated as a molecular manipulating medium. In addition to the direct observation of the diffusing molecules, fluorescence correlation spectroscopy (FCS) [35–37] and fluorescence recovery after photobleaching (FRAP) [38, 39] are also useful alternative techniques to characterize the molecular diffusivity in the lipid bilayer. However, we have to note that the data acquired from FCS and FRAP are essentially based on the ensemble characteristics derived from a number of molecules.

In any experimental approach, the artificial lipid bilayer on the solid substrate demonstrated sufficient fluidity for molecular diffusion if the temperature was above a phase transition temperature. Two-dimensional fluidity enables the manipulation of molecules in the bilayer by applying an appropriate bias. A typical example of this is electrophoresis, a method in which an electric field is applied to the bilayer [40–42]. Charged molecules present in the bilayer drift along the electric field lines. The drift velocity is dependent on the electric field strength, the size and charge of the molecule, bilayer fluidity, and so on. Molecular dependence of the drift velocity is a basic principle of an electrophoretic molecular separation (Figure 13.4) [43]. Recent investigations have revealed that electrophoresis on the artificial lipid bilayer can separate Texas Red DHPE isomers [44]. A POPC bilayer doped with 20 mol% cholesterol was used. The addition of cholesterol reduces band broadening during the electrophoresis. After applying an electric field of 100 V for 30 min, two Texas Red DHPE isomers were separated into two distinct bands in the POPC bilayer. Other than the lipid molecules, DNA-tethered vesicles and GPI-tethered proteins can also be separated via electrophoresis [45, 46].

As mentioned above, the charged molecules drift along the electric field direction in the electrophoretic manipulation. In principle, the drift direction is determined by the electric field direction. Differences in the drift velocity are what act to separate

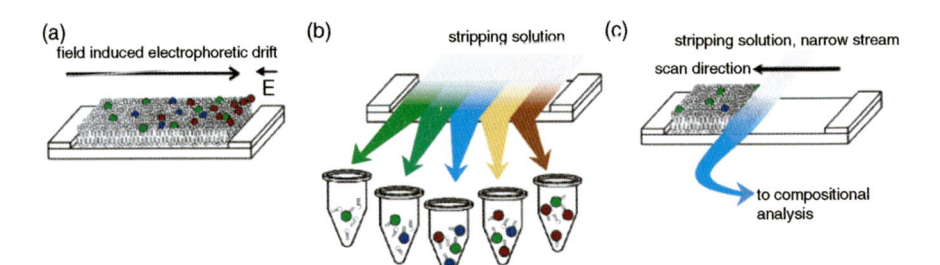

Figure 13.4 Schematic illustration of the electrophoretic molecular separation. (a) The charged molecules drift according to the electric field direction. (b) Separation of each fraction by applying several separate laminar flows. (c) Alternatively, each fraction can be separated by scanning the stripping laminar flow across the sample channel. Adapted from Ref. [43] with permission.

molecules into isolated bands. However, constructing an asymmetric obstacle on the substrate can modify and control the drift direction. This method of drift manipulation is known as the process based on the Brownian ratchet mechanism [47]. The deviation angle is determined by the original drift velocity, the diffusion coefficient, and the structure of the obstacle. This system is capable of separating molecules even if they have the same drift velocity but different diffusion constants. All of these electrophoretic manipulations in the lipid bilayer can be used to purify or separate biomaterials in a micro-fluid. However, electrophoresis can manipulate only charged molecules, which is a severe limitation of the electrophoretic system. To develop a versatile system that enables the manipulation of any molecule irrespective of its charge, a new concept must be introduced. One possibility is to exploit the self-spreading nature of a lipid bilayer, a macroscopic fluidic phenomenon, that will be introduced in the next section.

13.1.3
Self-Spreading of a Lipid Bilayer or Monolayer

The lipid bilayer grows spontaneously from a lipid aggregate on a hydrophilic substrate on immersing the substrate in an aqueous solution (Figure 13.5) [48, 49]. This phenomenon is called the "self-spreading" of a lipid bilayer. Any molecules in the lipid bilayer can be collectively transported by the molecular flow of the self-spreading. Since the self-spreading is a thermodynamically driven phenomenon, no input energy such as an electric field is needed for the molecular transportation. By taking advantage of this aspect, we can transport and manipulate any collection of molecules in the lipid bilayer, irrespective of their charge. By constructing a micro-channel on the substrate, we can control the direction of the self-spreading. Furukawa et al. fabricated photo-lithographic micro-channels with a width of 1–20 µm [50].

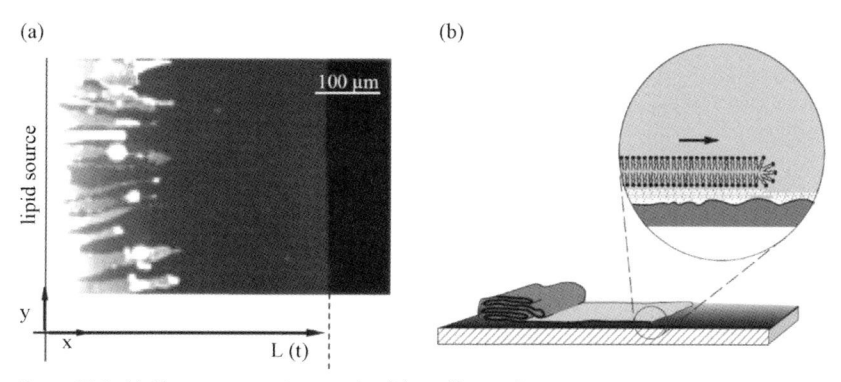

Figure 13.5 (a) Fluorescence micrograph of the self-spreading lipid bilayer doped with a dye molecule. The lipid bilayer spread on an oxidized silicon wafer from a deposited lipid aggregate illustrated on the left. (b) A schematic drawing of the self-spreading lipid bilayer from the lipid aggregate. Adapted from Ref. [48] with permission.

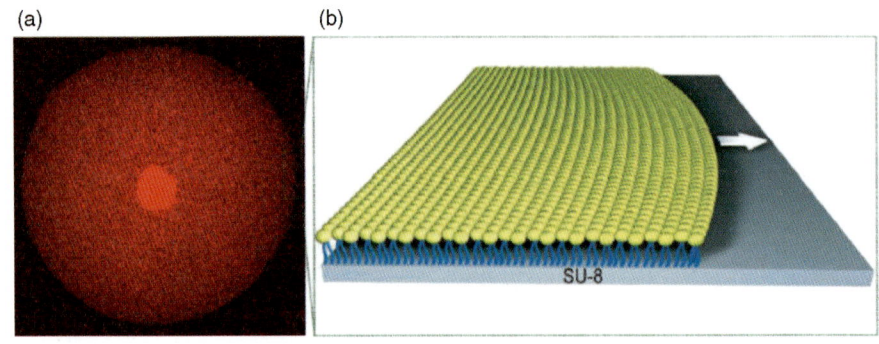

Figure 13.6 (a) Confocal micrograph of a circularly self-spreading lipid monolayer. A rhodamine-labeled lipid is doped to visualize the spreading behavior. (b) A schematic illustration of the front edge of the self-spreading lipid monolayer [51].

By placing the lipid aggregate near the micro-channel, the bilayer spreads into the micro-channel without changing the spreading dynamics. This system has several advantages compared with a conventional three-dimensional microfluidic system. First, input energy for the molecular transport is no longer needed, whereas the conventional system relies on an external bias such as a syringe pump or an electric field. This enables the miniaturization and simplification of the device. The second advantage is the low dimensionality of the lipid bilayer. The conventional system manipulates molecules in three-dimensional free space. However, the lipid bilayer confines the molecules into a quasi-two-dimensional plane. A small amount of molecule can be concentrated in the bilayer, leading to improvement in the manipulation and detection limits. Additionally, the lipid bilayer offers a field to manipulate a biomolecule in its native environment.

In addition to the self-spreading lipid bilayer, it was also found that a lipid monolayer showed similar spreading behavior on a hydrophobic surface (Figure 13.6) [51]. By fabricating an appropriate hydrophobic surface pattern, the spreading area and direction can be easily controlled. For both the self-spreading bilayer and monolayer, non-biased molecular transportation is an important key concept for the next generation of microfluidic devices.

13.1.4
Controlling the Self-Spreading Dynamics

To utilize the self-spreading bilayer in microfluidic devices, an understanding of their dynamics and other characteristics is necessary. The driving energy for the self-spreading has been explained as a gain in free energy via bilayer–substrate interaction [49]. Similar to the supported bilayer prepared by vesicle fusion, a hydration water layer is inserted under the spreading bilayer. Through this hydration layer, several bilayer–substrate interactions are imposed, such as the van der Waals interaction. Interactions involving lipid molecules and lipid membranes have long been investigated and discussed from both experimental and theoretical

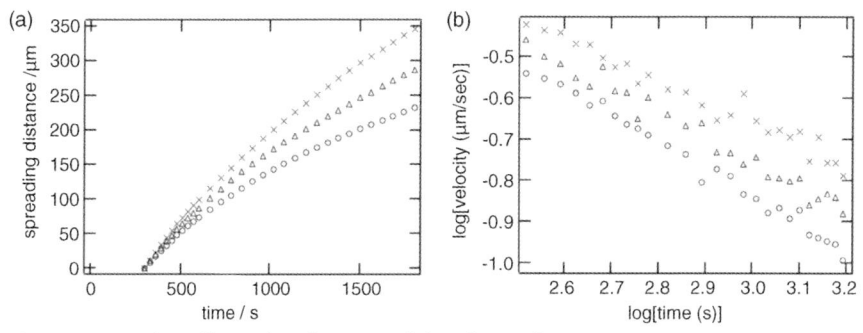

Figure 13.7 (a) The self-spreading distance and (b) velocity of egg-PC lipid bilayer in NaCl aqueous solutions with different concentrations. (×) 100 mM, (○) 10 mM, and (◆) 1 mM. Adapted from Ref. [53] with permission.

viewpoints [52]. Owing to this vast knowledge, the bilayer–substrate interaction energy can be controlled via several experimental conditions. For example, a dependence on the electrolyte concentration has been investigated by experimental and theoretical approaches [53]. Figure 13.7 shows the experimental results for the dependence of the spreading distance and the velocity on the NaCl concentration. It was clearly demonstrated that the spreading velocity increased on increasing the NaCl concentration. The self-spreading velocity as a function of time is given by

$$\log v(t) = \frac{1}{2}\log \beta - \frac{1}{2}\log t \tag{13.2}$$

where v, β, t are the spreading velocity, spreading coefficient, and time, respectively. Fitting the experimental date to Eq. (13.2) gives the value of β as 48, 33, and 21 $\mu m^2 s^{-1}$ in 100 mM, 10 mM, and 1 mM NaCl solutions, respectively. The spreading coefficient β is given by

$$\beta = \frac{Ed}{2\eta} \tag{13.3}$$

where E, d, η are the bilayer–substrate interaction energy, thickness of the water layer, and water viscosity, respectively. As is clear from Eq. (13.3), the spreading dynamics are closely correlated with the bilayer–substrate interaction energy. Assuming $d = 2$ nm and $\eta = 10^{-3}$ N s m^{-2}, we can expect the value of E to be 48, 33, and 21 μJ m^{-2} in 100 mM, 10 mM, and 1 mM NaCl solutions, respectively. For a theoretical estimation of E, three interaction energies were considered as the dominant components, that is, the van der Waals, the electrostatic double layer, and the hydration energies. Figure 13.8 shows the calculated interaction energy curves considering these three interaction energies on an egg-PC bilayer and hydrophilic glass substrate system. The minimum at around 2 nm corresponds to E in Eq. (13.3). The theoretically estimated values for E were calculated as 35, 33, 22 μJ m^{-2} in 100 mM, 10 mM, and 1 mM NaCl solutions, respectively, which is in good agreement with the experimental values.

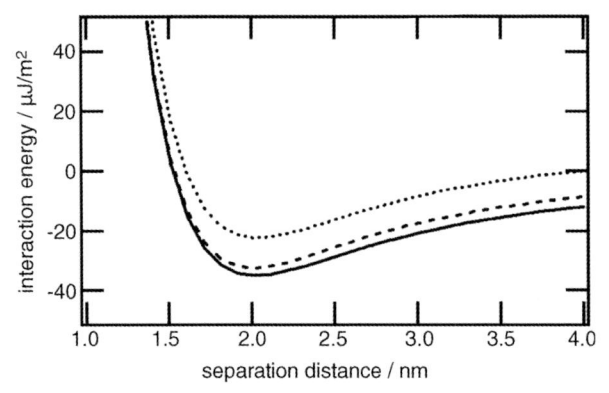

Figure 13.8 Calculated interaction energy curves for egg-PC on a surface-oxidized silicon substrate system. The NaCl concentration is 100 mM (solid line), 10 mM (broken line), and 1 mM (dotted line). Adapted from Ref. [53] with permission.

This result demonstrates that the self-spreading dynamics are controllable by tuning the bilayer–substrate interactions. The above-mentioned electrolyte dependence is an example of this fact. Considering that there are many parameters that alter the bilayer–substrate interaction, a diverse approach can be proposed. For example, Nissen et al. investigated the spreading dynamics on the substrate coated with polymetic materials [48]. They found that insertion of a hydrophilic and inert polymer layer under the self-spreading lipid bilayer strongly attenuated the bilayer–substrate interaction.

Of course, other physical and chemical conditions also affect the self-spreading dynamics. Figure 13.9 shows the dependences of β on the temperature and lipid

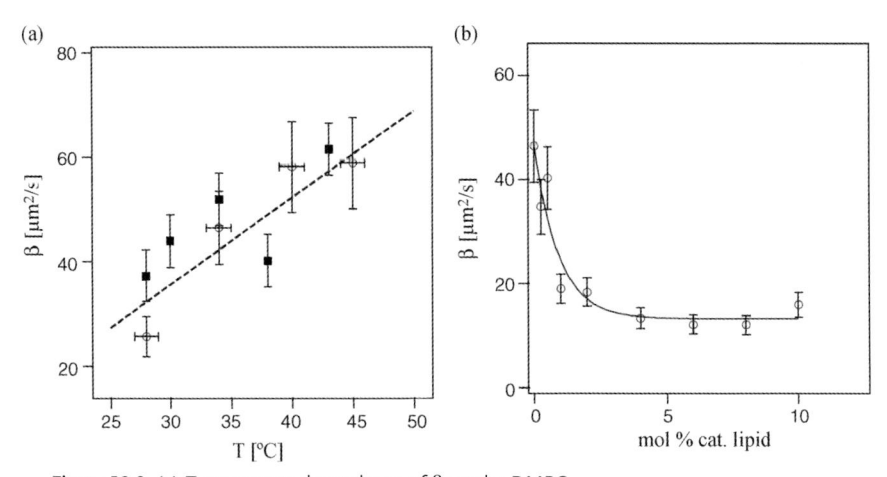

Figure 13.9 (a) Temperature dependence of β on the DMPC bilayer on glass (white) and silicon (black). (b) Dependence of β on the molar fraction of a cationic DMTAP additive in the DMPC bilayer. Adapted from Ref. [48] with permission.

composition. A gradual increase in β with temperature has been explained as being due to a change in the viscosity of water. However, the observed increase was more rapid than the change in the water viscosity. This discrepancy was attributed to the formation of water clusters in the hydration layer. It should also be taken into consideration that the viscosity of the lipid assembly itself is dependent on the temperature. This will change the frictional force present at the transfer of lipid molecules from the aggregate to the spreading bilayer, which Nabika *et al.* assumed to be negligible. Figure 13.9a demonstrates the bilayer composition dependence of the DMPC/DMTAP binary system. The bilayer composition is a critical parameter that determines the bilayer–substrate interaction energy, the bending energy of the bilayer, the viscosity, domain formation, and so on. DMTAP is a cationic lipid, with which the bilayer can interact more strongly with the negatively charged glass substrate. Thus, the spreading velocity was expected to be increased with DMTAP content. However, the experimental results showed the completely opposite behavior. This fact strongly suggests that it is not sufficient to consider only the bilayer–substrate interaction for a comprehensive understanding of the self-spreading dynamics. By comparing the structures of DMPC and DMTAP, it is clear that the structure of the head group is different and that DMTAP has a smaller head group. Thus, addition of DMTAP disturbs the formation of a thermodynamically stable bilayer structure. This energy cost reduces the self-spreading driving energy, which could be one of the reasons why the addition of DMTAP led to a decrease in β.

In addition to the spreading dynamics, the stacking structure of the self-spreading lipid bilayer is also controllable via the NaCl concentration [54, 55]. Further experimental and theoretical investigations regarding the control of self-spreading are required before we will be able to easily control the self-spreading behavior in microfluidic devices.

13.1.5
Molecular Manipulation on the Self-Spreading Lipid Bilayer

The most intriguing aspect of the self-spreading lipid bilayer is that any molecule in the bilayer can be transported without any external bias. The unique characteristic of the spreading layer offers the chance to manipulate molecules without applying any external biases. This concept leads to a completely non-biased molecular manipulation system in a microfluidic device. For this purpose, the use of nano-space, which occasionally offers the possibility of controlling molecular diffusion dynamics, would be a promising approach.

The first successful example was the use of a metallic nano-gate, in which a periodic array of nano-gates was constructed on the self-spreading substrate [56]. Due to its simplicity, the nano-gate substrate was prepared via nano-sphere lithography (NSL) (Figure 13.10b). However, to carry out a systematic and quantitative experiment on the structural dependence, the fabrication process was shifted to electron-beam lithography (EBL) (Figure 13.10b) [57, 58]. Figure 13.11a shows a fluorescence microscope image of the self-spreading egg-PC bilayer doped with TR-DHPE. The fluorescence intensity directly reflects the molecular concentration of TR-DHPE in

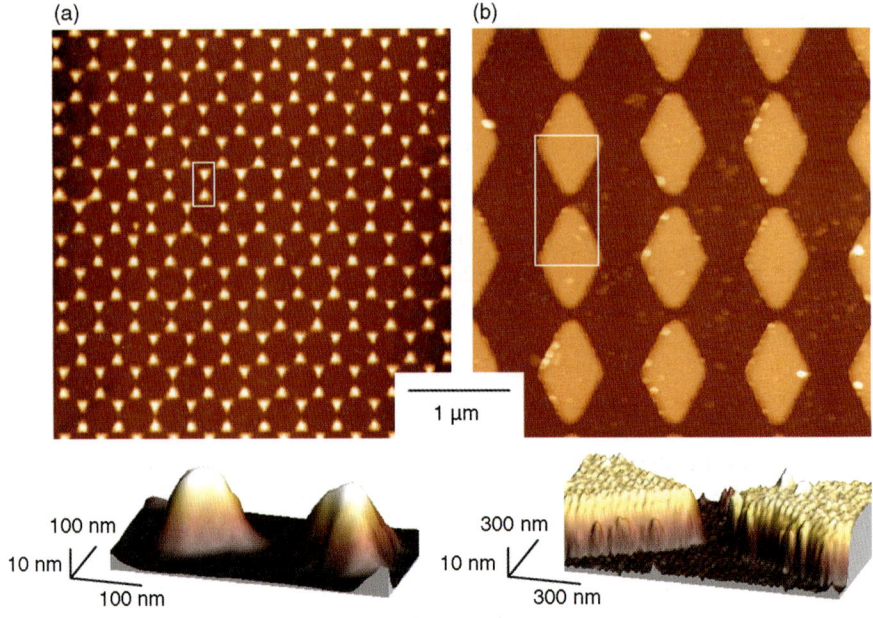

Figure 13.10 AFM image of (a) NSL and (b) EBL substrates. Lower panels show a enlarged three-dimensional images of the regions highlighted with the white squares with permission.

the self-spreading bilayer, under the condition that the dye content is below its self-quenching concentration. It is clear that the fluorescence intensity is not constant within the bilayer. The intensity gradually decreases from the spreading edge inwards. This is a characteristic aspect of the self-spreading lipid bilayer. During the self-spreading on a hydrophilic substrate, an elastic tension is imposed on the bilayer, causing area dilation [48]. This effect becomes more significant in the spreading edge

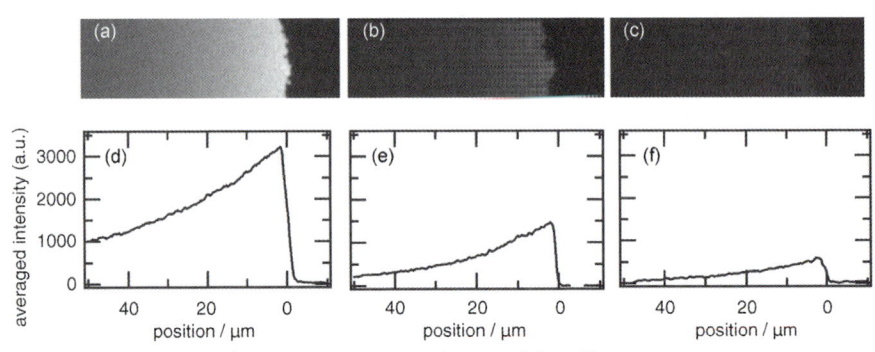

Figure 13.11 (a–c) Fluorescence microscope images of the self-spreading lipid bilayer on (a) flat glass, (b) EBL nano-gate with width 500 nm, and (c) EBL nano-gate with width 100 nm. (d–f) Averaged and area calibrated line profiles of the self-spreading bilayer on (d) flat glass, (e) EBL nano-gate with width 500 nm, and (f) EBL nano-gate with width 100 nm. Adapted from Ref. [57] with permission.

region. For a bulky molecule such as TR-DHPE, the less dense region is thermodynamically more favorable. Thus, TR-DHPE accumulates at the less dense spreading edge. This distribution gradient differs from molecule to molecule, depending on the molecular size and configuration. When the bilayer spreads on the nano-gate substrate, the fluorescence intensity appears to be reduced, indicating a molecular filtering effect. Periodic dark spots observed in the fluorescence micrographs on the nano-gate channel correspond to the non-wetting metallic architectures.

The observed molecular filtering effect can be explained along the same lines with a partitioning phenomenon on the lipid bilayer, in which doped molecules such as dye-labeled lipids exhibit inhomogeneous distribution when the bilayer has more than two coexisting phases [59]. The inhomogeneous distribution is the result of a difference in chemical potential among different phases. In the present case, the lipid density is thought to increase in the nano-gate region, judging from the attenuation in spreading dynamics [58]. This creates a similar situation to the above-mentioned phase coexisting system. In the case of TR-DHPE with a bulky head group, the solubility at the dense phase is known to be reduced compared to that of a non-compressed phase [60]. Therefore, the penetration ability of the TR-DHPE molecule into the densely packed nano-gate region is reduced. This is the proposed mechanism for the observed molecular filtering effect. The chemical potential is highly sensitive to any structural parameter of both the doped molecules and the medium lipid molecules. Based on the suggested mechanism, the nano-gate system can filter any molecule by recognizing any structural parameter, such as molecular size, charge, polarity, hydrophilicity, chirality, and configuration.

13.2
Summary

In this chapter, we have introduced a novel molecular filtering system using a self-spreading lipid bilayer and a periodic array of metal nano-gates. The filtering effect could be the result of the formation of a local chemical potential barrier in the nano-gate region during spreading. Since the self-spreading is a thermodynamically driven collective molecular flow, any molecule including non-charged molecules can be manipulated in this system, which is completely different from other ordinary separation systems such as those based on a conventional electrophoretic approach. The present system could be applied in micro- and nano-scopic device technologies, as it provides a versatile and completely non-biased filtering methodology.

Acknowledgment

This work was supported in part by Grants-in-Aid for Scientific Research on Priority Area "Molecular Nano Dynamics" (Area No. 432, No. 16205026) and for Grant-in-Aid for Young Scientist (B) (No. 18750001) from MEXT, Japan. We would like to thank Prof. H. Misawa and Prof. K. Ueno for the fabrication of the substrate with electron-beam litography.

References

1 Fu, J., Schoch, R. B., Stevens, A. L., Tannenbaum, S. R. and Han, J. (2007) A patterned anisotropic nanofluidic sieving structure for continuous-flow separation of DNA and proteins. *Nature Nanotechnol.*, **2**, 121–128.
2 Abgrall, P. and Nguyen, N. T. (2008) Nanofluidic devices and their applications. *Anal. Chem.*, **80**, 2326–2341.
3 Yuan, Z., Garcia, A. L., Lopez, G. P. and Petsev, D. N. (2007) Electrokinetic transport and separations in fluidic nanochannels. *Electrophoresis*, **28**, 595–610.
4 Sackmann, E. (1996) Supported membranes: scientific and practical applications. *Science*, **271**, 43–48.
5 Binder, W. H., Barragan, V. and Menger, F. M. (2003) Domains and rafts in lipid membranes. *Angew. Chem. Int. Ed.*, **42**, 5802–5827.
6 Groves, J. T. (2005) Molecular organization and signal transduction at intermembrane junctions. *Angew. Chem. Int. Ed.*, **44**, 3524–3538.
7 Tanaka, M. and Sackmann, E. (2005) Polymer-supported membranes as models of the cell surface. *Nature*, **437**, 656–663.
8 Brian, A. A. and McConnell, H. M. (1984) Allogeneic stimulation of cytotoxic T cells by supported planar membranes. *Proc. Natl. Acad. Sci. USA*, **81**, 6159–6163.
9 Richter, R. P., Bérat, R. and Brisson, A. R. (2006) Formation of solid-supported lipid bilayers: an integrated view. *Langmuir*, **22**, 3497–3505.
10 Tamm, L. K. and McConnell, H. M. (1985) Supported phospholipid bilayers. *Biophys. J.*, **47**, 105–113.
11 Parthasarathy, R. and Groves, J. T. (2004) Protein patterns at lipid bilayer junctions. *Proc. Natl. Acad. Sci. USA*, **101**, 12798–12803.
12 Dewa, T., Sugiura, R., Suemori, Y., Sugimoto, M., Takeuchi, T., Hiro, A., Iida, K., Gardiner, A. T., Cogdell, R. J. and Nango, M. (2006) Lateral organization of a membrane protein in a supported binary lipid domain: direct observation of the organization of bacterial light-harvesting complex 2 by total internal reflection fluorescence microscopy. *Langmuir*, **22**, 5412–5418.
13 Mossman, K. and Groves, J. (2007) Micropatterned supported membranes as tools for quantitative studies of the immunological synapse. *Chem. Soc. Rev.*, **36**, 46–54.
14 Ihalainen, P. and Peltonen, J. (2004) Immobilization of streptavidin onto biotin-functionalized Langmuir–Schaefer binary monolayers chemisorbed on gold. *Sens. Actuators, B*, **102**, 207–218.
15 Majd, S. and Mayer, M. (2005) Hydrogel stamping of arrays of supported lipid bilayers with various lipid compositions for the screening of drug-membrane and protein-membrane interactions. *Angew. Chem. Int. Ed.*, **44**, 6697–6700.
16 Lei, S. B., Tero, R., Misawa, N., Yamamura, S., Wan, L. J. and Urisu, T. (2006) AFM characterization of gramicidin-A in tethered lipid membrane on silicon surface. *Chem. Phys. Lett.*, **429**, 244–249.
17 Smith, E. A., Coyn, J. W., Cowell, S. M., Tokimoto, T., Hruby, V. J., Yamamura, H. I. and Wirth, M. J. (2005) Lipid bilayers on polyacrylamide brushes for inclusion of membrane proteins. *Langmuir*, **21**, 9644–9650.
18 Zhang, L., Longo, M. L. and Stroeve, P. (2000) Mobile Phospholipid Bilayers Supported on a Polyion/Alkylthiol Layer Pair. *Langmuir*, **16**, 5093–5099.
19 Munro, J. C. and Frank, C. W. (2004) In situ formation and characterization of poly (ethylene glycol)-supported lipid bilayers on gold surfaces. *Langmuir*, **20**, 10567–10575.
20 Lahiri, J., Kalal, P., Frutos, A. G., Jonas, S. J. and Schaeffler, R. (2000) Method for fabricating supported bilayer lipid membranes on gold. *Langmuir*, **16**, 7805–7810.

21 Dorvel, B. R., Keizer, H. M., Fine, D., Vuorinen, J., Dodabalapur, A. and Duran, R. S. (2007) Formation of tethered bilayer lipid membranes on gold surfaces: QCM-Z and AFM study. *Langmuir*, **23**, 7344–7355.

22 Yamazaki, V., Sirenko, O., Schafer, R. J., Nguyen, L., Gutsmann, T., Brade, L. and Groves, J. T. (2005) Cell membrane array fabrication and assay technology. *BMC Biotechnol.*, **5**, 18/1–18/11.

23 Kim, P., Lee, S. E., Jung, H. S., Lee, H. Y., Kawai, T. and Suh, K. Y. (2006) Soft lithographic patterning of supported lipid bilayers onto a surface and inside microfluidic channels. *Lab Chip*, **6**, 54–59.

24 Jackson, B. L. and Groves, J. T. (2004) Scanning probe lithography on fluid lipid membranes. *J. Am. Chem. Soc.*, **126**, 13878–13879.

25 Jackson, B. L. and Groves, J. T. (2007) Hybrid protein-lipid patterns from aluminum templates. *Langmuir*, **23**, 2052–2057.

26 Kohli, N., Vaidya, S., Ofoli, R. Y., Worden, R. M. and Lee, I. (2006) Arrays of lipid bilayers and liposomes on patterned polyelectrolyte templates. *Colloid Interface Sci.*, **301**, 461–469.

27 Carlson, J. W., Bayburt, T. and Sligar, S. G. (2000) Nanopatterning phospholipid bilayers. *Langmuir*, **16**, 3927–3931.

28 Wu, M., Holowka, D., Craighead, H. G. and Baird, B. (2004) Visualization of plasma membrane compartmentalization with patterned lipid bilayers. *Proc. Natl. Acad. Sci. USA*, **101**, 13798–13803.

29 Lenhert, S., Sun, P., Wang, Y., Fichs, H. and Mirkin, C. A. (2007) Massively parallel dip-pen nanolithography of heterogeneous supported phospholipid multilayer patterns. *Small*, **3**, 71–75.

30 Shi, J., Chen, J. and Cremer, P. S. (2008) Sub-100 nm patterning of supported bilayers by nanoshaving lithography. *J. Am. Chem. Soc.*, **130**, 2718–2719.

31 Lee, G. M., Ishikawa, A. and Jacobson, K. (1991) Direct observation of brownian motion of lipids in a membrane. *Proc. Natl. Acad. Sci. USA*, **88**, 6274–6278.

32 Saxton, M. J. and Jacobson, K. (1997) Single-particle tracking: Applications to membrane dynamics. *Annu. Rev. Biophys. Biomol. Struct.*, **26**, 373–371.

33 Schmidt, T., Schutz, G. J., Baumgartner, W., Gruber, H. J. and Schindler, H. (1996) Imaging of single molecule diffusion. *Proc. Natl. Acad. Sci. USA*, **93**, 2926–2929.

34 Denicourt, N., Tancrède, P., Brullemans, M. and Teissié, J. (1989) The liquid condensed diffusional transition of dipalmitoylphosphoglycerocholine in monolayers. *Biophys. Chem.*, **33**, 63–70.

35 Schwille, P., Korkach, J. and Webb, W. W. (1999) Fluorescence correlation spectroscopy with single-molecule sensitivity on cell and model membranes. *Cytometry*, **36**, 176–182.

36 Korlach, J., Schwille, P., Webb, W. W. and Feigenson, G. W. (1999) Characterization of lipid bilayer phases by confocal microscopy and fluorescence correlation spectroscopy. *Proc. Natl. Acad. Sci. USA*, **96**, 8461–8466.

37 Burns, A. R., Frankel, D. J. and Buranda, T. (2005) Local mobility in lipid domains of supported bilayers characterized by atomic force microscopy and fluorescence correlation spectroscopy. *Biophys. J.*, **89**, 1081–1093.

38 Lalchev, Z. I. and Mackie, A. R. (1999) Molecular lateral diffusion in model membrane systems. *Colloid Surf. B*, **15**, 147–160.

39 Fragata, M., Ohnishi, S., Asada, K., Ito, T. and Takahashi, M. (1984) Lateral diffusion of plastocyanin in multilamellar mixed-lipid bilayers studied by fluorescence recovery after photobleaching. *Biochemistry*, **23**, 4044–4051.

40 Groves, J. T., Ulman, N. and Boxer, S. G. (1997) Micropatterning fluid lipid bilayers on solid supports. *Science*, **275**, 651–653.

41 Groves, J. T., Boxer, S. G. and McConnell, H. M. (1997) Electric field-induced reorganization of two-component supported bilayer membranes. *Proc. Natl. Acad. Sci. USA*, **94**, 13390–13395.

42 Groves, J. T., Boxer, S. G. and McConnell, H. M. (1998) Electric field-induced critical demixing in lipid bilayer membranes. *Proc. Natl. Acad. Sci. USA*, **95**, 935–938.

43 Kam, L. and Boxer, S. G. (2003) Spatially selective manipulation of supported lipid bilayers by laminar flow: steps toward biomembrane microfluidics. *Langmuir*, **19**, 1624–1631.

44 Daniel, S., Diaz, A. J., Martinez, K. M., Bench, B. J., Albertorio, F. and Cremer, P. S. (2007) Separation of membrane-bound compounds by solid-supported bilayer electrophoresis. *J. Am. Chem. Soc.*, **129**, 8072–8073.

45 Ishii, C. Y. and Boxer, S. G. (2006) Controlling two-dimensional tethered vesicle motion using an electric field: interplay of electrophoresis and electro-osmosis. *Langmuir*, **22**, 2384–2391.

46 Groves, J. T., Wülfing, C. and Boxer, S. G. (1996) Electrical manipulation of glycan-phosphatidyl inositol-tethered proteins in planar supported bilayers. *Biophys. J.*, **71**, 2716–2723.

47 van Oudenaarden, A. and Boxer, S. G. (1999) Brownian ratchets: molecular separations in lipid bilayers supported on patterned arrays. *Science*, **285**, 1046–1048.

48 Nissen, J., Gritsch, S., Wiegand, G. and Rädler, J. O. (1999) Wetting of phospholipid membranes on hydrophilic surfaces – Concepts towards self-healing membranes. *Eur. Phys. J. B*, **10**, 335–344.

49 Rädler, J., Strey, H. and Sackmann, E. (1995) Phenomenology and kinetics of lipid bilayer spreading on hydrophilic surfaces. *Langmuir*, **11**, 4539–4548.

50 Furukawa, K., Nakashima, H., Kashimura, Y. and Torimitsu, K. (2006) Microchannel device using self-spreading lipid bilayer as molecule carrier. *Lab on a Chip*, **6**, 1001–1006.

51 Czolkos, I., Erkan, Y., Dommersnes, P., Jesorka, A. and Orwar, O. (2007) Controlled Formation and Mixing of Two-Dimensional Fluids. *Nano Lett.*, **7**, 1980–1984.

52 Israelachvili, J. N. (1991) *Intermolecular and Surface Forces*, 2nd edn, Academic Press, London and New York.

53 Nabika, H., Fukasawa, A. and Murakoshi, K. (2008) Tuning the dynamics and molecular distribution of the self-spreading lipid bilayer. *Phys. Chem. Chem. Phys.*, **10**, 2243–2248.

54 Nabika, H., Fukasawa, A. and Murakoshi, K. (2006) Control of the structure of self-spreading lipid membrane by changing electrolyte concentration. *Langmuir*, **22**, 10927–10931.

55 Suzuki, K. and Masuhara, H. (2005) Groove-spanning behavior of lipid membranes on microfabricated silicon substrates. *Langmuir*, **21**, 6487–6494.

56 Nabika, H., Sasaki, A., Takimoto, B., Sawai, Y., He, S. and Murakoshi, K. (2005) Controlling molecular diffusion in self-spreading lipid bilayer using periodic array of ultra-small metallic architecture on solid surface. *J. Am. Chem. Soc.*, **127**, 16786–16787.

57 Nabika, H., Takimoto, B., Iijima, N. and Murakoshi, K. (2008) Observation of self-spreading lipid bilayer on hydrophilic surface with a periodic array of metallic nano-gate. *Electrochim. Acta*, **53**, 6278–6283.

58 Nabika, H., Jijima, N., Takimoto, B., Ueno, K., Misawa, H. and Murakoshi, K. (2009) Segregation of molecules in lipid bilayer spreading through metal nano-gates. *Anal. Chem.*, **81**, 699-704.

59 Vaz, W. L. C. and Melo, E. (2001) Fluorescence spectroscopic studies on phase heterogeneity in lipid bilayer membranes. *J. Fluoresc.*, **11**, 255–271.

60 Baumgard, T., Hunt, G., Farkas, E. R., Webb, W. W. and Feigenson, G. W. (2007) Fluorescence probe partitioning between Lo/Ld phases in lipid membranes. *Biochim. Biophys. Acta*, **1768**, 2182–2194.

14
Microspectroscopic Study of Self-Organization in Oscillatory Electrodeposition
Shuji Nakanishi

14.1
Introduction

Solid surfaces with ordered nanostructures composed of periodic layers, dots, holes, and grooves (or ridges), provide unique optical, electronic, magnetic, and mechanical properties [1]. Photo- and electron beam lithography and ion beam etching (top-down method) have been widely used for creating desired nanostructures at the surface. Recently, atomic-scale fabrication using surface probe microscopy (SPM), such as scanning tunneling microscopy (STM) and scanning near-field optical microscopys (SNOM), has been developed as a powerful technique (bottom-up method) for designed, well-controlled nanostructure fabrication. However, these techniques now face serious problems, including the challenge of mass production and cost increases due to the expensive specialized apparatus required. With the goal of overcoming problems associated with conventional techniques, self-organization (bottom-up method) has recently attracted much attention.

In general, self-organization can be categorized into two different types, that is, static and dynamic. The former (static) is self-organization under conditions of thermodynamic equilibrium, in which ordered structures are formed on the basis of specific properties of the intermolecular forces. These structures, which can have regularity with almost the same size as the system components, are simple, rigid and stable. Self-assembled structures, such as lipid bilayers, close-packed crystals of nanospheres, and monolayers of thiol molecules on gold surfaces, are representative examples of this type. Numerous studies have been carried out on this type of self-organization, as summarized in a number of reviews [2–9]. The critical issues to be tackled next for these methods are to improve the regularity and to place the nanostructures with the desired sizes at specific desired locations.

In the other type of self-organization (dynamic self-organization), spontaneous ordering of the systems occurs under thermodynamically non-equilibrium conditions, in which various ordered structures with wavelengths tens to hundreds of thousands times larger than the size of the system components are formed by spatiotemporal synchronization of various factors [10–12]. The spatiotemporal order appearing in dynamic self-organization phenomena has unique and attractive properties for producing materials with ordered structures: (i) complex patterns appear spontaneously without any external control, (ii) the observed patterns have long-range order, and (iii) various ordered patterns are obtained simply by changing the experimental parameters. Since dynamic self-organized ordering in chemical systems vanishes when reactions stop and the systems are back in their equilibrium state, some sort of strategy to solidify the dynamic patterns is required for the fabrication of structures [2, 10, 11]. One of the strategies for achieving this is to use electrodeposition, in which the histories of ever-changing spatiotemporal orders caused by reactions can be recorded in a form of ordered architecture of the electrodeposited material. In this chapter, self-organization studies in oscillatory electrodeposition are reviewed, the focus being on nanostructure formation.

14.2
Dynamic Self-Organization in Electrochemical Reaction Systems

Chemical oscillation is a typical example of dynamic self-organization under non-equilibrium conditions [13–16]. Chemical reactions in oscillating systems proceed spatio-synchronically, and, in certain cases, produce a range of spatiotemporal patterns including dots, target patterns, and spirals. However, due to the paucity of examples of such pattern development, chemical oscillations have, in general, been regarded as specific or discrete phenomena. In contrast, a large number of oscillatory reactions have been observed in an electrochemical reaction at an electrode (solid) surface [17–20], indicating that electrochemical oscillations are not specific but general phenomena. Electrochemical systems more frequently display oscillatory behavior due to the effect of an autocatalytic (positive feedback) mechanism derived from the coupling of the electrical factor with chemical dynamics. Such a mechanism, which is rarely active in pure chemical systems, contributes considerably to the appearance of oscillation. Electrochemical oscillations have been reported for a variety of systems, including anodic metal dissolution, cathodic metal deposition, oxidation of hydrogen molecules and small organic compounds, and reduction of hydrogen peroxide and persulfate ions.

In comparison with other systems, electrochemical systems have strong advantages for the study of dynamic self-organization phenomena. For example, (i) the Gibbs energies for reactions can be regulated continuously and reversibly by tuning the electrode potential, and (ii) the oscillations can be observed via electric signals such as current or potential. (iii) The diffusion process can also be controlled by changing the sizes and geometrical arrangements of the electrodes in electrochemical cells, and (iv) the mode, period, and amplitude of the spatiotemporal patterns can be tuned easily by changing the geometrical arrangements of the electrodes and the applied potential or current. Based on the above, in the 1990s, detailed mathematical models for oscillations and spatial patterns in electrochemical systems were successfully constructed, which have enabled electrochemical oscillations and patterns to be controlled and designed.

According to the literature [21], all reported electrochemical oscillations can be classified into four classes depending on the roles of the true electrode potential (or Helmholtz-layer potential, E). Electrochemical oscillations in which E plays no essential role and remains essentially constant are known as "strictly potentiostatic" (Class I) oscillations, which can be regarded as chemical oscillations containing electrochemical reactions. Electrochemical oscillations in which E is involved as an essential variable but not as the autocatalytic variable are known as S-NDR (Class II) oscillations, which arise from an S-shaped negative differential resistance (S-NDR) in the current density (j) versus E curve. Oscillations in which E is the autocatalytic variable are known as N-NDR (Class III) oscillations, which have an N-shaped NDR. Oscillations in which the N-NDR is obscured by a current increase from another process are known as hidden N-NDR (HN-NDR; Class IV) oscillations. It is known that N-NDR oscillations are purely current oscillations, whereas HN-NDR oscillations occur in both current and potential. The HN-NDR oscillations can be further divided into three or four subcategories, depending on how the NDR is hidden.

14.3
Oscillatory Electrodeposition

As mentioned in the introduction, oscillatory electrodeposition is an interesting target from the point of view of the production of micro- and nanostructured materials because it has the possibility to produce ordered electrodeposits by recording ever-changing self-organized spatiotemporal patterns during the oscillation. Schlitte et al. were the first to report the formation of ordered architecture via an oscillatory electrodeposition [22]. They showed that the electrodeposition of Cu with a potential oscillation gave layered deposits. Krastev et al. reported that the oscillatory electrodeposition of Ag–Sb alloy gave similar layered deposits [23]. Interestingly, in this system, spiral and stripe patterns appeared at the surface of the deposits during the oscillation. The thickness of the layers in these examples was rather large, of the order of 100 µm or more. On the other hand, Switzer et al. reported that the oscillatory electrodeposition of Cu in alkaline solutions produced alternate Cu and Cu_2O multilayers with thickness of about 90 nm [24]. We have also reported that oscillatory electrodeposition of Cu–Sn alloy [25] and iron-group alloys [26] produced nano-period layered deposits. Thus, for all the examples shown above, layered deposits are formed in synchronization with electrochemical oscillations.

Another example is dendritic crystal growth under diffusion-limited conditions accompanied by potential or current oscillations. Wang et al. reported that electrodeposition of Cu and Zn in ultra-thin electrolyte showed electrochemical oscillation, giving beautiful nanostructured filaments of the deposits [27, 28]. Saliba et al. found a potential oscillation in the electrodeposition of Au at a liquid/air interface, in which the Au electrodeposition proceeds specifically along the liquid/air interface, producing thin films with concentric-circle patterns at the interface [29, 30]. Although only two-dimensional ordered structures are formed in these examples because of the quasi-two-dimensional field for electrodeposition, very recently, we found that three-dimensional metal latticeworks lying vertical to the electrode surface were spontaneously produced in synchronization with a potential oscillation in a normal electrochemical condition [31, 32]. Thus, these types of deposits are very attractive from the point of view of two- or three-dimensional structurization.

14.3.1
Formation of a Layered Nanostructure of Cu–Sn Alloy

Electrochemical oscillation during the Cu–Sn alloy electrodeposition reaction was first reported by Survila et al. [33]. They found the oscillation in the course of studies of the electrochemical formation of Cu–Sn alloy from an acidic solution containing a hydrosoluble polymer (Laprol 2402C) as a brightening agent, though the mechanism of the oscillatory instability was not studied. We also studied the oscillation system and revealed that a layered nanostructure is formed in synchronization with the oscillation in a self-organizational manner [25, 26].

Figure 14.1a shows a j vs. U curve in $Cu^{2+} + Sn^{2+} + H_2SO_4$ with (solid curve) and without (dashed curve) cationic surfactant. The addition of the surfactant causes a drastic change in the j vs. U curve. Namely, an NDR appears in a narrow potential region of about 5 mV near -0.42 V, where the Cu–Sn alloy is electrodeposited. Another notable point in the surfactant-added solution is that a current oscillation appears when the U is kept constant in (and near) the potential region of this NDR, as shown in Figure 14.1b. It was also revealed that both the NDR and current oscillation appeared only in the presence of cationic surfactant and not in the presence of anionic surfactant.

The structure of alloy films deposited during the current oscillation was investigated by scanning electron microscopy (SEM) and scanning Auger electron microscopy (AEM). Figure 14.1c illustrates schematically the procedure of sample preparation. The deposited film was etched with an Ar^+-ion beam, with the film being rotated. This procedure gave a bowl-shaped hollow of about 1 mm in diameter at the bottom, together with a slanting cross section of the deposited film. Figure 14.1d shows an SEM image (top view) of a sample thus prepared. Uniform concentric rings of gray and black colors in the region of the slanting cross-section clearly indicate the formation of a quite uniform layered structure spreading over a macroscopically wide range of 1×1 mm. It was confirmed that the number of sets of the gray and black layers (one period of the multilayer) agreed with the number of cycles of the current oscillation during which the deposit was formed, indicating that one oscillation cycle produced one layer of the deposit. Figure 14.1e compares the expanded SEM image in the region of the slanting cross-section with the profile (white curve) of the atomic ratio [Cu/(Cu + Sn)] in this region, in which we can see that Cu is rich in the black layer, whereas Sn is rich in the gray layer. The average thickness of one period of the multilayer was estimated from the sputter time to be about 38 nm. The thickness can be controlled via tuning of the oscillation period.

The NDR and the oscillation appear only in the presence of cationic surfactant and not in the presence of anionic surfactant, suggesting that the NDR arises from electrostatic adsorption of a cationic surfactant on a (negatively polarized) Cu–Sn

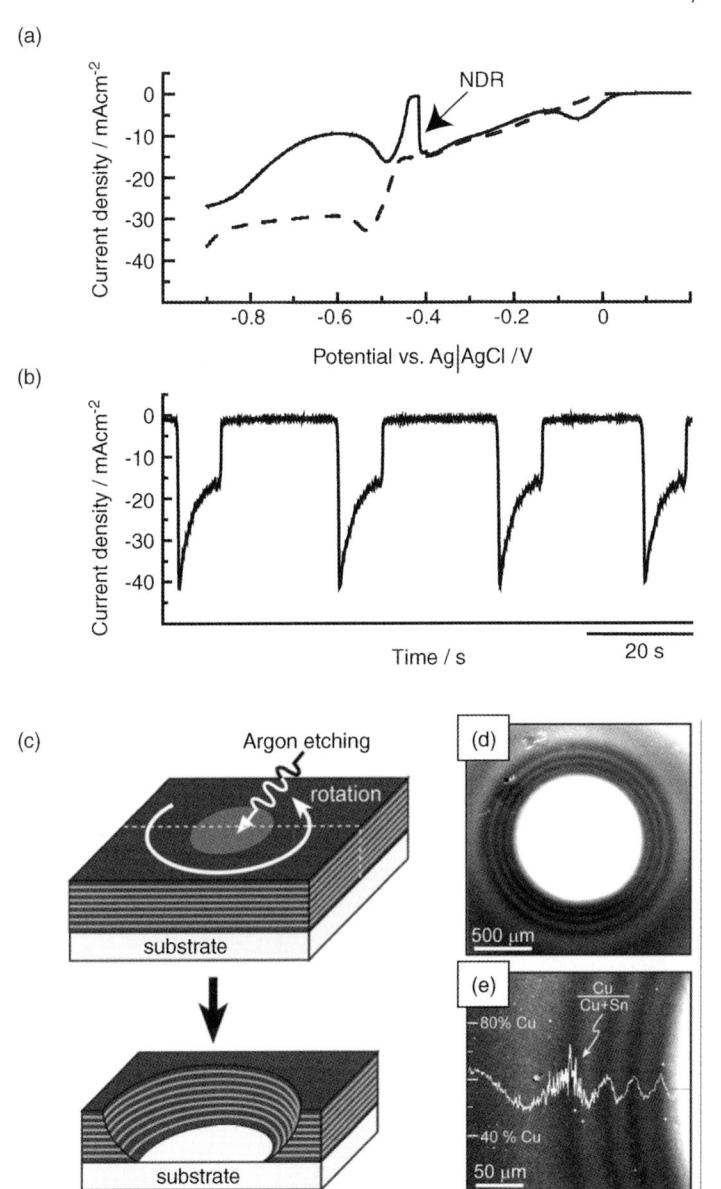

Figure 14.1 (a) Current density vs. potential curves obtained for the electrodeposition of Cu–Sn alloy in the presence (solid curve) and absence (dashed curve) of cationic surfactant. (b) Typical example of the current oscillation observed for Cu–Sn alloy electrodeposition. (c) Schematic illustration of sample preparation for SEM and AEM analyses. (d) SEM (top view) of a bowl-shaped hollow with a slanting cross-section, prepared in the deposited alloy film by Ar^+ ion etching. (e) Expanded SEM image, compared with the distribution of the atomic ratio [Cu/(Cu + Sn)] obtained with scanning AEM. (Reprinted from Ref. [25] with permission from the American Chemical Society.)

alloy (deposit) surface because the adsorption will retard the diffusion of electroactive metal ions to the alloy surface and thus decrease j. The negative polarization at the surface with negative potential shift will increase the amount of cationic surfactant, resulting in more decrease in j, that is, the appearance of the NDR (Figure 14.2).

On the basis of this argument, the mechanism for the current oscillation and the multilayer formation can be explained as follows. First note that U is kept constant externally with a potentiostat in the present case. In the high-current stage of the current oscillation, the true electrode potential (or Helmholtz double layer potential), E, is much more positive than U because E is given by $E = U - jAR$, where A is the electrode area, R is the resistance of the solution between the electrode surface and the reference electrode, and j is taken as negative for the reduction current. This implies that, even if U is kept constant in the region of the NDR, E is much more

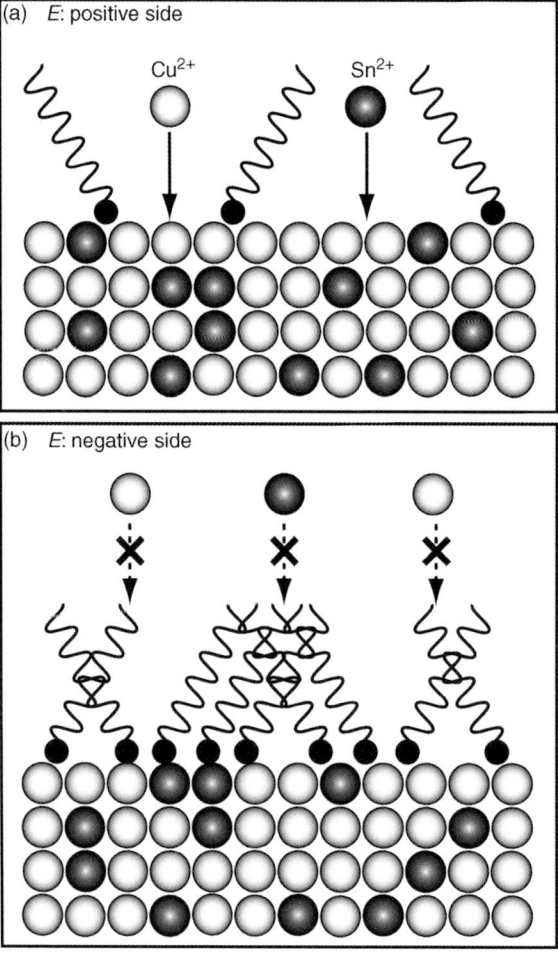

Figure 14.2 Schematic representation of surface reactions in the (a) low- and (b) high-current states, respectively.

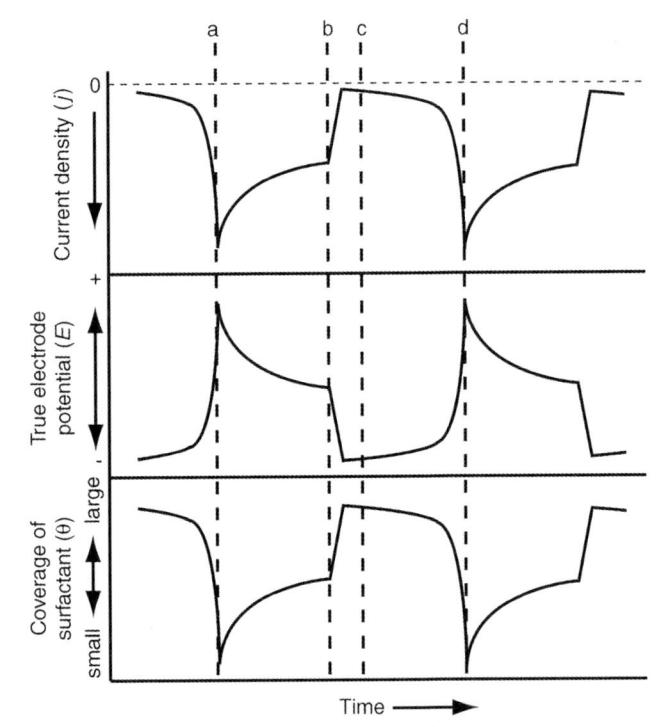

Figure 14.3 Schematic illustrations of expected current density vs. time, true electrode potential vs. time, and coverage of the surfactant vs. time curves.

positive than U, and hence the coverage (θ) of the adsorbed surfactant in this stage is small. Thus, effective diffusion of electroactive metal ions to the electrode surface occurs without retardation, which leads to a high j value due to active electrodeposition of Cu–Sn alloy (stage a in Figure 14.3).

The active alloy deposition, however, causes a decrease in the surface concentration of electroactive metal ions (hereafter denoted as C) owing to their slow diffusion from the solution bulk. This leads to a gradual decrease in j (in the absolute value) and thus to a decrease in the ohmic drop and a negative shift in E. The negative shift in E, in turn, leads to an increase in θ (Figure 14.2a). Thus, j decreases (stage b) owing to a decrease in the diffusion of metal ions to the electrode surface, and the system goes to a low-current stage, accompanied by a negative shift in E.

In the low-current stage (stage c), only slow deposition (or slow reduction of metal ions) occurs at vacant sites (atomic pinholes); thus, C gradually increases by diffusion from the solution bulk. The increase in C induces an increase in j and thus causes a positive shift in E (and a decrease in θ). When E is shifted to the positive, j increases (stage d) owing to an increase in the diffusion of metal ions, and the high-current stage is restored again.

The alternate-multilayer formation in the alloy deposit can also be explained on the basis of the above mechanism. First, we have to note that the j value in the low-current stage is very low; hence, this stage hardly contributes to the alloy formation. In the high-current stage, the E shifts gradually to the negative with time, as seen in Figure 14.3, and the Sn content in the alloy deposit increases with this negative shift in E. Accordingly, one period of the multilayer is formed by one cycle of the current oscillation.

14.3.2
Layered Nanostructures of Iron-Group Alloys

The electrodeposition of the iron-group alloys has an interesting aspect in that it leads to self-organized formation of layered structures in which the iron-group metals and the incorporated elements change their contents periodically. The formation of the layered structures was commonly observed in various electrodeposition systems of iron-group alloys [34–36]. It was also reported that an electrochemical oscillation was observed when the layered Ni–P alloy was deposited [37], implying that the layered structures of iron-group alloys are also formed by electrochemical oscillation.

Figure 14.4a shows the j vs. U curve in an electrolyte of $NiSO_4 + H_3BO_3 + NaCl$ with (solid-curve) and without (dashed-curve) NaH_2PO_2, which is the P-source of the Ni–P alloy. In the presence of NaH_2PO_2 (solid curve), the current started to flow at about -0.5 V and an NDR appeared in the potential region from -0.87 to -1.12 V. On the other hand, in the absence of NaH_2PO_2 (dashed curve), the current started to flow at a more negative potential than in the presence of NaH_2PO_2 and showed no NDR. It is to be noted that the j values in the presence and the absence of NaH_2PO_2 became nearly the same in a U range more negative than -1.0 V. This implies that the added NaH_2PO_2 has almost no effect on the deposition reaction in this U range.

Figure 14.4b shows a time course of a potential oscillation obtained in electrolyte with slightly different composition, which appeared spontaneously when j was kept at a constant value in the range $-55 < j < -75$ mA cm^{-2}, that is, in a range of j where U was in the NDR region of Figure 14.4a. It may be noted that the highest and lowest values of the oscillating potential in Figure 14.4b nearly coincide with the highest and lowest potentials of the NDR region, respectively. Figure 14.4c shows an Auger depth profile for a deposit film formed during the potential oscillation, in which the formation of the layered structure of Ni–P with different P-ratio in the alloy can be clearly seen. The thickness of one layer was estimated to be a few hundred nanometers, by dividing the thickness of the deposit, measured with an optical microscope, by the number of oscillation cycles.

Note that essentially the same behavior as for the Ni–P alloy deposition was observed in electrodeposition of other iron-group alloys, such as Co–W and Ni–W alloys. Namely, the deposition current in the presence of Na_2WO_4 (the W-source of the Co–W and Ni–W alloys) started to flow at a more positive potential than in the absence of Na_2WO_4, indicating that the electrodeposition of the Co–W and Ni–W alloys occurs by essentially the same mechanism as that of the Ni–P alloy, suggesting the presence of a general mechanism for the induced co-deposition of these alloys.

As mentioned above, the Ni–P (Co–W, Ni–W) alloy deposition current in the presence of the P-source (W-source) starts to flow at a more positive potential than in

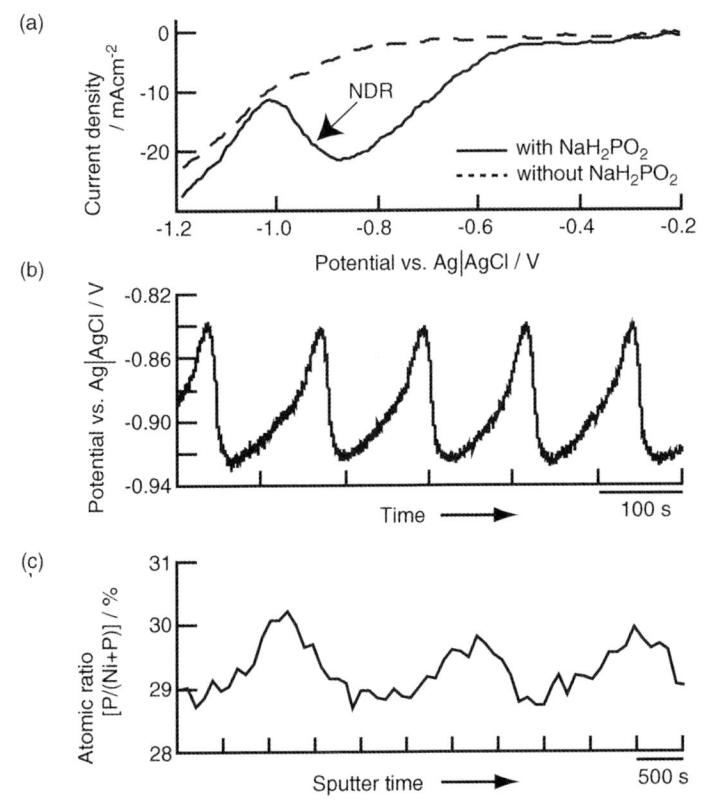

Figure 14.4 (a) Current density vs. potential curves obtained for electrodeposition of Ni–P alloy in the presence (solid curve) and absence (dashed curve) of NaH_2PO_2 (P-source of the alloy). (b) Typical example of the potential oscillation observed for Ni–P alloy electrodeposition. (c) An Auger depth profile for a deposit produced under the potential oscillation. (Reprinted from Ref. [26] with permission from American Chemical Society.)

its absence (Figure 14.4a). This fact indicates that P-source (W-source) or a species related to it acts as a promoter for the alloy deposition reaction. Sakai et al. attributed the origin of the NDR to desorption of the adsorbed promoter. The desorption of the adsorbed promoter (anionic species) may be caused by an increase in negative charges at the electrode surface by the negative potential shift (Figure 14.5) [26]. It was also revealed that the oscillation was caused by a positive feedback mechanism originating from the NDR, in a similar way to that in the Cu–Sn alloy electrodeposition system (Section 3.1.1).

14.3.3
Layered Nanostructure of Cu/Cu$_2$O

Switzer et al. found that Cu/Cu$_2$O layered nanostructures are electrodeposited with spontaneous potential oscillations from alkaline Cu(II)-lactate solution in a self-

(a) High-potential state

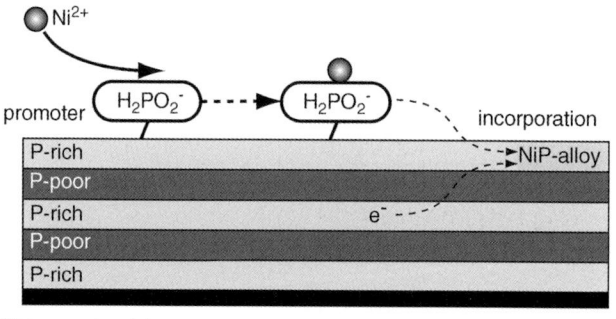

promoter H$_2$PO$_2^-$ ----> H$_2$PO$_2^-$ ----. incorporation
Ni^{2+}
P-rich → NiP-alloy
P-poor
P-rich   e$^-$
P-poor
P-rich

(b) Low-potential state

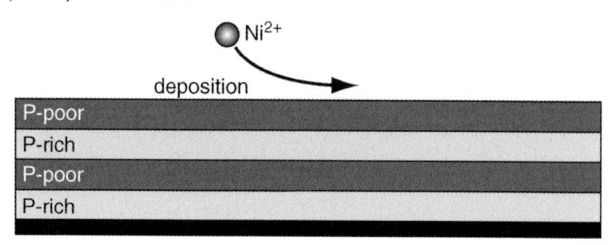

Ni^{2+}
deposition →
P-poor
P-rich
P-poor
P-rich

Figure 14.5 Schematic illustrations to explain the promotion effect of adsorbed H$_2$PO$_2^-$.

organized manner [24, 38]. Since the resultant deposits with layered nanostructures show interesting electronic [24, 39, 40] and optical [41] properties, this system has attracted a lot of attention from the point of view of self-organization of nanofunctional materials.

Figure 14.6a shows an example of the potential oscillation that is observed when the electrodeposition is performed in a solution with pH $\approx$ 8.3. Direct evidence for layering was obtained by Auger depth profile with Ar$^+$ ion sputtering [24], SEM [38], and STM [24]. Figure 14.6b shows an SEM cross-section of a film, from which the thickness of one layer is estimated to be about 62 nm. The phase composition, layer thicknesses, and resistivity of the films can be tuned by varying the applied current density of the solution pH. Essentially the same behavior can be observed in tartrate and citrate solutions [42–44]. Very interestingly, it was revealed that the resultant deposits can work as resonant tunneling devices, which show sharp NDR signatures at room temperature in perpendicular transport measurement [40]. The bias for the NDR maximum can be controlled simply by tuning the oscillation period (Figure 14.6c).

Leopold et al. and Nyholm et al. have investigated this oscillatory system by *in situ* confocal Raman spectroscopy [43], and *in situ* electrochemical quartz crystal microbalance [44], and *in situ* pH measurement [45] with the focus being on clarification of the oscillation mechanism. Based on the experimental results, a mechanism for the oscillations was proposed, in which variations in local pH close to the electrode surface play an essential role. Cu is deposited at the lower potentials of the oscillation followed by a simultaneous increase in pH close to the surface due to the protonation

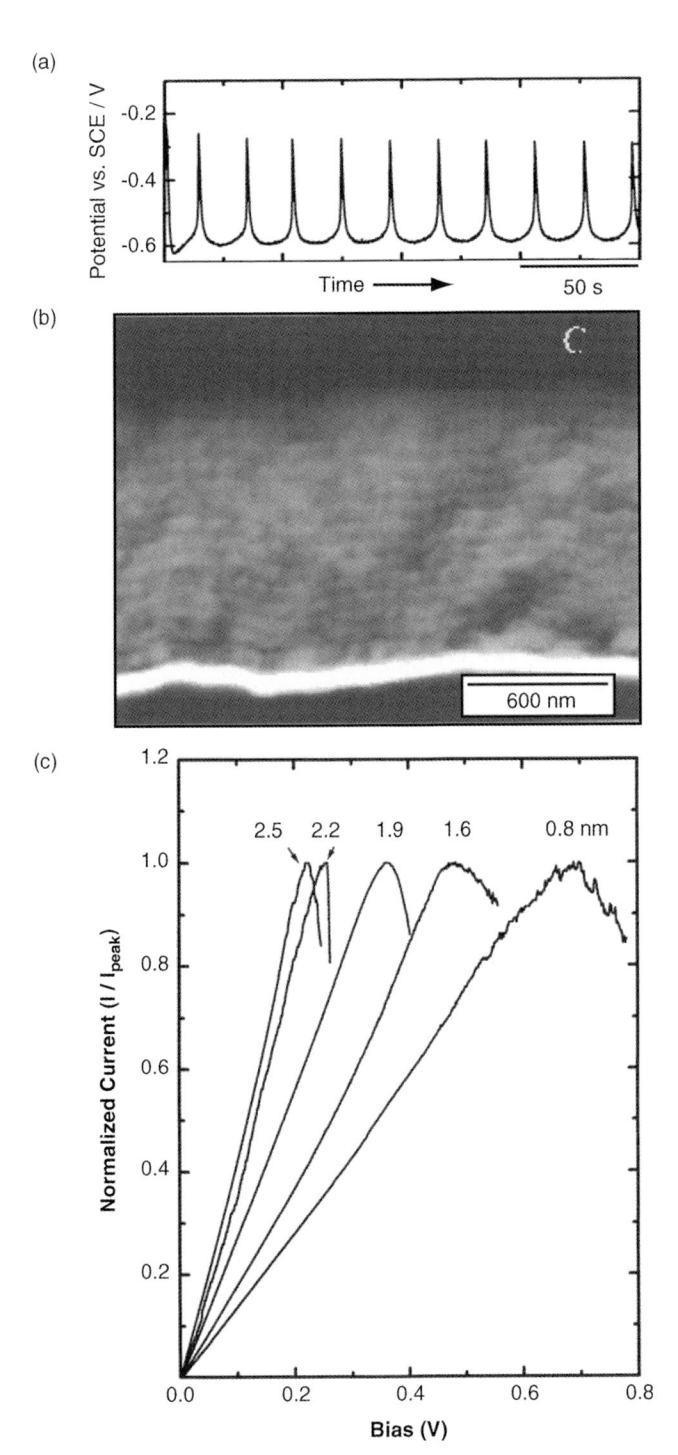

Figure 14.6 (a) Typical example of the potential oscillation observed for Cu/Cu$_2$O electrodeposition. (b) SEM image of a cross-section of the Cu/Cu$_2$O film grown under the oscillatory condition. (c) NDR curves for layered Cu/Cu$_2$O nanostructures as a function of the Cu$_2$O layer thickness. The NDR maximum shifts to higher applied bias for samples with thinner Cu$_2$O layers. (Reprinted from Refs. [38, 40] with permission from American Chemical Society.)

of the lactate (or tartrate) liberated from the Cu(II) complex. As the pH is increased, the deposition of Cu_2O becomes more favorable and a positive shift in potential is observed. When Cu_2O starts to electrodeposit, the production of OH^- stops and the pH decreases. The rate of the pH decrease is significantly increased by the presence of comproportionation between Cu(II) and copper as this reaction consumes OH^-. As a result of the decreasing local pH, the rate of electrodeposition of Cu_2O decreases, which has to be compensated for by a negative shift in the potential.

On the other hand, Switzer et al. proposed a different model for the oscillation. They attributed the oscillation to repetitive build-up and breakdown of a thin Cu_2O layer, which is a *p*-type semiconductor and acts as a thin rectifying (passivating) layer [24]. Disappearance of the oscillation under irradiated condition supports this model. Light will generate electron–hole pairs in the Cu_2O and lower the rectifying barrier at the semiconductor/solution interface.

14.3.4
Nanostructured Metal Filaments

More than 20 years ago, Matsushita et al. observed macroscopic patterns of electrodeposit at a liquid/air interface [46, 47]. Since the morphology of the deposit was quite similar to those generated by a computer model known as diffusion-limited aggregation (DLA) [48], this finding has attracted a lot of attention from the point of view of morphogenesis in Laplacian fields. Normally, thin cells with quasi 2D geometries are used in experiments, instead of the use of liquid/air or liquid/liquid interfaces, in order to reduce the effect of convection.

Recently, Wang et al. found electrodeposition in an "ultra-thin" layer of $CuSO_4$ showed spontaneous electrochemical oscillations and formation of straight Cu filaments with periodic corrugated nanostructures [28, 49–52]. Figure 14.7a shows schematically the experimental set-up with an "ultra-thin" layer of an electrolyte. To generate the ultra-thin electrolyte, the $CuSO_4$ solution was solidified by decreasing the temperature (The thickness of the electrolyte layer is about 200 nm). A typical example of the oscillation is shown in Figure 14.7b. Figure 14.7c shows the macroscopic morphology of the deposit obtained under the oscillation, in which finger-like branches developing outwards are seen. It can be seen by optical microscopy that the finger-like branches grown on the glass plate consist of long, narrow filaments, as shown in Figure 14.7d. AFM inspection of the microstructure of the deposit revealed that periodic corrugated structures exist on the filament (Figure 14.7e). The periodicity varies from several tens of nanometers to a few microns, depending on temperature, voltage or current applied across the electrodes, and the pH of the electrolyte. Analysis of the deposits by TEM diffraction [52] and scanning near-field optical microscopy (SNOM) [51] revealed that the periodic nanostructures on the electrodeposits correspond to the alternating growth of Cu and Cu_2O. Similar electrochemical oscillation and formation of nano-filaments were also observed in Zn electrodeposition (Figure 14.8) [27].

A simple model was proposed on the basis of the experimental findings, in which the solution pH plays a key role for the oscillatory instability [51]. The theoretical

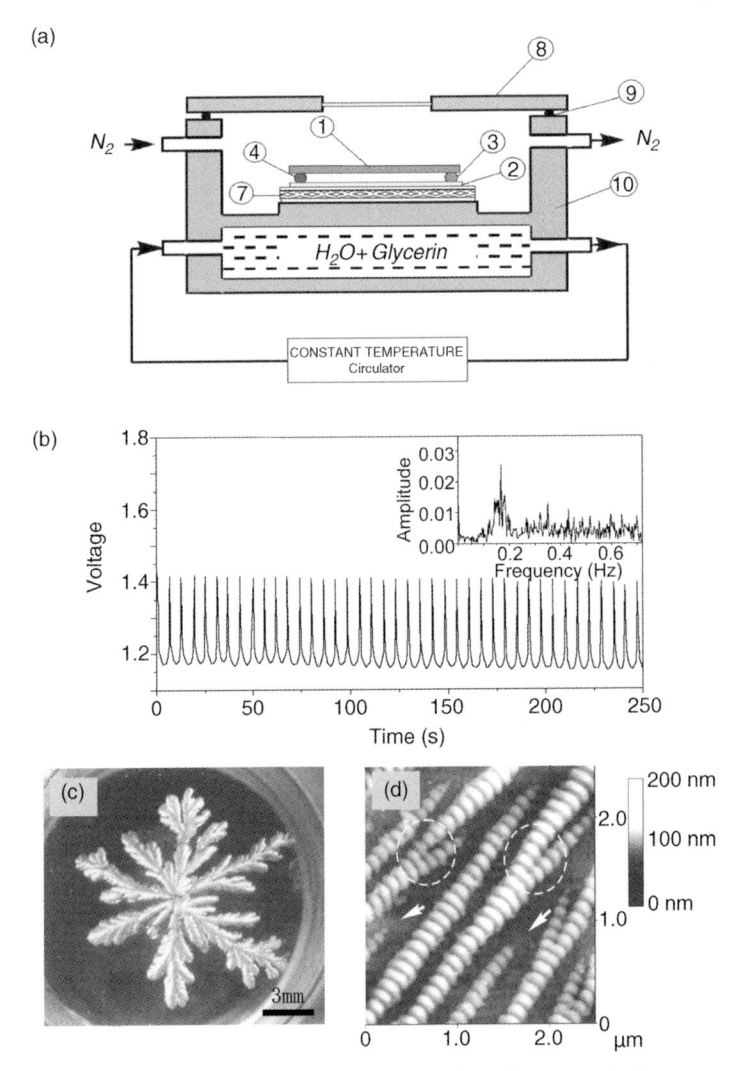

Figure 14.7 (a) A schematic diagram of the experimental set-up for the generation of an ultrathin electrolyte film and for electrodeposition. The cell for electrodeposition shown here has two parallel electrodes. (b) Voltage oscillation during the electrodeposition of Cu in the ultrathin electrolyte. Inset: the Fourier transform of the voltage oscillation. (c) The macroscopic view of the electrodeposits of copper grown from a circular electrodeposition cell. (d) The AFM view of the copper filaments. (Reprinted from Refs. [28] with permission from the American Physical Society.)

approach suggested that the oscillating local concentration of Cu^{2+} and H^+ triggers an alternating deposition of Cu and Cu_2O. Monte Carlo simulation of the spontaneous oscillation was also performed by the same authors [49]. The simulated layered structure of the deposited Cu/Cu_2O, as well as the correspondence between the pH oscillation and phase composition agreed qualitatively with the experimental results.

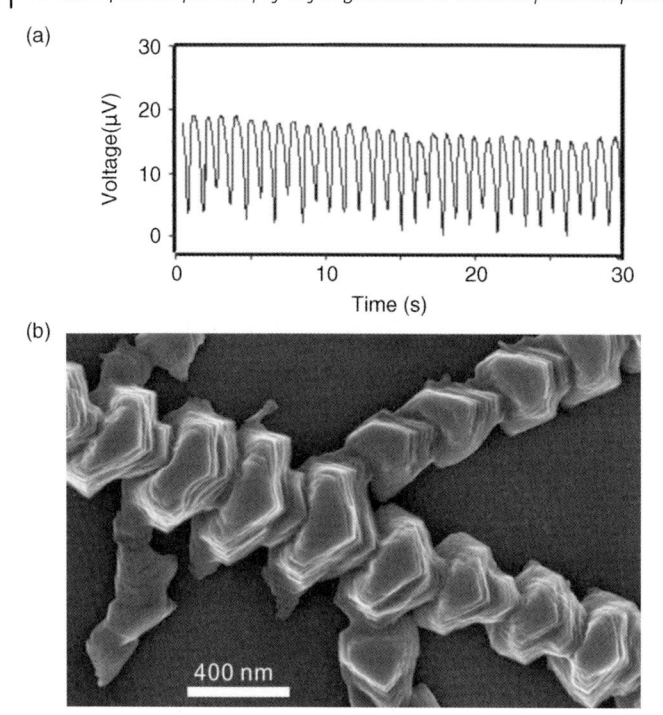

Figure 14.8 (a) A voltage oscillation obtained for Zn electrodeposition from an ultra-thin electrolyte. (b) SEM images of the Zn deposit obtained under the voltage oscillation. (Reprinted from Ref. [27] with permission from John Wiley & Sons Ltd.)

14.4
Raman Microspectroscopy Study of Oscillatory Electrodeposition of Au at an Air/Liquid Interface

As has been shown above, oscillatory electrodeposition is interesting from the point of view of the production of micro- and nanostructured materials. However, *in situ* observation of the dynamic change of the deposits had been limited to the micrometer scale by use of an optical microscope. Inspections on the nanometer scale were achieved only by *ex situ* experiments. Thus, information with regard to dynamic nanostructural changes of deposits in the course of the oscillatory growth was insufficient, although it is very important to understand how the macroscopic ordered structures are formed with their molecular- or nano-components in a self-organized manner.

Surface-enhanced Raman scattering (SERS) is a candidates for resolving this issue. Since the SERS effect is observed only at metal surfaces with nanosized curvature, this technique can also be used to investigate nanoscale morphological structures of metal surfaces. It is thus worth investigating SERS under oscillatory electrodeposition conditions. The author of this chapter and coworkers recently reported that

14.4 Raman Microspectroscopy Study of Oscillatory Electrodeposition of Au at an Air/Liquid Interface

in situ SERS from a gold film formed at a liquid/air interface by oscillatory electrodeposition can probe the dynamic nanostructural change of the deposits during the oscillatory growth [53]. Figure 14.9a is a schematic drawing of the experimental set-up for electrochemical formation of Au film. The working electrode (WE) was positioned at the center of the Pt-ring counter-electrode, and its tip was located just at the liquid/air interface. In the present electrodeposition system, the potential oscillates spontaneously (Figure 14.9b) and the Au deposition proceeds specifically along the liquid/air interface, resulting in the formation of an Au film with a concentric-circle pattern. Figure 14.10 shows SEM images of the Au film, taken by pulling it out from the electrolyte at stage 2 in Figure 14.9b (positive-end) of the potential oscillation. From this procedure, the growing front, namely the area denoted

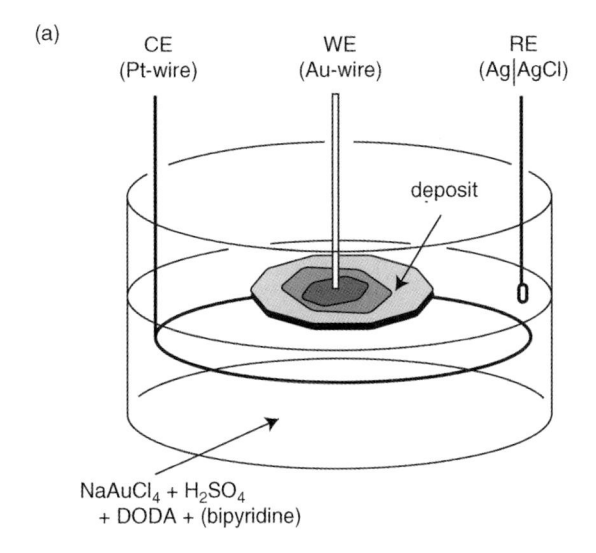

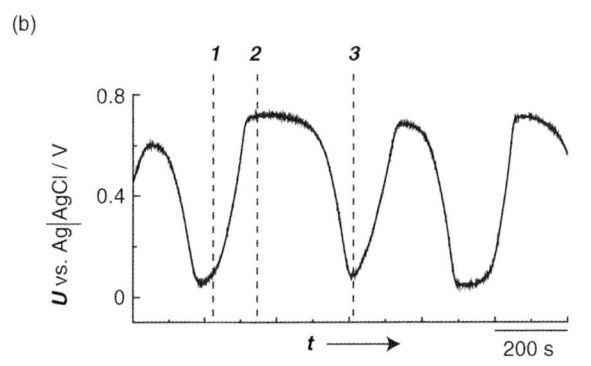

Figure 14.9 (a) Schematic illustration of the experimental set-up used for Au electrodeposition at a liquid/air interface. (b) Typical example of the potential oscillation during Au electrodeposition. (Reprinted from Ref. [53] with permission from the American Chemical Society.)

Figure 14.10 (a) SEM image of the growing front of the Au film at a stage where the potential is at the positive end of the potential oscillation (e.g., at stage-2 of Figure 14.9b). (b–d) Expansions of a part of b–d with increased magnification, respectively. The areas denoted by B, can be regarded as presenting a newly deposited part during the transition from A and B represent deposited parts in the preceding stages of the oscillation and newly deposited part, respectively (see text for details). The dotted lines in panels a and b are the border between areas A and B. (Reprinted from Ref. [53] with permission from the American Chemical Society.)

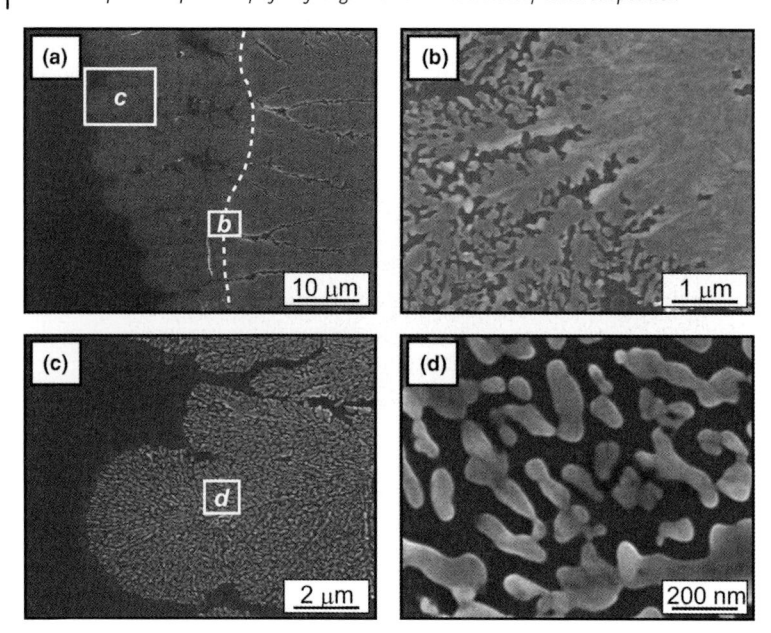

by B, can be regarded as presenting a newly deposited part during the transition from stage 1 to 2, that is, from the negative end to the positive end of the oscillation, whereas the area denoted by A represents a deposited part in the preceding stages of the oscillation. Figure 14.10b–d are the expanded images, from which we can see clearly that the deposit in area B is composed of numerous numbers of nanoneedles with width 20–80 nm, whereas the continuous film is formed in area A.

The SERS activity of the deposited Au film was investigated by measuring the Raman scattering intensity from bipyridine (bpy) adsorbed on the surface. Figure 14.11a is a schematic illustration to explain how the Raman signals were obtained under *in situ* conditions. The laser beam was focused at a position close to the growing front of the Au film, so that it could move and pass the laser spot at the liquid/air interface during one cycle of the oscillation. The Raman spectrum observed is shown as a function of t in Figure 14.11b, in which the intensity is expressed by the gray scale. All peaks at 1020, 1076, 1227, and 1293 cm^{-1} are assignable to vibrational modes of adsorbed bpy. The change in the Raman spectrum with t was measured concurrently with the measurement of the potential oscillation shown in Figure 14.11c The SERS intensity becomes stronger while U moves from the negative end to the positive end of the potential oscillation (e.g., from stage 1 to stage 2), whereas the intensity is weakened when U shifts from the positive end to the negative

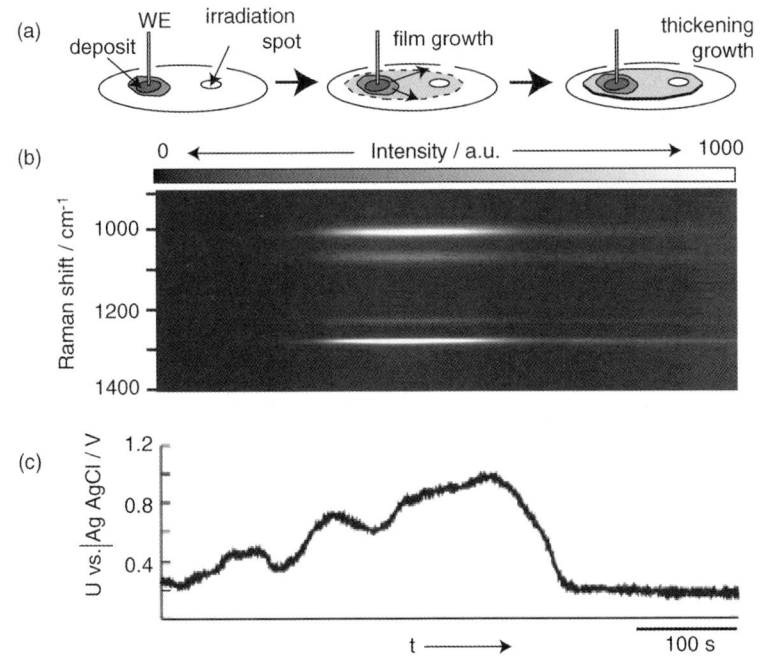

Figure 14.11 (a) Schematic illustrations to explain how the SERS signal was measured under *in situ* conditions. (b) Raman spectrum as a function of time. (c) The potential oscillation measured concurrently with the measurement of panel (b). (Reprinted from Ref. [53] with permission from the American Chemical Society.)

end (e.g., from stage 2 to stage 3). As mentioned earlier, when U moves from the negative end to the positive end of the oscillation, a 2D film composed of numerous numbers of nanoneedles is formed (Figure 14.10) and the SERS intensity increases, indicating that the nanoneedles are SERS-active. On the other hand, when U moves from the positive end to the negative end, the SERS intensity is weakened because the nanoneedles change the morphology into a continuous film via thickening and coalescing. These results clearly show that the SERS under *in situ* conditions probes the dynamic nanostructural change in the deposits proceeding under the oscillatory electrodeposition.

14.5
Summary

We have reviewed studies of the self-organized formation of ordered nanostructures by oscillatory electrodeposition. Although the mechanism is totally different in different cases and the structures of the resultant deposits vary greatly, they agree in that a unit structure is formed with one cycle of the oscillation. Periodic ordered structures are formed by periodic oscillation, and modulated oscillations give rise to the formation of modulated periodic structures. The important point with regard to the self-organized formation of ordered structures by oscillatory electrodeposition is that all the processes are spatiotemporally synchronized under non-equilibrium, nonlinear electrochemical dynamics. This principle is quite unique and is never realized by other methods. In order to design and tune the ordered structures, it is very important to understand the mechanism of the self-organization on a molecular or nano-level. From this point of view, it is very interesting to perform *in situ* microspectroscopic studies. In the present review, we described how the dynamic nanostructural change of a gold film formed by oscillatory electrodeposition was probed by *in situ* SERS. We can thus expect that further study along this line will greatly contribute to the preparation of designed and controlled ordered nanostructures at solid surfaces.

References

1 Geissler, M. and Xia, Y. N. (2004) Patterning: principles and some new developments. *Adv. Mater.*, **16**, 1249–1269.
2 Barth, J. V., Costantini, G. and Kern, K. (2005) Engineering atomic and molecular nanostructures at surfaces. *Nature*, **437**, 671–679.
3 Kunitake, T. (1992) Synthetic bilayer-membranes – molecular design, self-organization, and application. *Angew. Chem. Int. Ed.*, **31**, 709–726.
4 Lehn, J. M. (1990) Perspective in supramolecular chemistry – from molecular recognition towards molecular information-processing and self-organization. *Angew. Chem. Int. Ed.*, **29**, 1304–1319.
5 Love, J. C., Estroff, L. A., Kriebel, J. K., Nuzzo, R. G. and Whitesides, G. M. (2005) Self-assembled monolayers of thiolates on metals as a form of nanotechnology. *Chem. Rev.*, **105**, 1103–1169.
6 Murray, C. B., Kagan, C. R. and Bawendi, M. G. (2000) Synthesis and characterization of monodisperse nanocrystals and close-packed nanocrystal assemblies. *Annu. Rev. Mater. Sci.*, **30**, 545–610.
7 Ringsdorf, H., Schlarb, B. and Venzmer, J. (1988) Molecular architecture and function of polymeric oriented systems – models for the study of organization, surface recognition, and dynamics of biomembranes. *Angew. Chem. Int. Ed.*, **27**, 113–158.
8 Storhoff, J. J. and Mirkin, C. A. (1999) Programmed materials synthesis with DNA. *Chem. Rev.*, **99**, 1849–1862.
9 Ulman, A. (1996) Formation and structure of self-assembled monolayers. *Chem. Rev.*, **96**, 1533–1554.
10 Teichert, C. (2002) Self-organization of nanostructures in semiconductor heteroepitaxy. *Phys. Rep. -Rev. Sec. Phys. Lett.*, **365**, 335–432.
11 Grzybowski, B. A., Bishop, K. J. M., Campbell, C. J., Fialkowski, M. and Smoukov, S. K. (2005) Micro- and nanotechnology via reaction-diffusion. *Soft Matter*, **1**, 114–128.
12 Imbihl, R. and Ertl, G. (1995) Oscillatory kinetics in heterogeneous catalysis. *Chem. Rev.*, **95**, 697–733.
13 Epstein, I. R. (1998) *An Introduction to Nonlinear Chemical Dynamics; Oscillations, Waves, Patterns, and Chaos*, Oxford University Press, New York.
14 Scott, S. K. (1991) *Chemical Chaos*, Clarendon Press, Oxford.

References

15 Kapral, R. and Showalter, K. (1994) *Chemical Waves and Patterns*, Kluwer, London.
16 Briggs, T. S. and Rauscher, W. C. (1973) Oscillating iodine clock. *J. Chem. Educ.*, **50**, 496–496.
17 Fafiday, T. Z. and Hudson, J. L. (1995) *Modern Aspects of Electrochemistry*, vol. 27 (eds B. E. Conway, J. O'. M. Bockris and R. E. White), Plenum, New York, p. 383.
18 Koper, M. T. M. (1996) *Advances in Chemical Physics*, **92** (eds I. Prigogine and S. A. Rice), John Wiley & Sons, New York, p. 161.
19 Krischer, K. (2003) *Advances in Electrochemical Science and Engineering*, vol. 8 (eds R. C. Alkire, C. W. Tobias, H. Gerischer and D. M. Kolb), Wiley-VCH, Weinheim, p. 90.
20 Hudson, J. L. and Tsotsis, T. T. (1994) Electrochemical reaction dynamics – A review. *Chem. Eng. Sci.*, **49**, 1493–1572.
21 Strasser, P., Eiswirth, M. and Koper, M. T. M. (1999) Mechanistic classification of electrochemical oscillators – operational experimental strategy. *J. Electroanal. Chem*, **478**, 50–66.
22 Schlitte, F., Eichkorn, G. and Fischer, H. (1968) Rhythmic lamerllar crystal growth in electrolytic copper deposition. *Electrochim. Acta*, **13**, 2063–2075.
23 Krastev, I. and Koper, M. T. M. (1995) Pattern-formation during the electrodeposition of a silver antinomy alloy. *Phys. A*, **213**, 199–208.
24 Switzer, J. A., Hung, C. J., Huang, L. Y., Switzer, E. R., Kammler, D. R., Golden, T. D. and Bohannan, E. W. (1998) Electrochemical self-assembly of copper/cuprous oxide layered nanostructures. *J. Am. Chem. Soc.*, **120**, 3530–3531.
25 Nakanishi, S., Sakai, S., Nagai, T. and Nakato, Y. (2005) Macroscopically uniform nanoperiod alloy multilayers formed by coupling of electrodeposition with current oscillations. *J. Phys. Chem. B*, **109**, 1750–1755.
26 Sakai, S., Nakanishi, S. and Nakato, Y. (2006) Mechanisms of oscillations and formation of nano-scale layered structures in induced co-deposition of some iron-group alloys (Ni-P, Ni-W, and Co-W), studied by an in situ electrochemical quartz crystal microbalance technique. *J. Phys. Chem. B*, **110**, 11944–11949.
27 Liu, T., Wang, S., Wang, M., Peng, R. W., Ma, G. B., Hao, X. P. and Ming, N. B. (2006) Self-organization of periodically structured single-crystalline zinc branches by electrodeposition. *Surf. Interface Anal.*, **38**, 1019–1023.
28 Zhong, S., Wang, Y., Wang, M., Zhang, M. Z., Yin, X. B., Peng, R. W. and Ming, N. B. (2003) Formation of nanostructured copper filaments in electrochemical deposition. *Phys. Rev. E*, **67**, 061601.
29 Saliba, R., Mingotaud, C., Argoul, F. and Ravaine, S. (2001) Electroless deposition of gold films under organized monolayers. *J. Electrochem. Soc.*, **148**, C65–C69.
30 Saliba, R., Mingotaud, C., Argoul, F. and Ravaine, S. (2002) Spontaneous oscillations in gold electrodeposition. *Electrochem. Commun.*, **4**, 629–632.
31 Fukami, K., Nakanishi, S., Yamasaki, H., Tada, T., Sonoda, K., Kamikawa, N., Tsuji, N., Sakaguchi, H. and Nakato, Y. (2007) General mechanism for the synchronization of electrochemical oscillations and self-organized dendrite electrodeposition of metals with ordered 2D and 3D microstructures. *J. Phys. Chem. C*, **111**, 1150–1160.
32 Nakanishi, S., Fukami, K., Tada, T. and Nakato, Y. (2004) Metal latticeworks formed by self-organization in oscillatory electrodeposition. *J. Am. Chem. Soc.*, **126**, 9556–9557.
33 Survila, A., Mockus, Z. and Juskenas, R. (1998) Current oscillations observed during codeposition of copper and tin from sulfate solutions containing Laprol 2402C. *Electrochim. Acta*, **43**, 909–917.
34 Brenner, A. (1963) *Electrodeposition of Alloys*, Academic Press, New York.
35 Podlaha, E. J. and Landolt, D. (1996) Induced codeposition.1. An experimental investigation of Ni-Mo alloys. *J. Electrochem. Soc.*, **143**, 885–892.

36 Ogburn, F. and Johnson, C. E. (1975) Mechanical properties of electrodeposited brass. *Plating*, **62**, 142–147.

37 Lee, W. G. (1971) Improvement of solder connections by gold alloy plating. *Plating*, **58**, 997–1001.

38 Bohannan, E. W., Huang, L. Y., Miller, F. S., Shumsky, M. G. and Switzer, J. A. (1999) In situ electrochemical quartz crystal microbalance study of potential oscillations during the electrodeposition of Cu/Cu_2O layered nanostructures. *Langmuir*, **15**, 813–818.

39 Switzer, J. A., Hung, C. J., Bohannan, E. W., Shumsky, M. G., Golden, T. D. and VanAken, D. C. (1997) Electrodeposition of quantum-confined metal semiconductor nanocomposites. *Adv. Mater.*, **9**, 334–338.

40 Switzer, J. A., Maune, B. M., Raub, E. R. and Bohannan, E. W. (1999) Negative differential resistance in electrochemically self-assembled layered nanostructures. *J. Phys. Chem. B*, **103**, 395–398.

41 Mishina, E. D., Nagai, K. and Nakabayashi, S. (2001) Self-assembled Cu/Cu_2O multilayers: Deposition, structure and optical properties. *Nano Lett.*, **1**, 401–404.

42 Eskhult, J., Herranen, M. and Nyholm, L. (2006) On the origin of the spontaneous potential oscillations observed during galvanostatic deposition of layers of Cu and Cu_2O in alkaline citrate solutions. *J. Electroanal. Chem.*, **594**, 35–39.

43 Leopold, S., Arrayet, J. C., Bruneel, J. L., Herranen, M., Carlsson, J. O., Argoul, F. and Serant, L. (2003) In situ CRM study of the self-oscillating Cu-(II)-lactate and Cu-(II)-tartrate systems. *J. Electrochem. Soc.*, **150**, C472–C477.

44 Leopold, S., Herranen, M. and Carlsson, J. O. (2001) Spontaneous potential oscillations in the Cu(II)/tartrate and lactate systems, aspects of mechanisms and film deposition. *J. Electrochem. Soc.*, **148**, C513–C517.

45 Leopold, S., Herranen, M., Carlsson, J. O. and Nyholm, L. (2003) In situ pH measurement of the self-oscillating Cu(II)-lactate system using an electropolymerised polyaniline film as a micro pH sensor. *J. Electroanal. Chem.*, **547**, 45–52.

46 Matsushita, M., Hayakawa, Y. and Sawada, Y. (1985) Fractal structure and cluster statistics of zinc-metal trees deposited on a line electrode. *Phys. Rev. A*, **32**, 3814–3816.

47 Matsushita, M., Sano, M., Hayakawa, Y., Honjo, H. and Sawada, Y. (1984) Fractal structures of zinc metal leaves grown by electrodeposition. *Phys. Rev. Lett.*, **53**, 286–289.

48 Witten, T. A. and Sander, L. M. (1981) Diffusion-limited aggregation, A kinetic critical phenomenon. *Phys. Rev. Lett.*, **47**, 1400–1403.

49 Ha, M. J., Fang, F., Liu, J. M. and Wang, M. (2005) Monte carlo simulation of the spontaneous oscillation in electrochemical deposition. *Eur. Phys. J. D*, **34**, 195–198.

50 Wang, M., Feng, Y., Yu, G. W., Gao, W. T., Zhong, S., Peng, R. W. and Ming, N. B. (2004) Self-organization of nanostructured copper filament array by electrochemical deposition. *Surf. Interface Anal.*, **36**, 197–198.

51 Wang, Y., Cao, Y., Wang, M., Zhong, S., Zhang, M. Z., Feng, Y., Peng, R. W., Hao, X. P. and Ming, N. B. (2004) Spontaneous formation of periodic nanostructured film by electrodeposition: Experimental observations and modeling. *Phys. Rev. E*, **69**, 021607.

52 Zhang, M. Z., Wang, M., Zhang, Z., Zhu, J. M., Peng, R. W. and Ming, N. B. (2004) Periodic structures of randomly distributed Cu/Cu_2O nanograins and periodic variations of cell voltage in copper electrodeposition. *Electrochim. Acta*, **49**, 2379–2383.

53 Fukami, K., Nakanishi, S., Sawai, Y., Sonoda, K., Murakoshi, K. and Nakato, Y. (2007) In situ probing of dynamic nanostructural change of electrodeposits in the course of oscillatory growth using SERS. *J. Phys. Chem. C*, **111**, 3216–3219.

15
Construction of Nanostructures by use of Magnetic Fields and Spin Chemistry in Solid/Liquid Interfaces

Hiroaki Yonemura

15.1
Introduction

Applying the strong magnetic fields (>6 T) of a superconducting magnet to materials induces huge magnetic field effects (MFEs), in comparison with applying the magnetic fields (~1 T) of an electromagnet. It is expected that highly functional nanomaterials with new properties will be created, since new interfaces or nanostructures are constructed by strong magnetic fields [1].

The magnetic orientation of crystals [2–4], polymers [5, 6], fibrin fiber [7, 8], and carbon nanotubes [9, 10] by the strong magnetic fields of a superconducting magnet has been widely investigated. Two- or three-dimensional patterns of silver dendrites in a strong magnetic field were reported by Mogi *et al.* [11, 12] and Katsuki *et al.* [13, 14]. Significant morphological changes induced by magnetic fields were interpreted by the magnetohydrodynamic (MHD) mechanism in which the motions of ions in a magnetic field are influenced by the Lorentz force, and/or by the magnetic force of a gradient magnetic field. Magnetic field effects on the growth morphology in electropolymerization, photoelectrochemical reactions, and the redox behavior of polypyrrole as a conducting polymer, have been reported by Mogi *et al.* [15–17]. They were explained by the diamagnetic orientation of the polymers. Recently, Tanimoto *et al.* reported attractive 3D morphological chirality in membrane tubes prepared by a silicate garden reaction using a strong magnetic field [18, 19].

Magnetic field effects on the reaction kinetics or yields of photochemical reactions in the condensed phase have been studied [20–23]. They have proved powerful for verifying the mechanism of photochemical reactions including triplet states. Previously, we obtained photogenerated triplet biradicals of donor–acceptor linked compounds, and found that the lifetimes of the biradicals were remarkably extended in the presence of magnetic fields up to 1 T [24]. It has been reported that C_{60} and its derivatives form optically transparent microscopic clusters in mixed solvents [25, 26]. The clustering behavior of fullerene (C_{60}) is mainly associated with the strong three-dimensional hydrophobic interactions between the C_{60} units. Photoinduced

Molecular Nano Dynamics, Volume I: Spectroscopic Methods and Nanostructures
Edited by H. Fukumura, M. Irie, Y. Iwasawa, H. Masuhara, and K. Uosaki
Copyright © 2009 WILEY-VCH Verlag GmbH & Co. KGaA, Weinheim
ISBN: 978-3-527-32017-2

electron-transfer and photoelectrochemical reactions using C_{60} clusters have been extensively reported because of the interesting properties of C_{60} clusters [25, 26]. However, MFEs on the photoinduced electron-transfer reactions using the C_{60} cluster in mixed solvents had not until now been studied.

Magnetic field effects on the photoelectrochemical reactions of photosensitive electrodes are very important for practical applications of the MFEs in controlling the photoelectronic functions of molecular devices. Previously, we have examined MFEs on the photoelectrochemical reactions of photosensitive electrodes modified with zinc-tetraphenylporphyrin-viologen linked compounds [27, 28] and semiconductor nanoparticles [29, 30]. However, MFEs on the photoelectrochemical reactions of photosensitive electrodes modified with nanoclusters have not yet been reported.

First, we attempted to create highly functional nanomaterials containing new physical or chemical properties using single-walled carbon nanotubes (SWNTs) (Section 15.2.1), a C_{60} derivative (Section 15.2.2), and diluted magnetic semiconductor $(Zn_{1-x}Mn_xS)$ nanoparticles (Section 15.2.3), since new interfaces or nanostructures are constructed by strong (>6 T) or normal magnetic fields due to an electromagnet (~1 T). Secondly, we attempted to obtain new information to examine spin chemistry at the solid/liquid interface or the nanostructures using electrodes modified with C_{60} nanoclusters (Section 15.3.2).

15.2
Construction of Nanostructures by the use of Magnetic Fields

15.2.1
Magnetic Orientation and Organization of SWNTs or their Composite Materials Using Polymer Wrapping

Since the discovery of SWNTs, they have been expected to become the building blocks of the next generation of functional nanomaterials. However, their strong cohesive property and poor solubility have restricted the use of SWNTs for fundamental and applied research fields. One method to overcome these problems is to make the SWNTs more soluble by wrapping them with polymers [31]. At the same time, the fabrication of high-performance carbon nanotube (CNT)-based composites is driven by the ability to create anisotropy at the molecular level to obtain appropriate functions.

Many groups have reported the orientation of CNTs by various methods, such as spinning CNTs [32], condensing viscous flow-aligned polymer-CNT composites [33], electric field [34, 35], and the use of the cooperative reorientation of a liquid crystal-CNT suspension in an electric field [36]. Several groups have reported magnetic orientation of CNTs by using strong magnetic fields [9, 10, 37, 38]. However, the magnetic orientation of composites between CNT and a conducting polymer have been scarcely reported.

In addition, SWNTs have been expected to act as acceptors or molecular wires in molecular photoelectric conversion since they have attractive electron-accepting properties and a one-dimensional nanowire structure. Recently, the photoelectrochemical reactions of electrodes modified with SWNT composites with a photosensitizer such as a semiconductor or donor molecules have been reported [39–41]. For example, novel SWNT composites due to electrostatic interaction using poly(sodium 4-styrenesulfonate) (PSS) or pyrene derivatives have been reported [39, 40, 42–44]. Recently, we examined the photoelectrochemical reactions of a modified electrode of composite materials consisting of poly[2-methoxy-5-(2'-ethylhexyloxy)-1,4-phenylene vinylene] (MEHPPV) as a conjugated polymer or ruthenium tris(2,2'-bipyridine) $(Ru(bpy)_3^{2+})$-PSS complex and SWNT [45]. The magnetic orientation of these composites on the electrodes is expected to improve the photoelectrochemical properties of the electrodes.

We examined the magnetic orientation or organization of the SWNTs or the polymer-wrapped SWNTs using MEHPPV by measurements of AFM images and polarized absorption spectra [46–48].

SWNTs (HiPco, Carbon Nanotechnologies Incorporated) were shortened by ultrasonication with a probe-type sonicator in mixed acids (H_2SO_4 and HNO_3) under ice-cooling. After diluting the mixture with water (MiliQ), the shortened SWNTs were purified by filtration through a PTFE membrane filter (pore size: 1 µm or 0.2 µm) or by chromatography (Sepadex G-50).

The SWNT/MEHPPV composites were prepared by the following procedure. The shortened SWNTs were added to a DMF solution of MEHPPV. The suspension was then sonicated with a bath-type sonicator. Centrifugation (6000 rpm) of the suspension for 15 min gave a DMF solution of the SWNT/MEHPPV composite.

The shortened SWNTs were added to the aqueous solution in the absence and the presence of $NaHCO_3$. These suspensions were then sonicated with a bath-type sonicator. Centrifugation (6000 rpm) of the suspensions gave aqueous solutions of the shortened SWNTs with and without $NaHCO_3$.

The formation of SWNT/MEHPPV composites was confirmed by absorption and fluorescence spectra. The DMF solution of SWNT/MEHPPV composites or the aqueous solution of the shortened SWNTs was then dropped onto a mica or glass plate. The magnetic processing of the composites or the SWNTs was carried out by using a superconducting magnet (8 T) in the horizontal direction, as described below.

The magnetic field was applied by using two superconducting magnets (horizontal and vertical direction of magnetic field). In the horizontal direction, a superconducting magnet (Oxford Instruments Spectromag-1000) was used, as reported in the previous papers [9, 10]. In the superconducting magnet a bore tube (50 mm diameter) was installed horizontally. Distribution of the magnetic field was approximated by a Gaussian distribution. The maximum strength of the magnetic field was 8.0 T at the center position.

In the case of SWNT/MEHPPV composites, after drying at 283 K under the magnetic field of 8 or 0 T (control), AFM images of the SWNT/MEHPPV composites on the mica were measured (Figure 15.1). The heights of the top of the SWNT/MEHPPV composites were 6–15 nm, indicating that they consist of bundles of 4–21 SWNTs, since the diameter of the SWNTs is 0.7–1.5 nm.

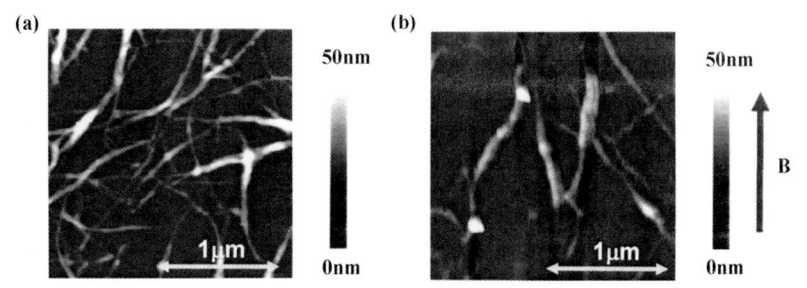

Figure 15.1 AFM images of SWNT/MEHPPV composites on mica placed in (a) 0 and (b) 8 T.

The AFM images strongly indicate that the SWNT/MEHPPV composites are oriented randomly in the absence of a magnetic field (0 T) (Figure 15.1a), while they are oriented with the tube axis of the composites parallel to the magnetic field in the presence of a magnetic field (8 T) (Figure 15.1b). We examined the effect of the length of the shortened SWNT on the magnetic orientation of the SWNT/MEHPPV composites. In the long SWNTs (average length; 2.2 μm), the magnitude of the magnetic orientation (Figure 15.1b) was greater in comparison with that in the short SWNTs (average length; 1.3 μm) in the AFM measurement.

In the case of the shortened SWNTs in the absence and the presence of NaHCO$_3$, after drying at 283 K under a magnetic field of 8 or 0 T (control), the AFM images of the shortened SWNTs on the mica were also measured. The images strongly indicate that the SWNTs are oriented randomly in the absence of a magnetic field, while the SWNTs oriented with the tube axis of the composites parallel to the magnetic field in the presence of a magnetic field (8 T).

On the basis of comparison of the AFM images between the SWNT/MEHPPV composites and the SWNT in the absence and presence of NaHCO$_3$, the magnetic orientation of SWNT/MEHPPV composites can most likely be ascribed to the anisotropy in susceptibilities of the SWNTs.

Polarized absorption spectra of the SWNT/MEHPPV composites on glass plates were measured in the near-IR region (900–1600 nm) with a UV–VIS–NIR spectrometer. In the absence of a magnetic field, the absorbance of the band (at about 1150 nm) due to the semiconducting SWNT was almost the same in both polarization directions (horizontal and vertical) (Figure 15.2a). On the other hand, in the presence of a magnetic field (8 T), the absorption band (at about 1150 nm) of the SWNT/MEHPPV composites on the glass plates in the horizontal polarization direction ($B(//); 0°$) was larger than that in the vertical polarization direction ($B(\perp); 90°$) (Figure 15.2b).

Next, the polarized absorption spectra of the shortened SWNTs in the absence and presence of NaHCO$_3$ on the glass plates were measured in the near-IR region (1000~1600 nm). In the absence of a magnetic field, the absorption band (at about 1450 nm) due to the semiconducting SWNT was observed in both polarization directions (horizontal and vertical). The absorption bands on the glass plate are similar to that in aqueous solution. In contrast, in the presence of a magnetic field (8 T), the absorbance (at about 1450 nm) due to the semiconducting SWNT in the

15.2 Construction of Nanostructures by the use of Magnetic Fields

(a)

(b)

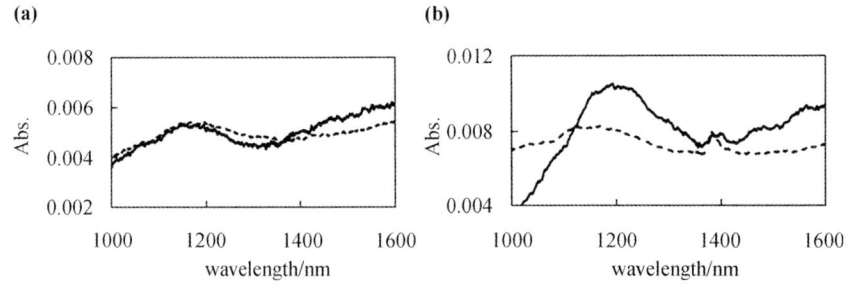

Figure 15.2 Polarized absorption spectra of SWNT/MEHPPV composites on glass plates in (a) the absence (0 T) and (b) the presence of magnetic processing (8 T). In the absence of magnetic processing (0 T), the polarization direction of the light against the longitudinal direction of the glass plates is horizontal (0°; solid line) or vertical (90°; broken line). In the presence of magnetic processing (8 T), the polarization direction of the light against the direction of the magnetic field is horizontal $B(//)$ (0°; solid line) or vertical $B(\perp)$ (90°; broken line). AFM images of SWNT/MEHPPV composites on mica placed in (a) 0 and (b) 8 T.

horizontal polarization direction ($B(//)$; 0°) was much larger than that in the vertical polarization direction ($B(\perp)$; 90°).

The above results of polarized absorption spectra on glass plates support the magnetic orientation of the SWNT/MEHPPV composites and the shortened SWNTs in the absence and presence of $NaHCO_3$, where the composites or the SWNTs were oriented with the tube axis of the composites or the SWNTs parallel to the magnetic field (8 T), suggested by the results of the AFM images (Figure 15.1).

On the basis of the polarized absorption spectra (Figure 15.2), the magnetic orientation of SWNT/MEHPPV composites can also most likely be ascribed to the anisotropy in susceptibilities of SWNTs, similar to the comparison of the AFM images (Figure 15.1), as described above.

In the vertical direction, a superconducting magnet (Japan Superconductor Technology, JMTD-LH15T40) was used in the present study, as reported in the previous papers [18, 19]. It has a room temperature bore tube width of 40 mm. The distribution of the magnetic field is reported in the previous papers [18, 19]. The maximum field (B_{max}(vertical)) and field (B) ×gradient field (dB/dz) were 15 T and 1500 $T^2 m^{-1}$, respectively, where z is the distance from the center of the bore tube along the tube. Three samples were placed at positions in the bore tube, for which B and BdB/dz were 5.6 T and $-940 T^2 m^{-1}$ for the top position, 15 T and $0 T^2 m^{-1}$ for the middle position, and 9.8 T and $+1070 T^2 m^{-1}$ for the bottom position and one was placed outside the tube as a control.

In the case of the vertical direction of magnetic field, after drying at ambient temperature under the magnetic field at the three positions (top, middle, and bottom) and in the absence of a magnetic field (outside the bore tube) as the control, the AFM images of the SWNTs on the mica were measured (Figure 15.3).

An organized network of bundles consisting of a certain amount of nanofibers, several nanometers in width, was observed at the top position, as shown in Figure 15.3a and b. Interesting nanostructures were not observed at the other positions. The heights of the top of the nanofibers were 2–3 nm. The results indicate

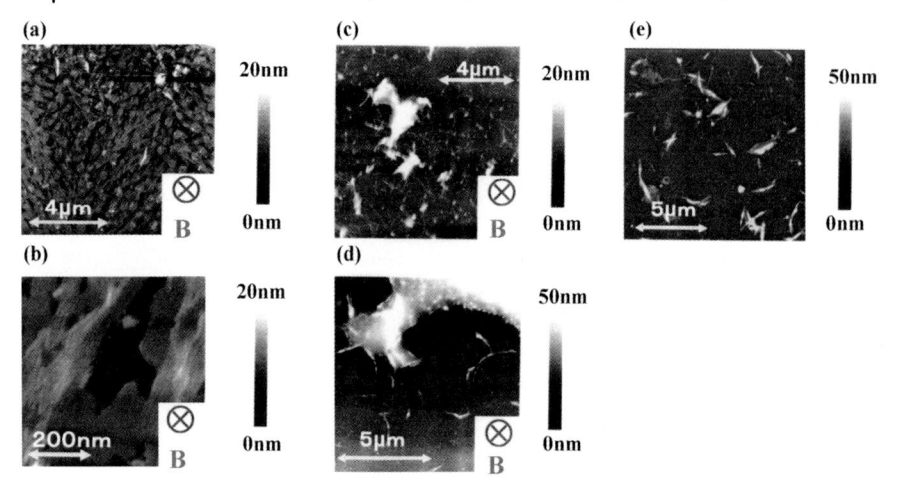

Figure 15.3 AFM images of SWNTs on mica placed at the top ((a) and (b)), the middle (c), and the bottom (d) positions using magnet apparatus (vertical direction of magnetic field) and outside position (e) of the bore tube (0 T).

that the nanofibers consist of individual SWNT or bundles of some SWNTs, since the diameter of the SWNTs is 0.7–1.5 nm.

On the basis of these observations, an interesting formation of nanostructures consisting of SWNTs was probably achieved by magnetic force, magnetic orientation, interaction of induced magnetic moment of SWNTs due to strong magnetic fields, and self-assembly of SWNTs due to hydrophobic interaction in aqueous solution and so on [46, 48].

15.2.2
Effects of Magnetic Processing on the Morphological, Electrochemical, and Photoelectrochemical Properties of Electrodes Modified with C_{60}-Phenothiazine Nanoclusters

We examined the effects of magnetic processing on the morphological, electrochemical, and photoelectrochemical properties of electrodes modified with nanoclusters of $C_{60}N^+$ and MePH (Figure 15.4) using a strong magnetic field [49].

Clusters of $C_{60}N^+$ and MePH were prepared by dissolving $C_{60}N^+$ and MePH in THF-H_2O (2 : 1) mixed solvent using first injection methods [50]. $C_{60}N^+$ and MePH form optically transparent clusters. The formation of nanoclusters of $C_{60}N^+$–MePH (diameter about 100 nm) was verified from absorption measurements and AFM.

The magnetic processing of the samples was carried out using a superconducting magnet (Oxford Instruments Spectromag-1000) as described in Section 15.2.1. The THF–H_2O mixed solution of $C_{60}N^+$ and MePH was dropped onto a mica or indium tin oxide (ITO) electrode. The samples were placed at three positions in the strong magnetic field and one outside the magnetic field (control). The magnetic field was applied horizontally to the surface of the mica or ITO electrode at 283 K. After drying

15.2 Construction of Nanostructures by the use of Magnetic Fields

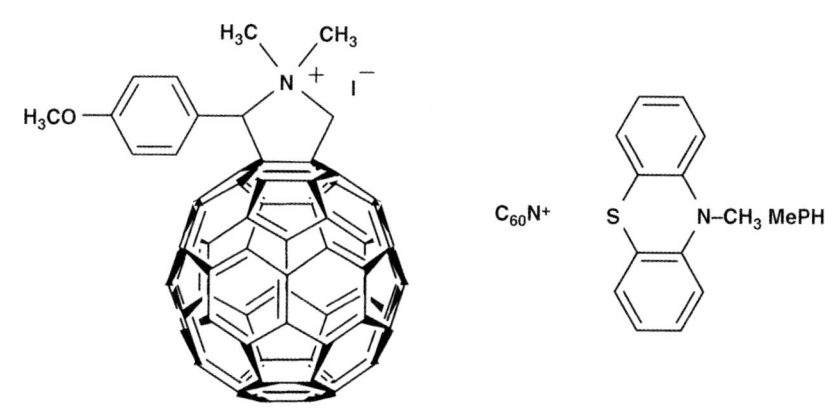

Figure 15.4 Chemical structures of a C_{60} derivative ($C_{60}N^+$) and methylphenothiazine (MePH).

the solvent at 283 K, AFM of the sample on the mica and electrochemical and photoelectrochemical measurements of the sample on the ITO electrode, that is, the electrodes modifed with $C_{60}N^+$–MePH clusters, were carried out.

The formations of clusters of $C_{60}N^+$–MePH were also examined by AFM. From the AFM image on a mica surface in the absence of magnetic field at 283 K (Figure 15.5a), many roughly spherical nanoclusters were clearly observed. The diameters of the clusters of $C_{60}N^+$ were estimated to be about 100 nm. The results were in fair agreement with those of the $C_{60}N^+$–MePH clusters prepared using a different method in previous studies [51, 52].

The morphological effects of applying magnetic fields to $C_{60}N^+$–MePH clusters on mica were also examined by AFM. The images indicate that the $C_{60}N^+$–MePH clusters in the presence of magnetic processing (Figure 15.5b) were smaller than those in the absence of magnetic processing (Figure 15.5a). In the AFM images obtained for clusters in the presence of magnetic processing (Figure 15.5b), non-spherical clusters were observed.

The morphological effects of applying magnetic fields can probably be ascribed to the MHD mechanism and/or the convective flow of the suspension produced by the

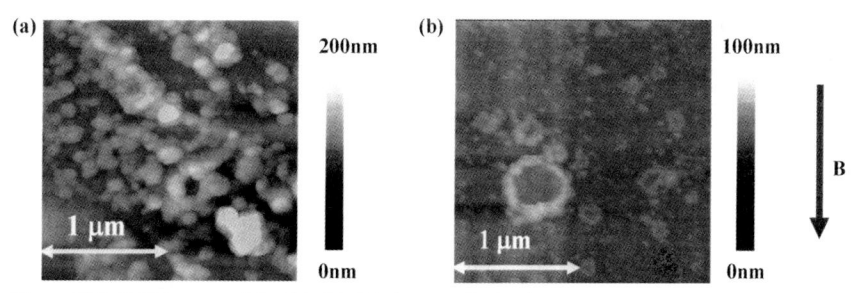

Figure 15.5 AFM images of $C_{60}N^+$–MePH clusters obtained from $C_{60}N^+$ and MePH in THF–H_2O (2:1) mixed solvent on a mica surface in the absence (a) and presence of magnetic processing at 8 T (b) at 283 K.

magnetic force. In other words, these phenomena can probably be ascribed to the effect of the Lorentz force since the $C_{60}N^+$–MePH clusters had positive charge. These morphological effects of applying magnetic fields can also probably be ascribed to the magnetic force due to the gradient magnetic fields.

The redox potentials of the ITO electrodes modified with $C_{60}N^+$–MePH clusters were measured by cyclic voltammetry and differential pulse voltammetry in the absence and presence of magnetic processing.

In the cyclic voltammetry, the oxidation peaks of PH were clearly observed in positive scans for all the modified electrodes. In contrast, reduction peaks of C_{60} were clearly observed in the absence of magnetic processing but not in the presence of magnetic processing.

Next, differential pulse voltammetry of the ITO electrodes modified with $C_{60}N^+$–MePH clusters was carried out to observe clearly the peak corresponding to the reduction of C_{60} in the presence of magnetic processing. The first reduction peaks corresponding to the $C_{60}N^+$ nanocluster in the presence of magnetic processing were negative-shifted compared with those in the absence of magnetic processing (Figure 15.6a, ($-1200 \leftarrow 0$ mV vs Ag/AgCl)). In contrast, with MePH, the first oxidation peaks were observed at about 600 mV vs Ag/AgCl in the absence and presence of magnetic processing.

Nakashima and coworkers reported that the DPVs due to the reduction of C_{60} were influenced by the type of cation electrolyte in the cast films of C_{60} lipids on electrodes [53]. The results showed the importance of the cationic charge of the lipids and the microenvironments in the films for the electrochemistry of C_{60}. Therefore, the Coulombic interaction between $C_{60}^{\bullet-}$ and the cation is most likely responsible for the marked negative shifts in the present study, since the interaction between the anion $C_{60}^{\bullet-}$ and the cation $C_{60}N^+$ was changed by the morphology of the $C_{60}N^+$ nanoclusters. These considerations are in fair agreement with the small shifts in the redox potential for MePH. The MFEs on the electrochemical properties of $C_{60}N^+$ clusters are in good agreement with the morphological effects of applying magnetic fields seen in the AFM images (Figure 15.5).

The photoelectrochemical properties of ITO electrodes modified with $C_{60}N^+$–MePH clusters were also measured in the presence and absence of magnetic

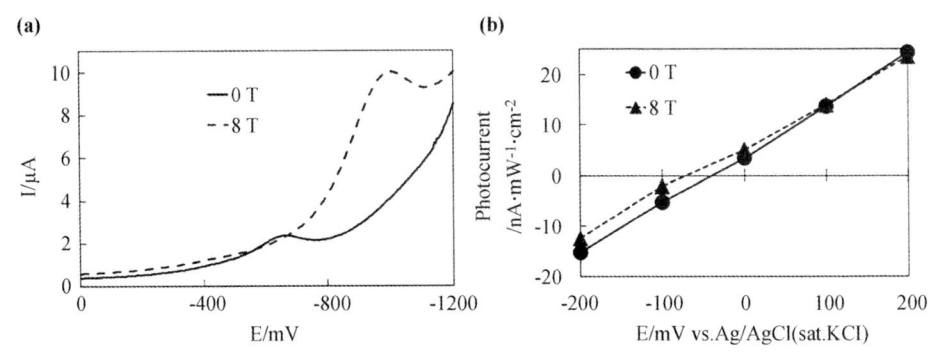

Figure 15.6 Effects of magnetic processing on (a) DPV curves (b) potential dependences of photocurrents of $C_{60}N^+$–MePH clusters on ITO electrodes.

processing. In both cases, photoirradiation of the modified electrodes afforded anodic photocurrents.

The photocurrent action spectrum in the absence of magnetic processing was in fair agreement with the absorption spectrum of the mixture of $C_{60}N^+$ and MePH in the THF:H_2O (2:1) mixed solvent. A similar photocurrent action spectrum was also observed in the presence of magnetic processing. These results indicate that the photocurrents are caused by the photoexcitation of $C_{60}N^+$ clusters in the presence and absence of magnetic processing.

The potential dependences of the photocurrents of the electrodes modified with $C_{60}N^+$–MePH nanoclusters in the presence of magnetic processing were also different from those in the absence of magnetic processing (Figure 15.6b). The magnetic field effects during AFM (Figure 15.4), and differences in electrochemical and photoelectrochemical measurements (Figure 15.5) can most likely be ascribed to the difference in the reduction potentials between the absence and the presence of magnetic processing due to the morphological change of the $C_{60}N^+$ nanoclusters.

The potential dependences of the photocurrents of the modified electrodes were examined in all the ITO electrodes modified with $C_{60}N^+$–MePH clusters in the absence and presence of magnetic processing. There were appreciable differences (Figure 15.6b). The potentials at zero photocurrent in the presence of magnetic processing were slightly negatively shifted compared with those in the absence of magnetic processing. The reaction scheme of the photoelectrochemical reactions of the ITO electrodes modified in the absence and presence of magnetic processing is shown in Figure 15.7. On the basis of the reaction scheme (Figure 15.7), the results can most likely be ascribed to the difference in the first reduction potentials of the $C_{60}N^+$ clusters in the presence and absence of magnetic processing and are in good agreement with those of the electrochemical properties.

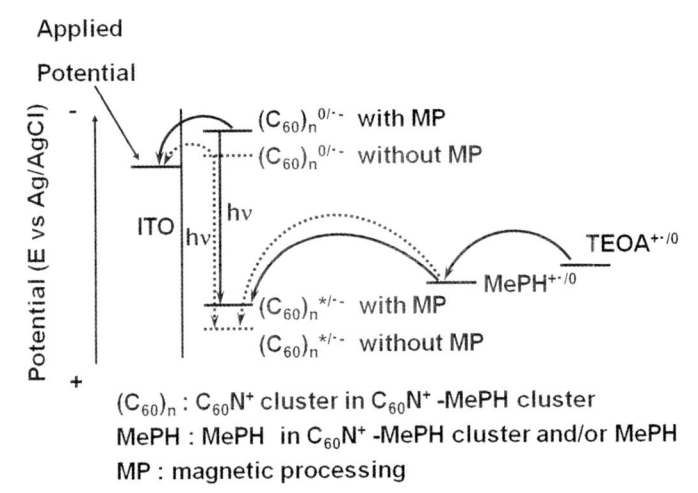

Figure 15.7 Schematic presentation of reaction scheme of photoelectrochemical reactions of ITO electrodes modified with $C_{60}N^+$–MePH clusters in the absence and presence of magnetic processing.

The MFEs on electrochemical and photoelectrochemical measurements can most likely be ascribed to the difference in the reduction potentials between the clusters in the absence and presence of magnetic processing using the morphological change of $C_{60}N^+$ nanoclusters.

We demonstrated that the morphology of nanostructures, electrochemical, and photoelectrochemical properties in the electrodes modified with nanoclusters of C_{60} can be controlled by applying a strong magnetic field. The present study provides useful information for designing novel nanodevices whose photofunctions can be controlled by a magnetic field.

15.2.3
Effects of Magnetic Processing on the Luminescence Properties of Monolayer Films with Mn^{2+}-Doped ZnS Nanoparticles

Semiconductor nanoparticles have been intensively studied because of their properties of quantum size effects [54]. A number of synthetic techniques have been reported and their characteristics have been studied by various spectroscopic methods [55, 56]. However, magnetic field effects (MFEs) on the photoelectrochemical properties of semiconductor nanocrystals had not until now been reported.

We reported, for the first time, MFEs on photocurrents of modified electrodes with cadmium sulfide nanoparticles (Q-CdS) ascribing them to their quantum size effects [29]. However, the MFEs were smaller than those on the modified electrodes with the porphyrin-viologen linked compounds as reported in previous papers [27, 28]. In a diluted magnetic semiconductor and its nanoparticles, a variety of unusual magnetic and magneto-optical properties due to exchange interaction between the band electrons and the magnetic ions have been reported [57, 58]. We found that the MFEs on the photocurrent responses from modified electrodes with diluted magnetic semiconductor nanoparticles (Q-$Cd_{1-x}Mn_xS$) were substantially enhanced by the presence of Mn^{2+} ions [30]. The MFEs on photocurrents observed in the modified electrodes with semiconductor nanoparticles (Q-CdS and Q-$Cd_{1-x}Mn_xS$) are most likely explained by an electron–hole pair mechanism.

We prepared monolayered films with Q-$Zn_{1-x}Mn_xS$ and examined the effects of magnetic processing during their preparation on their luminescence properties [59].

Q-$Cd_{1-x}Mn_xS$ was prepared using the bis(2-ethylhexyl) sulfosuccinate (AOT) reversed micelle method. The size of the nanoparticles was controlled by changing the $W(=[H_2O]/[AOT])$ values (2.5, 3.0, 5.0, 7.5, 10). Luminescence peaks were observed at 583~589 nm in the Q-$Zn_{1-x}Mn_xS$ or the alkanethiol-capped Q-$Zn_{1-x}Mn_xS$, except for the Q-$Zn_{1-x}Mn_xS$ ($W=2.5$) samples. Luminescence at 583~589 nm was observed in the Q-$Cd_{1-x}Mn_xS$ samples and is ascribed to Mn^{2+} ion in the nanoparticles due to energy transfer from ZnS. The luminescence was enhanced by capping with alkanethiol.

Monolayer films with the alkanethiol-capped ZnS:Mn nanoparticles were fabricated on quartz substrates by the layer-by-layer method using a self-assembled monolayer of 1,6-hexanedithiol.

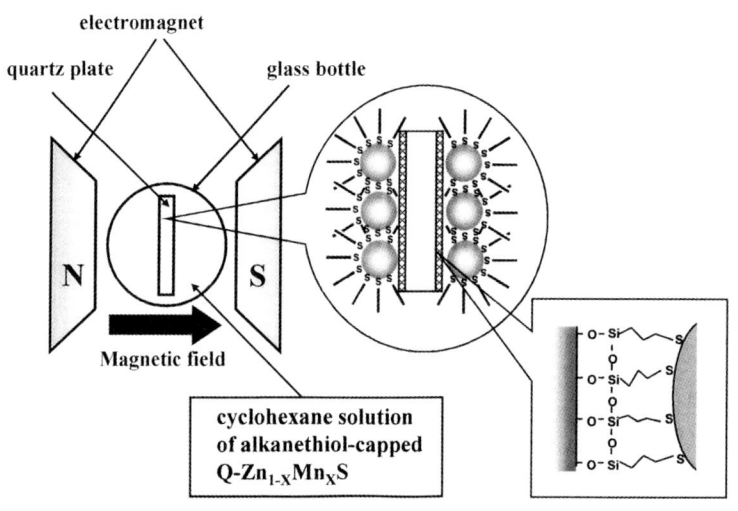

15.2 Construction of Nanostructures by the use of Magnetic Fields | 269

Figure 15.8 Experimental set-up for magnetic processing in the preparation of monolayer films consisting of Q-Zn$_{1-x}$Mn$_x$S.

Magnetic processing in the preparation of monolayer films was carried out as shown in Figure 15.8. Polarized luminescence spectra for monolayer films prepared in the absence and the presence of magnetic processing were measured without a magnetic field. The polarization degrees (p-values) of luminescence for the monolayer films consisting of Q-Zn$_{1-x}$Mn$_x$S ($W=5.0$, 10 and $x=0$, 0.05, 0.10) were compared in the absence and the presence of magnetic processing (0.2, 0.6, 0.8 T). The p-value of luminescence for a cyclohexane solution of alkanethiol-capped Q-Zn$_{1-x}$Mn$_x$S ($W=5.0$ and $x=0.10$) was estimated to be 0.007. The polarized luminescence spectra for the monolayer film consisting of Q-Zn$_{1-x}$Mn$_x$S ($W=5.0$ and $x=0.10$) in the absence of magnetic processing was recorded and the p-value was estimated to be 0.024. The increment in the p-value can probably be ascribed to the immobilization of the Q-Zn$_{1-x}$Mn$_x$S on the quartz plates.

The p-values for the monolayer films consisting of Q-Zn$_{1-x}$Mn$_x$S ($W=5.0$, 10 and $x=0.10$) increased with increasing magnetic field in the magnetic processing, as shown in Table 15.1. At the same magnetic field (0.8 T), the effect of x-values on the

Table 15.1 Effect of magnetic processing on polarization degrees (p-values) of luminescence for monolayer films consisting of Q-Zn$_{1-x}$Mn$_x$S ($x=0.10$).

	Magnetic Field (B)		
	0.2 T	0.6 T	0.8 T
$W=5$	0.028	0.050	0.075
$W=10$	0.029	0.060	0.076

Table 15.2 Effect of magnetic processing on polarization degrees (p-values) of luminescence for monolayer films consisting of Q-Zn$_{1-x}$Mn$_x$S ($x = 0.10$).

	x-value		
	0.02	0.05	0.10
$W = 5$	0.028	0.050	0.079

p-values for the monolayer films consisting of Q-Zn$_{1-x}$Mn$_x$S ($W = 5.0$) was examined (Table 15.2), the p-values increased with increasing x-value.

As a reference, magnetic processing in the preparation of monolayer films consisting of Q-ZnS without Mn^{2+} ion was also carried out. The luminescence at $\sim$400 nm was observed in the monolayer films, however, no magnetic field effects on the p-values for the monolayer films were observed.

On the basis of the observations, the enhancements of the p-values for the Q-Zn$_{1-x}$Mn$_x$S monolayer films can most likely be ascribed to the interaction between Mn^{2+} ion as a magnetic ion and the external magnetic field. The enhancements are probably caused by magnetic orientation of the Q-Zn$_{1-x}$Mn$_x$S on the quartz substrates.

15.3
Spin Chemistry at Solid/Liquid Interfaces

15.3.1
Magnetic Field Effects on the Dynamics of the Radical Pair in a C$_{60}$ Clusters–Phenothiazine System

Recently, photochemical and photoelectrochemical properties of fullerene (C$_{60}$) have been widely studied [60]. Photoinduced electron-transfer reactions of donor–C$_{60}$ linked molecules have also been reported [61–63] In a series of donor–C$_{60}$ linked systems, some of the compounds show novel properties, which accelerate photoinduced charge separation and decelerate charge recombination [61, 62]. These properties have been explained by the remarkably small reorganization energy in their electron-transfer reactions. The porphyrin–C$_{60}$ linked compounds, where the porphyrin moieties act as both donors and sensitizers, have been extensively studied [61, 62].

It has been reported that C$_{60}$ and its derivatives form optically transparent microscopic clusters in mixed solvents [25, 26]. Photoinduced electron-transfer and photoelectrochemical reactions using the C$_{60}$ clusters have been extensively reported because of the interesting properties of C$_{60}$ clusters [25, 26]. The MFEs on the decay of the radical pair between a C$_{60}$ cluster anion and a pyrene cation have been observed in a micellar system [63]. However, the MFEs on the photoinduced electron-transfer reactions using the C$_{60}$ cluster in mixed solvents have not yet been studied.

15.3 Spin Chemistry at Solid/Liquid Interfaces

A fullerene derivative containing a positive charge ($C_{60}N^+$) (Figure 15.4) forms optically transparent clusters (diameter about 100 nm) in THF/H_2O mixed solvent as seen from the results of the absorption spectra and AFM images, and dynamic light scattering as described above (Section 15.2.2). We examined photoinduced electron-transfer reaction and MFEs on the dynamics of the radical pair generated by the intermolecular electron transfer reaction between the $C_{60}N^+$ cluster $(C_{60}N^+)_n$ and MePH (Figure 15.4).

Transient absorption spectra of the $C_{60}N^+$ cluster $(C_{60}N^+)_n$–MePH system following laser excitation at 355 nm indicate that the photoinduced intermolecular electron-transfer from the triplet excited state of PH to the $C_{60}N^+$ cluster $(C_{60}N^+)_n$ occurs as shown in Figure 15.9a.

MFEs on the dynamics of the radical pair in $C_{60}N^+$ clusters $(C_{60}N^+)_n$–MePH system were examined in THF–H_2O (2:1) mixed solvent. MFEs on the decay profiles of the transient absorption at 520 nm due to the phenothiazine cation radical ($PH\cdot^+$) are shown in Figure 15.9b. The decay was retarded in the presence of the magnetic field. In addition, the absorbance at 10 μs after laser excitation increased with increasing magnetic field. The result indicated that the yield of the escaped $PH^{\bullet+}$ increased with the increase in magnetic field. Therefore, the MFEs on the decay profile were clearly observed.

The magnitudes of the MFEs were evaluated by the following equation; $\Delta = Abs(BT)/Abs(0T)$, where Abs(BT) and Abs(0T) represent the absorbance at 10 μs at 520 nm in the presence (BT) and the absence of magnetic field. The Δ-values increased with increasing magnetic field. The Δ-value became 2.7 at 1.2 T as shown in Figure 15.9b.

The MFEs on the $C_{60}N^+$ cluster–MePH are explained by a radical pair mechanism between the $C_{60}N^+$ cluster, $(C_{60}N^+)_n$ and MePH as shown in Figure 15.9a. MePH is mainly photoexcited by the 355 nm laser light, because of the excess amount of MePH in comparison with $C_{60}N^+$. The singlet excited state of PH ($^1PH^*$) is generated by the

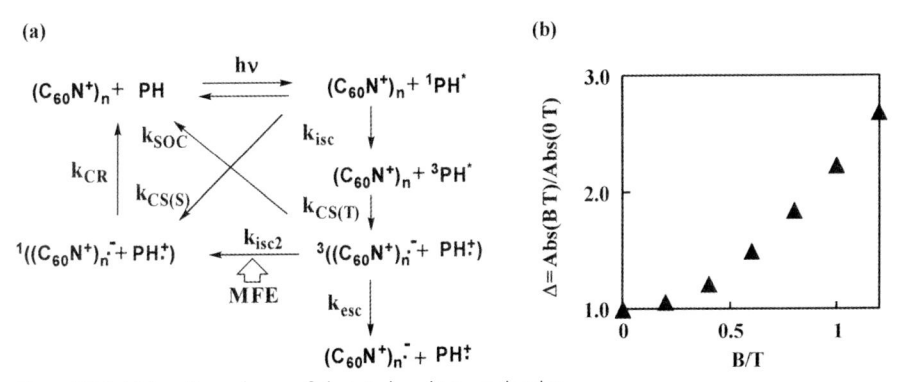

Figure 15.9 (a) Reaction scheme of photoinduced intermolecular electron-transfer in $C_{60}N^+$ cluster- MePH system. (b) MFEs on $\Delta = Abs(BT)/Abs(0T)$ in $C_{60}N^+$–MePH in THF–H_2O (2:1) mixed solvent.

laser excitation. The intersystem-crossing process (k_{isc}) occurs and ^{3}PH* is generated. The intermolecular electron-transfer process ($k_{CS(T)}$) from ^{3}PH* to the $C_{60}N^+$ cluster $(C_{60}N^+)_n$, occurs and generates the triplet radical pair, $^3((C_{60}N^+)_n^{\bullet -} + PH^{\bullet +})$. The triplet radical pair disappears partly via a spin–orbit coupling (SOC)-induced intersystem-crossing process (k_{SOC}) to the ground state. The process (k_{SOC}) is independent of the magnetic field. The triplet radical pair decays to the ground state via the singlet radical pair, $^1((C_{60}N^+)_n^{\bullet -} + PH^{\bullet +})$. The intersystem-crossing process (k_{isc2}) for the radical pair, but not the reverse electron-transfer process (k_{CR}) from the singlet radical pair to the ground state, becomes a rate-determining step for the radical pair to decay via a reverse electron-transfer reaction. The intersystem-crossing process (k_{isc2}) is influenced by the magnetic field. In the absence of magnetic field, the three sublevels of triplet radical pair are degenerate. In the presence of a magnetic field, Zeeman splitting of the triplet sublevels occurs. As a consequence of Zeeman splitting, the intersystem crossing process (k_{isc2}) is retarded in the presence of a magnetic field. As a result, the escape process (k_{esc}) from the triplet radical pair increases with increasing magnetic field and the yield of the escaped PH$^{\bullet +}$ increases with increasing magnetic field. In the presence of a magnetic field, the intersystem-crossing process (k_{isc2}) is controlled by the spin–lattice relaxation from triplet sublevels. Thus the MFEs are explained in terms of a spin–lattice relaxation mechanism [20–23].

15.3.2
Magnetic Field Effects on Photoelectrochemical Reactions of Electrodes Modified with the C_{60} Nanocluster-Phenothiazine System

As a study of spin chemistry at solid/liquid interfaces, we have examined MFEs on the photoelectrochemical reactions of photosensitive electrodes modified with nanoclusters containing $C_{60}N^+$ and MePH (Figure 15.4), intended for utilization of C_{60} as photofunctional nanodevices.

The nanoclusters of the mixture of $C_{60}N^+$ and MePH (Figure 15.4) were prepared by dissolving $C_{60}N^+$ and MePH in the THF–H$_2$O mixed solvent prepared using a different method to that described above (Section 15.2.2) [49]. Similar spherical nanoclusters (diameter about 100 nm) were also observed, similar to Figure 15.5a. Self-assembled monolayers of HS(CH$_2$)$_2$SO$_3^-$Na$^+$ (MPS) were prepared by immersing a gold electrode in an ethanol solution of MPS. Modified electrodes with clusters of $C_{60}N^+$ alone and of the mixture of $C_{60}N^+$ and MePH were fabricated by immersing the MPS-modified electrode in the THF–H$_2$O mixed solution containing the respective nanoclusters. The AFM image (Figure 15.10a) indicated that the modified electrodes with nanoclusters of $C_{60}N^+$ and MePH were fabricated by immersing the MPS-modified electrode in the THF–H$_2$O mixed solution containing the nanoclusters.

Photoelectrochemical measurements were carried out by using a three-electrode cell containing the modified electrode as a working electrode, a platinum electrode as a counter electrode, and an Ag/AgCl electrode as a reference electrode. Na$_2$SO$_4$ was used as the supporting electrolyte. Photocurrents from the modified electrode were

15.3 Spin Chemistry at Solid/Liquid Interfaces

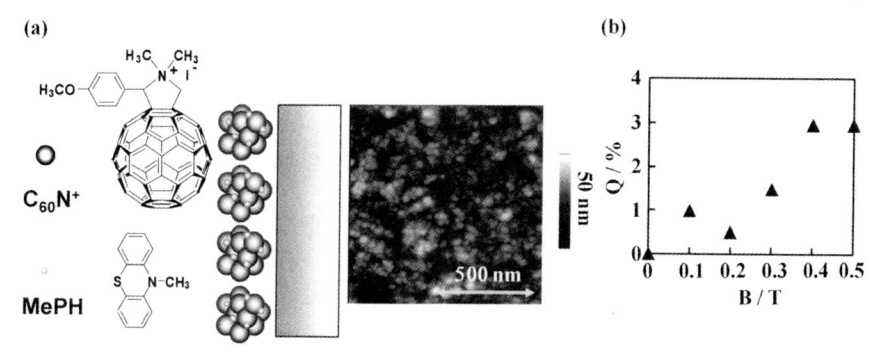

Figure 15.10 (a) AFM image and schematic illustration of electrode modified with nanocluster of $C_{60}N^+$ and MePH. (b) Magnetic field dependence on Q-values.

measured under a controlled potential at 0 V vs Ag/AgCl with a potentiostat in the presence of triethanolamine as a sacrificial electron donor under nitrogen atmosphere. The MFEs on the photoelectrochemical measurements were carried out according to the previous paper [29, 30]. The electrode cell was placed at the pole-gap of an electromagnet.

Photoirradiation of the modified electrode with nanoclusters of $C_{60}N^+$ alone or the mixture of $C_{60}N^+$ and MePH afforded anodic photocurrents. The photocurrent action spectrum was in fair agreement with the absorption spectrum of the THF–H_2O (2:1) mixed solution containing nanoclusters of the mixture of $C_{60}N^+$ and MePH or $C_{60}N^+$ alone. These results strongly indicate that the photocurrents can be ascribed to photoexcitation of the nanoclusters of $C_{60}N^+$.

MFEs on the photocurrents of the modified electrode with nanoclusters of the mixture of $C_{60}N^+$ and MePH were examined to verify the photocurrent generation mechanism. In the presence of a magnetic field, the photocurrents clearly increased.

The magnitude of the MFEs on the photocurrent is expressed as follow:

$$Q = (I(B) - I(0))/I(0) \times 100 \qquad (15.1)$$

where $I(B)$ and $I(0)$ are the photocurrent in the presence and absence of magnetic fields, respectively. The Q-value (%) increased gradually with increasing magnetic field up to 0.5 T (about 3% at 0.5 T), as shown in Figure 15.10b.

Though the photoexcited species of the $C_{60}N^+$ cluster, $(C_{60}N^+)_n$ in the present study is different from that of MePH, as described in Section 15.3.1, the formation of a triplet radical pair occurred in both cases. Therefore, the present MFEs observed in the photocurrents (Figure 15.10b) are expected to be explained by a similar mechanism to that shown in Figure 15.9a In fact, the magnetic field dependence on the yield of the escaped $PH^{\bullet+}$ (Δ-value) (Figure 15.9b) was in good agreement with that on the Q-value in Figure 15.10b. Accordingly, the MFEs on the photocurrents can be explained by a spin–lattice relaxation mechanism in a radical pair mechanism [20–23].

15.4
Summary

We first demonstrated that the morphology of nanostructures, electrochemical, photoelectrochemical, or luminescence properties in the functional materials consisting of SWNTs, their composites, diluted magnetic semiconductor ($Zn_{1-x}Mn_xS$) nanoparticles, and nanoclusters of C_{60} can be controlled by magnetic processing, including the use of a strong magnetic field. The magnetic processing in the present studies provides useful information for designing novel nanodevices whose photofunctions can be controlled by a magnetic field.

The MFEs on the photoelectrochemical reactions of photosensitive electrodes modified with nanoclusters containing $C_{60}N^+$ and MePH were examined as a study of spin chemistry at solid/liquid interfaces. The results can be expected to lead to an epochmaking means of reaction control involving photoelectrochemical processes. The results also provide useful information for designing novel nanodevices whose photofunctions can be controlled by a magnetic field.

Acknowledgments

The author is grateful to Mr. H. Horiuchi for the preparation of photoelectrochemical cells, Professor S. Yamada for discussion, and Professor Y. Tanimoto and Associate Professor Y. Fujiwara for using the superconducting magnets. The author also thanks The Center of Advanced Instrumental Analysis, Kyushu University, for ^{1}H-NMR measurements. This study was financially supported by the Grant-in-Aids for Scientific Research: Priority Areas (Area 767, No. 15085203) and (Area 432, No. 17034051), Scientific Research (C) (No. 17550131), and twenty first century COE Program "Function Innovation of Molecular Informatics" from MEXT of the Japan.

References

1 Yamaguchi, M (2006) *Magneto-Science Magnetic Field Effects on Materials: Fundamentals and Applications* (eds M. Yamaguchi and Y. Tanimoto), Kodansha-Springer, Chapter 1.1.
2 Fujiwara, M., Fukui, M. and Tanimoto, Y. (1999) Magnetic orientation of benzophenone crystals in fields up to 80.0 kOe. *J. Phys. Chem. B*, **102**, 2627–2630.
3 Fujiwara, M., Tokunaga, R. and Tanimoto, Y. (1998) Crystal growth of potassium nitrate in a magnetic field of 80 kOe. *J. Phys. Chem. B*, **102**, 5996–5998.
4 Fujiwara, M., Chidiwa, T., Tokunaga, R. and Tanimoto, Y. (1998) Crystal growth of trans-azobenzene in a magnetic field of 80 kOe. *J. Phys. Chem. B*, **108**, 3417–3419.
5 Kimira, T., Yamato, M., Koshimizu, W., Koike, M. and Kawai, T. (2000) Magnetic orientation of polymer fibers in suspension. *Langmuir*, **16**, 858–861.
6 Kimira, T., Yoshino, M., Yamane, T., Yamato, M. and Tobita, M. (2004) Uniaxial alignment of the smallest diamagnetic susceptibility axis using time-dependent magnetic fields. *Langmuir*, **20**, 5669–5672.

7 Torbet, J., Freyssinet, J.-M. and Hudry-Clergeon, G. (1981) Oriented fibrin gels formed by polymerization in strong magnetic fields. *Nature*, **289**, 91–93.

8 Yamagishi, A., Takeuchi, T., Higashi, T. and Date, M. (1989) Diamagnetic orientation of polymerized molecules under high magnetic field. *J. Phys. Soc. Jpn.*, **58**, 2280–2283.

9 Fujiwara, M., Oki, E., Hamada, M., Tanimoto, Y., Mukouda, I. and Shimomura, Y. (2001) Magnetic orientation and magnetic properties of a single carbon nanotube. *J. Phys. Chem. A*, **105**, 4383–4386.

10 Fujiwara, M., Kawakami, K. and Tanimoto, Y. (2002) Magnetic orientation of carbon nanotubes at temperatures of 231 K and 314 K. *Mol. Phys.*, **100**, 1085–1088.

11 Mogi, I., Okubo, S. and Nakagawa, Y. (1995) Dense radial growth of silver metal leavesin a high magnetic field. *J. Phys. Soc. Jpn.*, **60**, 3200–3202.

12 Mogi, I., Kamiko, M. and Okubo, S. (1995) Magnetic field effects on fractal morphology in electrochemical deposition. *Physica B*, **211**, 319–322.

13 Katsuki, A., Uechi, I. and Tanimoto, Y. (2004) Effects of a high magnetic field on the growth of 3-dimensional silver dendrites. *Bull. Chem. Soc. Jpn.*, **77**, 275–279.

14 Tanimoto, Y., Katsuki, A., Yano, H. and Watanabe, S. (1997) Effect of high magnetic field on the silver deposition from its aqueous solution. *J. Phys. Chem. A*, **101**, 7357363.

15 Mogi, I. (1996) Magnetic field effects on the dopant-exchange process in polypyrrole. *Bull. Chem. Soc. Jpn.*, **69**, 2661–2666.

16 Mogi, I., Watanabe, K. and Motooka, M. (1999) Effects of magnetoelectropolymerization on doping-undoping behavior of polypyrrole. *Electrochemistry*, **67**, 1051–1053.

17 Konno, A., Mogi, I. and Watanabe, K. (2001) Effect of strong magnetic fields on the photocurrent of a poly(N-methylpyrrole) modified electrode. *J. Electroanal. Chem.*, **507**, 202–205.

18 Uechi, I., Katsuki, A., Dunin-Barkovskiy, L. and Tanimoto, Y. (2004) 3D-morphological chirality induction in zinc silicate membrane tube using a high magnetic field. *J. Phys. Chem. B*, **108**, 2527–2530.

19 Duan, W., Kitamura, S., Uechi, I., Katsuki, A. and Tanimoto, Y. (2005) Three-dimensional morphological chirality induction using high magnetic fields in membrane tubes prepared by a silicate garden reaction. *J. Phys. Chem. B*, **109**, 13445–13450.

20 Steiner, U. E. and Ulrich, T. (1989) Magnetic field effects in chemical kinetics and related phenomena. *Chem. Rev.*, **89**, 51–147.

21 Nakagaki, R., Tanimoto, Y. and Mutai, K. (1993) Magnetic field effects upon photochemistry of bifuctional chain molecules. *J. Phys. Org. Chem.*, **6**, 381–392.

22 Tanimoto, Y. (1998) *Dynamic Spin Chemistry* (eds S. Nagakura H. Hayashi and T. Azumi), Kodansha-Wiley, Tokyo/New York, Chapter 3.

23 Tanimoto, Y. and Fujiwara, Y. (2003) Handbook of photochemistry and photobiology, in *Inorganic Photochemistry*, vol. **1** (ed. H. S. Nalwa), American Scientific Publishers, Chapter 10.

24 Yonemura, H., Noda, M., Hayashi, K., Tokudome, H., Moribe, S. and Yamada, S. (2002) Photoinduced intramolecular electron transfer reactions in fullerene-phenothiazine linked compounds: effects of magnetic field and spacer chain length. *Mol. Phys.*, **100**, 1395–1403.

25 Kamat, P. V., Barazzouk, S., Hotchandani, S. and Thomas, K. G. (2000) Nanostructured thin films of C_{60} – aniline dyad clusters: electrodeposition, charge separation, and photoelectrochemistry. *Chem. Eur. J.*, **6**, 3914–3921.

26 Biji, V., Barazzouk, S., Thomas, K. G., George, M. V. and Kamat, P. V. (2001) Photoinduced electron transfer between 1,2,5-triphenylpyrrolidinofullerene cluster aggregates and electron donors. *Langmuir*, **17**, 2930–2936.

27 Yonemura, H., Ohishi, K. and Matsuo, T. (1996) Magnetic field effects on photoelectrochemical responses of modified electrodes with porphyrin-viologen linked compound as a Langmuir–Blodgett film. *Chem. Lett.*, 661–662.

28 Yonemura, H., Ohishi, K. and Matsuo, T. (1997) Magnetic field effects on photoelectrochemical responses of modified electrodes with Langmuir-Blodgett monolayer containing porphyrin-viologen linked compounds. *Mol. Cryst. Liq. Cryst.*, **294**, 221–224.

29 Yonemura, H., Yoshida, M., Mitake, S. and Yamada, S. (1999) Magnetic field effects on photocurrent responses from modified electrodes with CdS nanoparticles. *Electrochemistry*, **67**, 1209–1210.

30 Yonemura, H., Yoshida, M. and Yamada, S. (2001) Reduction of photocurrents from modified electrodes with $Cd_{1-x}Mn_xS$ nanoparticles in the presence of magnetic fields. *Stud. Surf. Sci. Catal.*, **132**, 741–744.

31 O'Connell, M. J., Boul, P., Ericson, L. M., Huffman, C., Wang, Y., Haroz, E., Kuper, C., Tour, J., Ausman, K. D. and Smalley, R. E. (2001) Reversible water-solubilization of single-walled carbon nanotubes by polymer wrapping. *Chem. Phys. Lett.*, **342**, 265–271.

32 Ericson, L. M., Fan, H., Peng, H., Davis, V. A., Zhou, W., Sulpizio, J., Wang, Y., Booker, R., Vavro, J., Guthy, C., Nicholas, A., Parra-Vasquez, G., Kim, M. J., Ramesh, S., Saini, R. K., Kittrell, C., Lavin, G., Schmidt, H., Adams, W. W., Billups, W. E., Pasquali, M., Hwang, W.-F., Hauge, R. H., Fischer, J. E. and Smalley, R. E. (2004) Macroscopic, neat, single-walled carbon nanotube fibers. *Science*, **305**, 1447–1450.

33 Vigolo, B., Pénicaud, A., Coulon, C., Sauder, C., Pailler, R., Journet, C., Bernier, P. and Poulin, P. (2000) Macroscopic fibers and ribbons of oriented carbon nanotubes. *Science*, **290**, 1331–1334.

34 Kamat, P. V., Thomas, K. G., Barazzouk, S., Girishkumar, G., Vinodgopal, K. and Meisel, D. (2004) Self-assembled linear bundles of single wall carbon nanotubes and their alignment and deposition as a film in a dc field. *J. Am. Chem. Soc.*, **126**, 10757–10762.

35 Takahasi, T., Murayama, T., Higuchi, A., Awano, H. and Yonetake, K. (2006) Aligning vapor-grown carbon fibers in polydimethylsiloxane using dc electric or magnetic field. *Carbon*, **44**, 1180–1188.

36 Dierking, I., Scalia, G. and Morales, P. (2005) Liquid crystal-carbon nanotube dispersions. *J. Appl. Phys.*, **97**, 044309-1–044309-5.

37 Fischer, J. E., Zhou, W., Vavro, J., Llaguno, M. C., Guthy, C., Haggenmueller, R., Casavant, M. J., Walters, D. E. and Smalley, R. E. (2003) Magnetically aligned single wall carbon nanotube films: Preferred orientation and anisotropic transport properties. *J. Appl. Phys.*, **93**, 2157–2163.

38 Zaric, S., Ostojic, G. N., Kono, J., Shaver, J., Moore, V. C., Hauge, R. H., Smalley, R. E. and Wei, X. (2004) Estimation of magnetic susceptibility anisotropy of carbon nanotubes using magnetophotoluminescence. *Nano Lett.*, **4**, 2219–2221.

39 Guldi, D. M., Rahman, G. M. A., Prato, M., Jux, N., Qin, S. and Ford, W. (2005) Single-wall carbon nanotubes as integrative building blocks for solar-energy conversion. *Angew. Chem. Int. Ed.*, **44**, 2015–2018.

40 Rahman, G. M. A., Guldi, D. M., Cagnoli, R., Mucci, A., Schenetti, L., Vaccari, L. and Prato, M. (2005) Combining single wall carbon nanotubes and photoactive polymers for photoconversion. *J. Am. Chem. Soc.*, **127**, 10051–10057.

41 Robel, I., Bunker, B. A. and Kamat, P. V. (2005) Single-walled carbon nanotube-CdS nanocomposites as light-harvesting assemblies: Photoinduced charge-transfer interactions. *Adv. Mater.*, **17**, 2458–22463.

42 Guldi, D. M., Rahman, G. M. A., Ramey, J., Marcaccio, M., Paolucci, D., Paolucci, F., Qin, S., Ford, W. T., Balbinot, D., Jux, N., Tagmatarchis, N. and Prato, M. (2004) Donor-acceptor nanoensembles of soluble

carbon nanotubes. *Chem. Commun.*, **10**, 2034–2035.

43 Guldi, D. M., Rahman, G. M. A., Jux, N., Tagmatarchis, N. and Prato, M. (2004) Integrating single-wall carbon nanotubes into donor-acceptor nanohybrids. *Angew. Chem. Int. Ed.*, **43**, 5526–5530.

44 Paloniemi, H., Lukkarinen, M., Ääritalo, T., Areva, S., Leiro, J., Heinonen, M., Haapakka, K. and Lukkari, J. (2006) Layer-by-layer electrostatic self-assembly of single-wall carbon nanotube polyelectrolytes. *Langmuir*, **22**, 74–83.

45 Yonemura, H., Yamamoto, Y. and Yamada, S. (2008) Photoelectrochemical reactions of electrodes modified with composites between conjugated polymer or ruthenium complex and single-walled carbon nanotube. *Thin Solid Films*, **516**, 2620–2625.

46 Yonemura, H. (2006) High magnetic effects on nanostructures and photoproperties. *Kagaku to Kyoiku*, **54** (1), 20–23.

47 Yonemura, H., Yamamoto, Y., Yamada, S., Fujiwara, Y. and Tanimoto, Y. (2008) Magnetic orientation of single-walled carbon nanotubes or their composites using polymer wrapping. *Sci. Technol. Adv. Mater.*, **9** (024213), 1–6.

48 Yonemura, H., Yamamoto, Y., Yamada, S., Fujiwara, Y. and Tanimoto, Y., manuscript in preparation.

49 Yonemura, H., Wakita, Y., Kuroda, N., Yamada, S., Fujiwara, Y. and Tanimoto, Y. (2008) Effects of magnetic processing on electrochemical and photoelectrochemical properties of electrodes modified with C_{60}-phenothiazine nanoclusters. *Jpn. J. Appl. Phys.*, **47**, 1178–1183.

50 Sun, Y.-P. and Bunker, C. E. (1993) C_{70} in solvent mixtures [7]. *Nature*, **365**, 398–398.

51 Yonemura, H., Kuroda, N., Moribe, S. and Yamada, S. (2006) Photoindiced electron-transfer and magnetic field effects on the dynamics of the radical pair in a C_{60} clusters–phenothiazine system. *C. R. Chim.*, **9**, 254–260.

52 Yonemura, H., Kuroda, N. and Yamada, S. (2006) Magnetic field effects on photoelectrochemical reactions of modified electrodes with C_{60}-phenothiazine nanoclusters. *Sci. Technol. Adv. Mater*, **7**, 643–648.

53 Nakanishi, T., Ohwaki, H., Tanaka, H., Murakami, H., Sagara, T. and Nakashima, N. (2004) Electrochemical and chemical reduction of fullerenes C_{60} and C_{70} embedded in cast films of artificial lipids in aqueous media. *J. Phys. Chem. B*, **108**, 7754–7762.

54 Ogawa, S., Fan, F.-R.F. and Bard, A. J. (1995) Scanning tunneling microscopy, tunneling spectroscopy, and photoelectrochemistry of a film of Q-CdS particles incorporated in a self-assembled monolayer on a gold surface. *J. Phys. Chem.*, **99**, 11182–11189.

55 Nakanishi, T., Ohtani, B. and Uosaki, K. (1998) Fabrication and characterization of CdS-nanoparticle mono- and multilayers on a self-assembled monolayer of alkanedithiols on gold. *J. Phys. Chem. B*, **102**, 1571–1577.

56 Steigerwald, M. L., Alivisatos, A. P., Gibson, J. M., Harris, T. D., Kortan, R., Muller, A. J., Thayer, A. M., Duncan, T. M., Dougalss, D. C. and Brus, L. E. (1988) Surface derivatization and isolation of semiconductor cluster molecules. *J. Am. Chem. Soc.*, **110**, 3046–3050.

57 Furdya, J. K. and Kossut, J. (1988) Diluted Magnetic Semiconductors, in *Semiconductors and Semimetals*, vol. **25**, Academic, New York.

58 Yanata, Y., Suzuki, K. and Oka, Y. (1993) Magneto-optical properties of CdMnSe microcrystallites in SiO glass prepared by rf sputtering. *J. Appl. Phys.*, **73**, 4595–4598.

59 Yonemura, H., Yanagita, M., Horiguchi, M., Nagamatsu, S. and Yamada, S. (2008) Characterization of mono- and multilayered films with Mn^{2+}-doped ZnS nanoparticles and luminescence properties of the monolayered films prepared by applying magnetic fields. *Thin Solid Films*, **516**, 2432–2437.

60 Guldi, D. M. and Prato, M. (2000) Excited-state properties of C_{60} fullerene derivatives. *Acc. Chem. Res.*, **33**, 695–703.

61 Imahori, H., Mori, Y. and Matano, Y. (2003) Nanostructured artificial photosynthesis. *J. Photochem. Photobiol. C*, **4**, 51–83.

62 Imahori, H. and Sakata, Y. (1999) Fullerenes as novel acceptors in photosynthetic electron transfer. *Eur. J. Org. Chem.*, **10**, 2445–2457.

63 Haldar, M., Misra, A., Banerjee, A. K. and Chowdhury, M. (1999) Magnetic field effect on the micellar (C)-pyrene radical-pair system. *J. Photochem. Photobiol. A*, **127**, 7–12.

16
Controlling Surface Wetting by Electrochemical Reactions of Monolayers and Applications for Droplet Manipulation

Ryo Yamada

16.1
Introduction

Reversible control of surface physicochemical properties has received much attention [1, 2] because it enables researchers to construct novel fluidic devices such as pumps [3], micro-shutters [4] and variable focus lenses [5]. Organic monolayers are of particular interest for controlling surface characteristics because surface structural changes and chemical reactions are designed with an accuracy of a single molecular thickness. In this chapter, I would like to give a brief review of self-assembled monolayers and describe our studies on the droplet manipulation realized by controlling the wetting distribution of the surface.

16.1.1
Self-Assembled Monolayers

When appropriate molecular–molecular and molecular–surface interactions are present, an ordered monolayer is formed spontaneously on surfaces (Figure 16.1). This process is called self-assembly (SA) and monolayers formed in this manner are called self-assembled monolayers (SAMs).

SAMs of alkanethiols on an Au(1 1 1) surface are widely used to control surface properties, electron transfer processes and to stabilize nano-clusters [6, 7]. SAMs are formed by chemical bond formation between S and Au when an Au(1 1 1) substrate is immersed in a solution containing several mM of alkanethiols for hours to days. Various functions have been realized by using SAMs of alkanethiols on Au substrates as listed in Table 16.1.

Figure 16.2a and b show schematic drawings of top and side views of the SAMs of alkanethiols on the Au(111) surface, respectively [28]. The basic molecular arrangement is $(\sqrt{3} \times \sqrt{3})R30°$ with respect to the Au(111) surface. Closer inspection of the structure revealed the existence of a c(4 × 2) superlattice of $(\sqrt{3} \times \sqrt{3})R30°$. The alkyl chain is tilted from the surface normal by about 30° with all-trans conformation. This

Molecular Nano Dynamics, Volume I: Spectroscopic Methods and Nanostructures
Edited by H. Fukumura, M. Irie, Y. Iwasawa, H. Masuhara, and K. Uosaki
Copyright © 2009 WILEY-VCH Verlag GmbH & Co. KGaA, Weinheim
ISBN: 978-3-527-32017-2

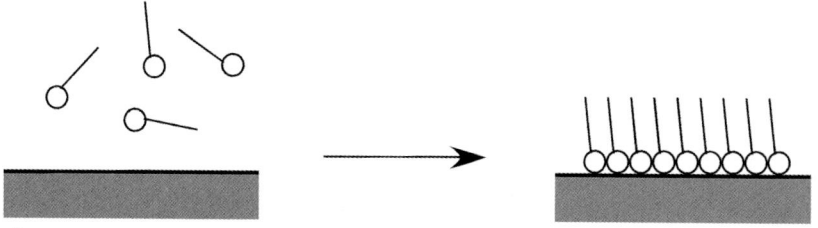

Figure 16.1 Spontaneous formation of a monolayer on a clean surface.

tilt angle comes from the conditions for close packing of alkyl chains. The plane defined by an all-trans carbon molecular skeleton alternatively changes its direction.

16.1.2
Preparation of Gradient Surfaces

A surface whose physicochemical properties gradually change as a function of the position is called a gradient surface. While the gradient surface has potential applications as a substrate for combinatorial studies [29–31], the gradient of surface wetting is of particular interest because spontaneous motions of a droplet are induced on the surface [32].

The gradient surface is easily prepared by using SA technique. For example, vapor phase deposition of the molecules is useful to change the spatial distribution of the coverage, as shown in Figure 16.3. In this method, a droplet containing molecules is put beside the substrate [32]. The vaporized molecules diffuse and are adsorbed on the substrate. The closer the position on the substrate to the droplet is, the faster the adsorption of the molecules is. As a result, a gradient of coverage is formed on the surface. SA in dilute solutions is also used to make a gradient [33]. When the substrate is slowly pulled up from the dilute solution of molecules, the bottom part of the substrate possesses higher coverage of molecules. Other

Table 16.1 Functions realized by SAMs and terminal functional groups.

Functions	Functional Groups
Redox	ferrocene [8], quinone [9], $Ru(NH_3)_6^{2+}$ [10]
Photochemistry	porphyrin [11], $Ru(bpy)_3^{2+}$ [12]
Catalysis	ferrocene [13], Ni-azamacrocyclic complex [14], metal-porphyrin [15]
Optical second harmonic generation	ferrocenyl-nitrophenyl ethylene [16]
Sensor	quinone [17], cyclodextrin [18], enzyme [19]
Structural transition	azobenzene [20], spiropyran [21]
Mediator	ferrocene [22], pyridine [23]
Wetting control	carboxyl group [6], hydroxyl group [6], sulfonic acid [6], methyl group [6]
Bonding	carboxyl group [24], amine [25], phosphonate [26], thiol [27]

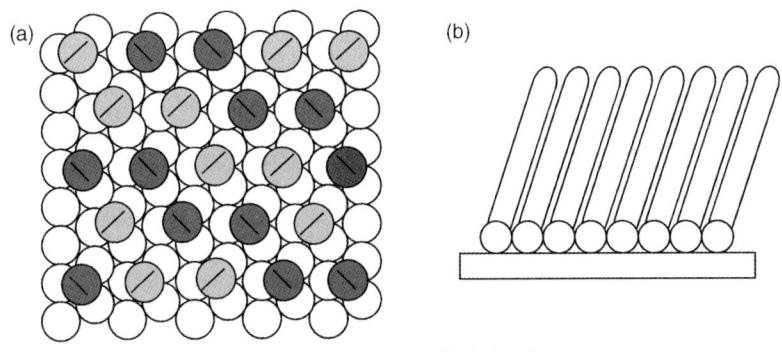

Figure 16.2 (a) Model of molecular arrangement (shaded circle) with respect to the Au(111) surface (open small circle). The diagonal slash indicates the azimuthal orientation of the plane defined by the C–C–C backbone of the all trans-hydrocarbon chain. (b) Side view of the molecules. Circles represent sulfur atoms.

methods using microfluidics [34] and partial decomposition of monolayers [35, 36] have been developed.

16.1.3
Spontaneous Motion of a Droplet on Wetting Gradients

Manipulation of a droplet on a solid surface is of growing interest because it is a key technology to construct lab-on-a-chip systems. The imbalance of surface tensions is known to cause spontaneous motion of a droplet on the surface, as mentioned above. The wetting gradient causing liquid motion has been prepared by chemical [32], thermal [37], electrochemical [3] and photochemical [38–40] methods.

Chaudhury and Whitesides prepared the gradient surface by vapor deposition of hexamethyldisilazane (HMDS) molecules on silicon oxide surface [32]. Because of the gradual change in the coverage of the molecules, gradients in the wetting by water were generated. When a water droplet was put on the gradient surface, the droplet moved spontaneously. The droplet moved on the gradient surface because of the difference in advancing and receding contact angles at the front and rear ends of the droplet [32] (Figure 16.4). Interestingly, the droplet can climb the slope without any external power source.

Abott et al. demonstrated that the flow in a microfluidic channel can be controlled by using the wetting gradient generated by electrochemical reactions of the surfactant

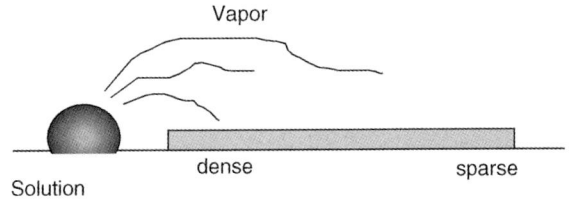

Figure 16.3 Vapor phase deposition of molecules and gradients of coverage of molecules formed on a surface.

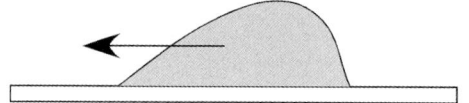

Figure 16.4 Spontaneous motion of a droplet on wetting gradients.

dissolved in the solution [3]. Two electrodes were placed in the micro-fluidic channel. The surfactant was oxidized at one electrode and reduced at the other electrode. As a result, the concentration of oxidized molecules gradually changed as a function of the distance between the two electrodes and a flow of liquid was generated. By using this method, they succeeded in moving the solid particles in micro-fluidic channels.

16.1.4
Surface Switching

Physicochemical properties of surfaces can be reversibly changed by using reversible reactions of monolayers and polymers [2]. This phenomenon is called surface switching and the surfaces possessing this characteristic are sometimes called smart surfaces. Surface switching using monolayers was proposed by Lahann et al. [41]. They prepared a loosely packed monolayer of 16-mercapto-hexadecanoic acid. The orientation of the molecule was altered by the surface charge controlled by an external voltage source. When the surface was negatively charged, the molecules were straightened due to the electrostatic repulsion. On the contrary, the molecules made a bow shape when the surface was positively charged. These conformational changes were confirmed by surface analysis techniques such as sum frequency generation spectroscopy. As a result of surface switching, the contact angle of the water droplet was reversibly changed as a function of the voltage applied to the surface.

Ichimura et al. used photoisomerization of the azobenzene molecule to create wetting gradients [38]. The silicon oxide surface was modified with a monolayer of calyx[4]resorcinarene derivative possessing azobenzene units. The azobenzene was in the trans-state under ambient conditions. When the surface was exposed to UV light (365 nm), photoisomerization took place and the cis-isomer, which has a high dipole moment, was generated on the surface. This isomerization resulted in a change in the surface wetting. The cis-isomer is converted to the trans-isomer by illumination with blue light (436 nm). The gradient of surface wetting was generated by asymmetrical illumination of light. The molecular shuttle was also used to generate a wetting gradient by photo-irradiation [40].

Electrochemical reactions of Fc-alkanethiol monolayers have been used to control the wetting of the gold surface [41–45]. We observed the change in surface wetting accompanying the electrochemical reactions by putting an oil (nitrobenzene) droplet on the surface, as shown in Figure 16.5. The surface covered with Fc-alkanethiol is hydrophobic and, thus, the oil was spread on the surface as shown in Figure 16.5a. When Fc was oxidized, the anions in the solution were trapped on the surface to compensate the positive charge of Fc^+ and determined the surface wetting. The surface became hydrophilic when the anions were hydrophilic, which

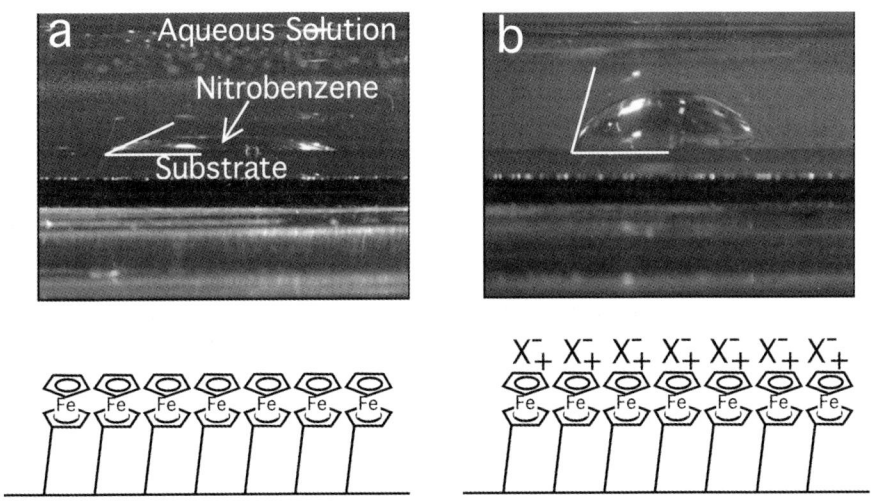

Figure 16.5 Change of contact angle of nitrobenezene droplet. The droplet was put on a gold substrate covered with Fc-alkanethiols in a 1 M aqueous solution of HClO$_4$. Fc was neutral (a) and positively charged (b).

is usual for aqueous solutions, and, thus, the oil droplet was repelled as shown in Figure 16.5b.

Figure 16.6a shows an experimental set-up to monitor the interfacial tension of a nitrobeneze/1 mM HClO$_4$ aqueous solution interface as a function of the potential of the electrodes by a hanging meniscus method. A gold plate modified with Fc-alkanethiol was first hung in the aqueous phase (the upper open area in Figure 16.6a). Then, the gold plate was lowered. When the gold plate reached the aqueous solution/nitrobenzene interface, the mass measured by the microbalance changed due to the interfacial force working on the plate. The change in measured mass, ΔW, is given by

$$\Delta W/p = \gamma \cos\theta$$

where p is the perimeter of the plate (4 cm), γ is the interfacial tension of the liquid/liquid interface, and θ is the contact angle [46]. Since γ was almost constant during the potential change, ΔW was attributed to the change in the contact angle θ.

Figure 16.6b shows the experimental results. Open and closed circles represent the values measured after the plate was moved downward and upward in Figure 16.6a, respectively, at each potential. In both processes, the contact angles increased as the potential was changed towards the positive direction. The observed behavior of the contact angles is attributed to the change in surface wetting of the gold-substrate due to the oxidation and reduction of Fc. The hysteresis of the contact angle during the upward and downward process was apparent from the measurement. Note that inserting and pulling steps are related to the advancing and receding processes of droplet motion on the surface, respectively.

The hysteresis in contact angles caused the hysteresis of the deformation of the droplet put on the substrate during the potential cycle. When the potential was

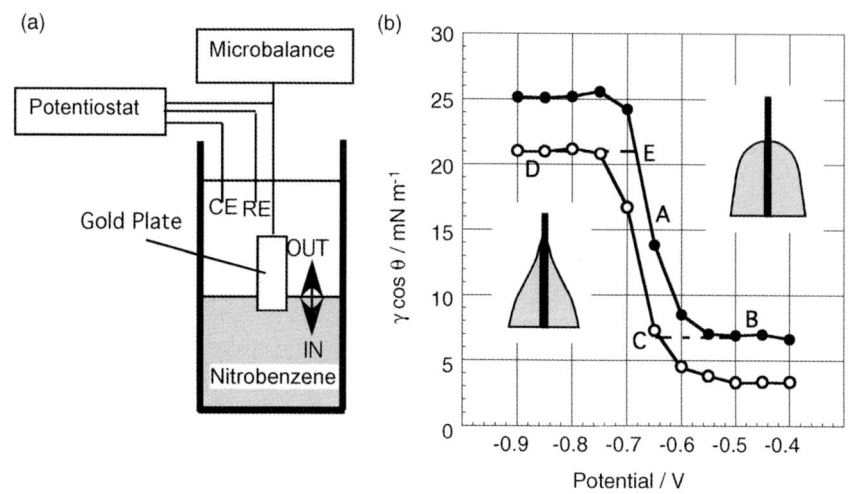

Figure 16.6 (a) Schematic drawing of experimental set-up for the evaluation of the interfacial tension under potential control. (b) Relative change in contact angle as a function of the potential after the substrate was inserted into (open circles) and pulled from the nitrobenzene phase. Insets are schematic drawings of the side views of the contact lines. The potential was described with respect to the Au/AuO$_x$ reference electrode.

scanned in a positive direction, the nitrobenzene droplet shrank on the substrate since the surface became repulsive to nitrobenzene. The contact line of the droplet receded in this process and, therefore, the change in the contact angle followed the curve A–B in Figure 16.6b. At the point B, when the potential was scanned in the negative direction, the droplet kept its shape until the potential reached −0.66 V (point C). At point D, the droplet spread. When the direction of the potential scan was changed from negative to positive, the droplet did not deform until the potential reached about −0.68 V (point E).

16.2
Ratchet Motion of a Droplet

16.2.1
Ratchet Motion of a Droplet on Asymmetric Electrodes

When a droplet is deformed asymmetrically, the ratchet motions of the droplet can be induced as demonstrated on the vibrated gradient surface and on a saw-shaped electrode on which the wetting was changed by electrowetting [48].

We showed that the ratchet motions of a droplet were induced by very simple asymmetric guides [45]. Figure 16.7 shows a schematic drawing of the electrode causing the ratchet motions of a droplet.

(a)

```
1mm
 ↓      Glass
        Gold electrode      4mm
 ↑      Glass
        ← 25mm →
```

(b)

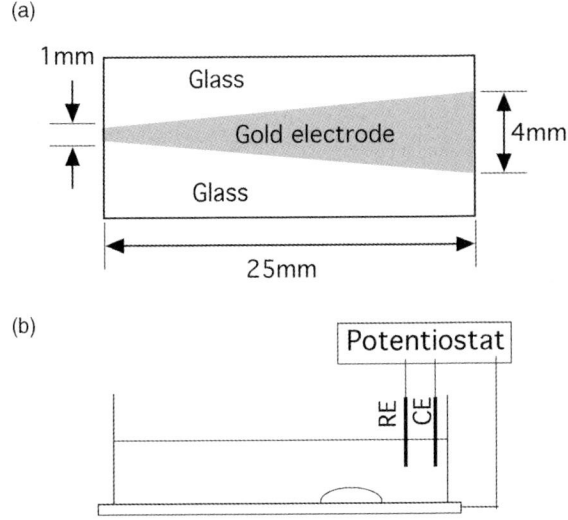

Potentiostat
RE | CE

Figure 16.7 Schematic drawing of the asymmetric electrode pattern. The gold electrode was covered with a Fc-alkanethiol monolayer. The wetting of the gold electrode was switched from wetting to repulsive and vice versa by changing the electrochemical potential of the electrode.

The surface of the V-shaped electrode was covered with Fc-alkanethiol monolayers. The droplet of nitrobenzene having a diameter larger that the width of the electrode was put on the electrode. The droplet of nitrobenzene was deformed following the electrode structure because it did not spill from the area of the electrode into the surrounding glass surface, which was hydrophilic. The contact line at the left side of the droplet, denoted A in Figure 16.8a, was longer than that at the right side, denoted B, on the V-shaped electrode. The difference in the length of the contact lines generated the imbalance in net forces acting on the droplet in the deformation process. When the surface wetting was changed from repulsive to wetting, only the contact line facing the wider side (A in Figure 16.8b) proceeded because of the stronger net-force generated for the longer contact line of the droplet. On the contrary, only the narrow contact line (denoted B in Figure 16.8c) retracted, as shown in Figure 16.8c, when the surface was changed from wetting to repulsive. The sequential spreading and shrinking of the droplet resulted in its net transport.

The presented results show that the simple asymmetric pattern caused directional deformations and transport of a droplet. This technique is applicable to generation of a flow in microfluidic devices.

16.2.2
Ratchet Motion of a Droplet Caused by Dynamic Motions of the Wetting Boundary

The ratchet motion of a droplet was induced by electrochemical control of the wetting distributions on the surface [44]. Figure 16.9 shows the principles for the generation

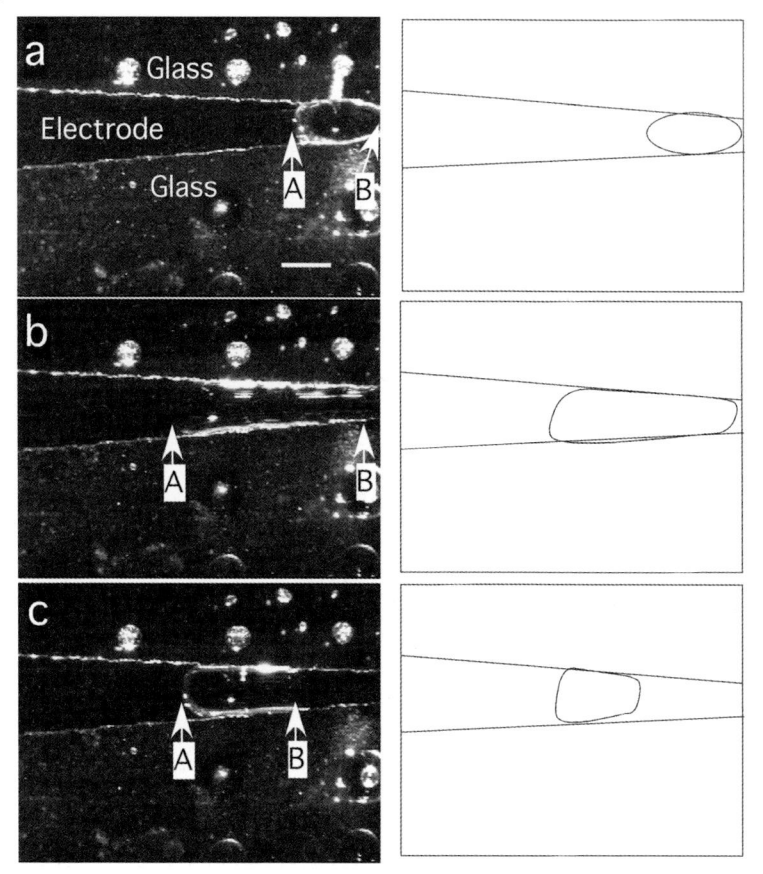

Figure 16.8 Directional deformations of the nitrobenzene droplet. The length of the scale bar in (a) is 3 mm. When the surface was changed from repulsive to wetting (a–b), the left side of the droplet (A) spread. When the surface was back to repulsive (b–c), the right side of the droplet (B) shrank. These directional deformations resulted in a net transport of the droplet.

and control of the wetting gradient by the electrochemical method. The gold thin-film covered with Fc-alkanethiol monolayer was immersed in the electrolyte solution and the potential of the gold substrate, E_{offset}, was controlled with respect to the reference electrode in the solution. A lateral bias voltage, V_{bias}, was applied to the gold substrate to generate in-plane gradients in the electrochemical potential [49]. The electrochemical potential between "A" and "B" denoted in Figure 16.9a is expected to change linearly when only a negligible current flows into the counter electrode in the electrochemical cell. Figure 16.9b schematically shows the relationships between these electrochemical parameters and the concentration of Fc^+ in the monolayer as a function of the position on the gold substrate, x. When the redox potential of Fc^+/Fc in the monolayer is in the electrochemical potential window generated in the gold substrate, the concentration of Fc^+ gradually changes around the point P shown in

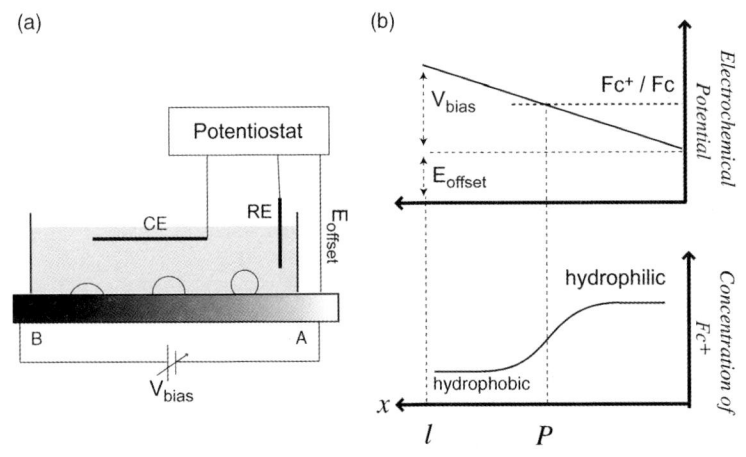

Figure 16.9 (a) Schematic drawing of the experimental configuration. (b) Potential profile and wetting distribution on the substrate under the biased condition.

Figure 16.9b. Since Fc- and Fc^+-covered surfaces are known to be hydrophobic and hydrophilic, respectively [42, 43], the wetting gradient is expected to be formed around P. Therefore, we call P the wetting boundary. The position of the wetting boundary and the magnitude of the wetting gradient can be reversibly controlled as functions of E_{offset} and V_{bias}/l, respectively, where l is the length of the substrate in the biased direction.

Figure 16.10 show pictures of the nitrobenzene droplets on the gold electrode covered with Fc-monolayer in aqueous solution when V_{bias} (0.35 V) was applied. The potential of the substrate was measured by two wires in contact with the substrate. It was clearly shown that the position of the wetting boundary moved when E_{offset} was changed, as shown in Figure 16.10b–d.

Figure 16.11 shows the motions of the droplet caused by the shift in the wetting boundary. Initially, E_{offset} and E_{bias} were set to -300 mV and -500 mV, respectively. The left side of the substrate in Figure 16.11 was more wetting to nitrobenzene than the right side of the substrate under this bias condition. The initial appearance of the droplet, shown in Figure 16.11a, indicated that the droplet was on the repulsive side of the wetting boundary at this stage. When E_{offset} was changed to more negative values to move the wetting boundary to the droplet, the droplet gradually spread into the wetting areas. Figure 16.11b shows the droplet when E_{offset} was -340 mV. The droplet was almost completely wet, indicating that the wetting boundary was at the right side of the droplet. When E_{offset} was changed to the initial value to move the wetting boundary, the droplet gradually shrunk from the right edge and returned to its original appearance, as shown in Figure 16.11c. The unidirectional spreading-and-shrinking cycle resulted in the net transport of the droplet by an inchworm motion.

The rate of the droplet motion was limited by the rate of the viscous flow of the liquid because electrochemical reaction took place much faster (in less than 1 s) than the deformation of the droplet ($\sim$10 s). Pinning of the droplet sometimes occurred at a defect on the surface. In most cases, the pinned droplet could be moved again by

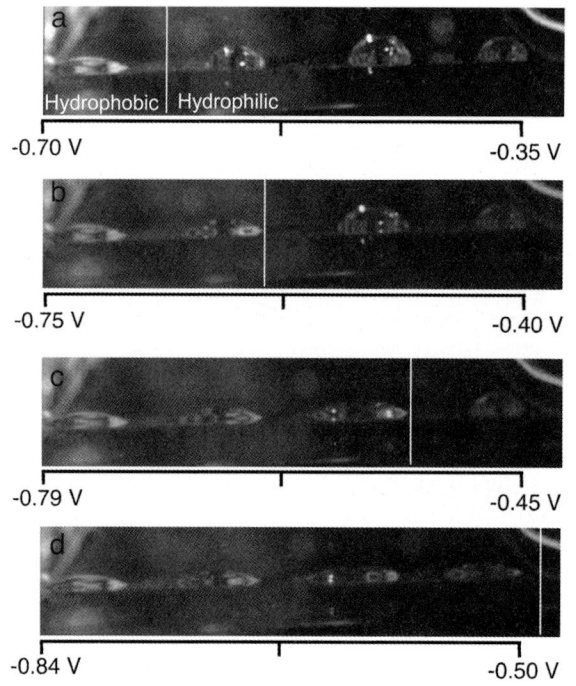

Figure 16.10 Photographs of nitrobenzene droplets. V_{bias} was fixed at 0.35 V and E_{offset} was varied (a–d). The line in the photograph represents the position of the wetting boundary estimated from the shape of the droplets. Note that the current peak for the Fc^+/Fc reaction was observed at about -0.5 V in the cyclic voltammogram.

making the wetting gradient stronger or changing the limits of the potential cycles. The inchworm motion could be repeated many times, as shown in Figure 16.11d, and the reverse motion was possible by changing the bias direction.

The positions of the contact lines moved as a result of the imbalance between the forces acting on the droplet edges during the deformation of the droplet. Since the surface had wetting gradient, the forces acting on the contact lines at the more wetting side were always larger than those acting on the other side during the deformation of the droplet. As a result, the rear and front ends were pinned when the droplet spread and shrank, respectively. It should be noted that the directional spreading of the droplet was caused by the imbalance of advancing contact angles acting on the opposite sides of the droplet while the directional shrinking was due to the imbalance of receding contact angles. These mechanisms differentiate the motion of the droplet presented here from the motion driven by the static wetting gradient originating from the difference in the contact angles between *advancing* and *receding* sides of the droplet.

The motion of droplets in solution was used to manipulate micro-particles on surfaces. The dichloromethane droplet pushed aside hydrophilic glass beads of about

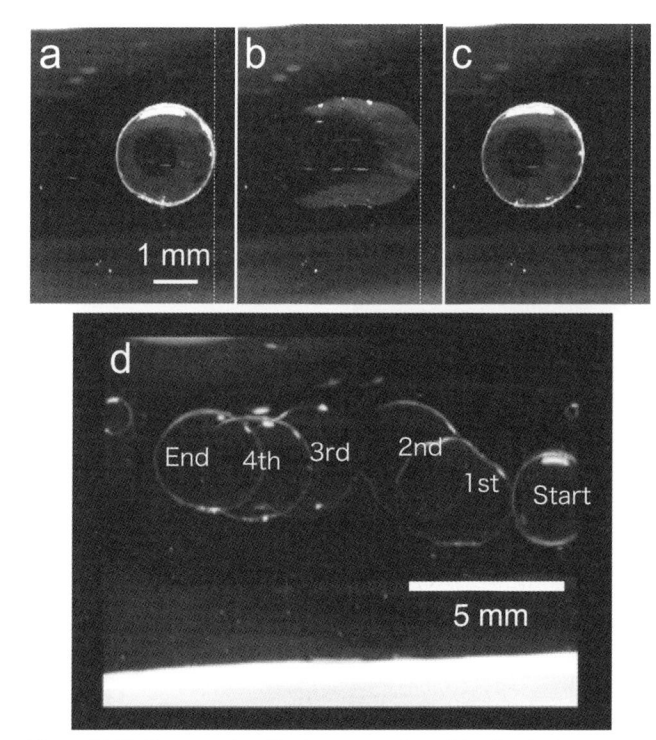

Figure 16.11 Photographs of inchworm motion of the droplet in solution. See text for details. $E_{BIAS} = 0.5$ V. $E_{offset} = -300$ mV (a), -340 mV (b) and -300 mV (c). (d) Trace of the multi-step inchworm motions of the droplet. Six photographs were superimposed.

40 μm diameter on the substrate in an aqueous solution, as shown in Figure 16.12a, because the hydrophilic glass beads could not get into the oil droplet. In contrast, the hydrophobic beads, which initially dispersed randomly in the dichloromethane droplet, were gathered at the receding end of the droplet during its motion, as shown in Figure 16.12b and c, because the hydrophobic beads could not get out of the droplet. In both cases, the liquid/liquid and/or solid/liquid/liquid interfaces acted as barriers to move the particles directly.

16.3 Conclusion

Surface switching coupled with geometric and potential asymmetry was used to cause directional motion of a droplet. Sophisticated design and active control of surface properties are important technology for motion control on the micro/nano-scales.

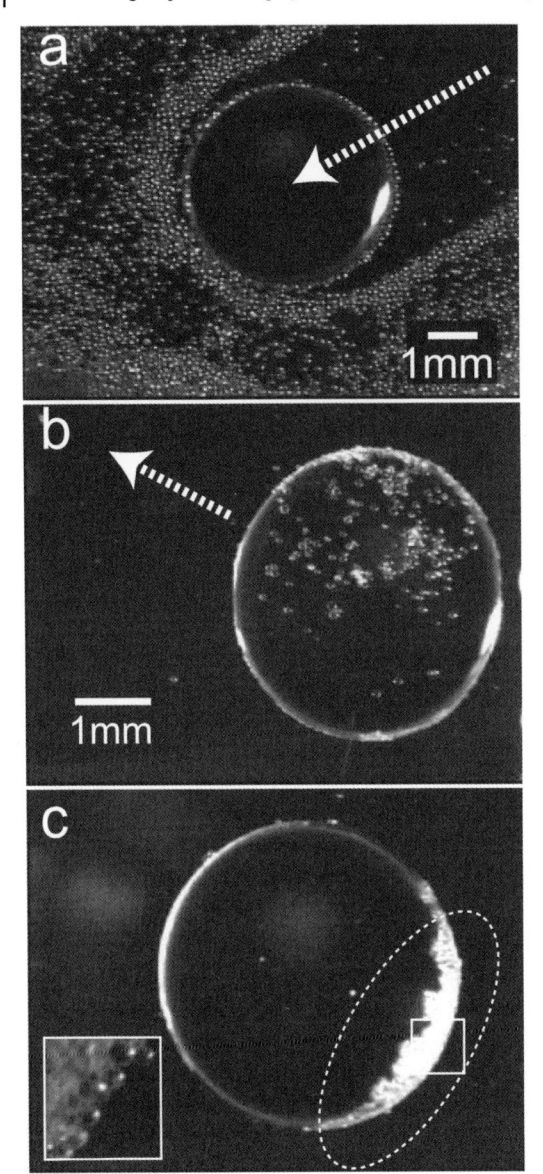

Figure 16.12 Transportation of glass-beads by the droplet. (a) Hydrophilic glass beads were pushed by the oil droplet. The droplet moved from the upper right of the figure. A magnified image of the region within the square is shown in the inset. Hydrophobic beads in the droplet (b) were carried with the motion of the oil droplet (c). The droplet shown in (c) was moved for several mm in the direction shown by the arrow in (b).

Acknowledgments

This work was supported by Grants-in-Aid for Scientific Research on Priority Areas "Molecular Nano Dynamics" from Ministry of Education, Culture, Sports, Science and Technology.

References

1 Russell, T. P. (2002) *Science*, **297**, 964–967.
2 Liu, Y., Li, Mu., Liu, B. and Kong, J. (2005) *Chem. Eur. J.*, **11**, 2622–2631.
3 Gallardo, B. S., Gupta, V. K., Eagerton, F. D., Jong, L. I., Craig, V. S., Shah, R. R. and Abbott, N. L. (1999) *Science*, **283**, 57–60.
4 Hayes, R. A. and Feenstra, B. J. (2003) *Nature*, **425**, 383–385.
5 Berge, B. and Peseux, J. (2000) *Eur. Phys. J. E*, **3**, 159–163.
6 Ulman, A. (1991) *An Introduction to Ultrathin Organic Films from Langmuir-Blodgett to Self-Assembly*, Academic Press, San Diego.
7 Love, J. C., Estroff, L. A., Kriebel, J. K., Nuzzo, R. G. and Whitesides, G. M. (2005) *Chem. Rev.*, **105**, 1103–1169.
8 Shimazu, K., Yagi, I., Sato, Y. and Uosaki, K. (1992) *Langmuir*, **8**, 1385–1387.
9 Sato, Y., Fujita, M., Mizutani, F. and Uosaki, K. (1996) *J. Electroanal. Chem.*, **409**, 145–154.
10 Finklea, H. O. and Hanshew, D. D. (1992) *J. Am. Chem. Soc.*, **114**, 3173–3181.
11 Uosaki, K., Kondo, T., Zhang, X.-Q. and Yanagida, M. (1997) *J. Am. Chem. Soc.*, **119**, 8367–8368.
12 Sato, Y. and Uosaki, K. (1995) *J. Electroanal. Chem.*, **384**, 57–66.
13 Sato, Y., Yabuki, S. and Mizutani, F. (2000) *Chem. Lett.*, 1330–1331.
14 Gobi, K., Kitamura, F., Tokuda, K. and Ohsaka, T. (1999) *J. Phys. Chem. B*, **103**, 83–88.
15 Shimazu, K., Takechi, M., Fujii, H., Suzuki, M., Saiki, H., Yoshimura, T. and Uosaki, K. (1996) *Thin Solid Films*, **273**, 250–253.
16 Kondo, T., Horiuchi, S., Yagi, I., Ye, S. and Uosaki, K. (1999) *J. Am. Chem. Soc.*, **121**, 391.
17 Ye, S., Yashiro, A., Sato, Y. and Uosaki, K. (1996) *J. Chem. Soc., Faraday Trans.*, **92**, 3813–3821.
18 Wang, Y. and Kaifer, A. E. (1998) *J. Phys. Chem. B*, **102**, 9922–9927.
19 Katz, E., Lötzbeyer, T., Schiereth, D. D., Schuhmann, W. and Schmidt, H.-L. (1994) *J. Electroanal. Chem.*, **373**, 189–200.
20 Caldwell, W. B., Campbell, D. J., Chen, K., Herr, B. R., Mirkin, C. A., Malik, A., Durbin, M. K., Dutta, P. and Huang, K. G. (1995) *J. Am. Chem. Soc.*, **117**, 6071–6082.
21 Blonder, R., Willner, I. and Buckmann, A. F. (1998) *J. Am. Chem. Soc.*, **120**, 9335–9341.
22 Sato, Y., Itoigawa, H. and Uosaki, K. (1993) *Bull. Chem. Soc. Jpn.*, **66**, 1032–1037.
23 Taniguchi, I., Tomosawa, K., Yamaguchi, H. and Yasukouchi, K. (1982) *J. Chem. Soc., Chem. Commun.*, 1032–1033.
24 Aizenberg, J., Black, A. J. and Whitesides, G. M. (1999) *Nature*, **398**, 495–498.
25 Keller, S. W., Kim, H.-N. and Mallouk, T. E. (1994) *J. Am. Chem. Soc.*, **116**, 8817–8818.
26 Lee, H., Kepley, L. J., Houg, H. G., Akhter, S. and Mallouk, T. E. (1988) *J. Phys. Chem.*, **92**, 2597–2601.
27 Bethell, D., Brust, M., Schiffrin, D. J. and Kiely, C. (1996) *J. Electroanal. Chem.*, **409**, 137–143.
28 Poirier, G. E. (1997) *Chem. Rev.*, **97**, 1117–1127, and references therein.
29 Zhao, B. (2004) *Langmuir*, **20**, 11748–11755.

30 Wu, T., Efimenko, K. and Genzer, J. (2002) *J. Am. Chem. Soc.*, **124**, 9394–9395.
31 Loos, K., Kennedy, S. B., Eidelman, N., Tai, Y., Zharnikov, M., Amis, E. J., Ulman, A. and Gross, R. A. (2005) *Langmuir*, **21**, 5237–5241.
32 Chaudhury, M. K. and Whitesides, G. M. (1992) *Science*, **256**, 1539–1541.
33 Morgenthaler, S., Lee, S., Zürcher, S. and Spencer, N. D. (2003) *Langmuir*, **19**, 10459–10462.
34 Caelen, I., Bernard, A., Juncker, D., Michel, B., Heinzelmann, H. and Delamarche, E. (2000) *Langmuir*, **16**, 9125–9130.
35 Ballav, N., Shaporenko, A., Terfort, A. and Zharnekov, M. (2007) *Adv. Mater.*, **19**, 998–1000.
36 Ballav, N., Weidner, T. and Zharnikov, M. (2007) *J. Phys. Chem. C*, **111**, 12002–12010.
37 Cazabat, A. M., Heslot, F., Troian, S. M. and Carles, P. (1990) *Nature*, **346**, 824–826.
38 Ichimura, K., Oh, S.-K. and Nakagawa, M. (2000) *Science*, **288**, 1624–1626.
39 Abbott, S., Ralston, J., Reynolds, H. and Hayes, R. (1999) *Langmuir*, **15**, 8923–8928.
40 Bern,á J., Leigh, D. A., Lubomska, M., Mendoza, S. M., Pérez, E. M., Rudolf, P., Teobaldi, G. and Zerbetto, F. (2005) *Nature Mater.*, **4**, 704–710.
41 Lahann, J., Mitragotri, S., Tran, T., Kaido, H., Sundaram, J., Choi, I. S., Hoffer, S., Somorjai, G. A. and Langer, R. (2003) *Science*, **299**, 371–374.
42 Abbott, N. L. and Whitesides, G. M. (1994) *Langmuir*, **10**, 1493–1497.
43 Sondag-Huethorst, J. A. M. and Fokkink, L. G. J. (1994) *Langmuir*, **10**, 4380–4387.
44 Yamada, R. and Tada, H. (2005) *Langmuir*, **21**, 4254–4256.
45 Yamada, R. and Tada, H. (2006) *Colloids Surf. A*, **276**, 203–206.
46 Adamson, A. and Gast, A. P. (1997) *Physical Chemistry of Surfaces*, 6th edn, Wiley-Interscience.
47 Daniel, S., Sircar, S., Gliem, J. and Chaudhury, M. K. (2004) *Langmuir*, **20**, 4085–4092.
48 Marquet, C., Buguin, A., Talini, L. and Silberzan, P. (2002) *Phys. Rev. Lett.*, **88**, 168301.
49 Terrill, R. H., Balss, K. M., Zhang, Y. and Bohn, P. W. (2000) *J. Am. Chem. Soc.*, **122**, 988–989.

17
Photoluminescence of CdSe Quantum Dots: Shifting, Enhancement and Blinking

Vasudevanpillai Biju and Mitsuru Ishikawa

17.1
Introduction

Quantum dots are nanocrystals of semiconductors, metals and organic materials in which excitons (electrons and holes) are three-dimensionally confined. The dimension of quantum dots is typically on the 1–25 nm scale (nanoscale). At this scale, the surface-to-volume ratios of materials become large and their electronic energy states become discrete. The large surface-to-volume ratios and the discrete energy states give unique electronic, optical, magnetic, and mechanical properties to nanomaterials. While nanoscience advances with the size-dependent properties of nanomaterials, large surface tension and friction limit their applications in the advancement of nanotechnology. In general, as the sizes of semiconductor, metal and organic materials are decreased towards the nanoscale, their optical and electronic properties become size- and shape-dependent and vary largely from those in the bulk and at the atomic/molecular levels. The size- and shape-dependent properties on the nanoscale are attributed to the quantum confinement effect, strong confinement of electrons and holes when the radii of nanoparticles are less than the exciton Bohr radius of the material. Remarkable advances in the field of semiconductor quantum dots emerged recently when the fundamental principles underlying light–matter interactions and the quantum confinement effect became clearly understood. A rationale between size and energy states in semiconductor quantum dots was developed by Luis Brus [1, 2] by applying the particle-in-a-sphere-model approximation to the bulk Wannier Hamiltonian. According to the approximation, the energy of the lowest excited state in a quantum confined system is given by Eq. 17.1.

$$E = \frac{\hbar^2 \pi^2}{2R^2}\left[\frac{1}{m_e} + \frac{1}{m_h}\right] - \frac{1.8e^2}{cR} \tag{17.1}$$

where, R is the radius of a nanocrystal, m_e is the effective mass of an electron, m_h is the effective mass of a hole, and e is the electronic charge. In simple words, as the size decreases the bandgap increases. This equation was proposed even before strong quantum confinement was experimentally observed in colloidal quantum dots.

Molecular Nano Dynamics, Volume I: Spectroscopic Methods and Nanostructures
Edited by H. Fukumura, M. Irie, Y. Iwasawa, H. Masuhara, and K. Uosaki
Copyright © 2009 WILEY-VCH Verlag GmbH & Co. KGaA, Weinheim
ISBN: 978-3-527-32017-2

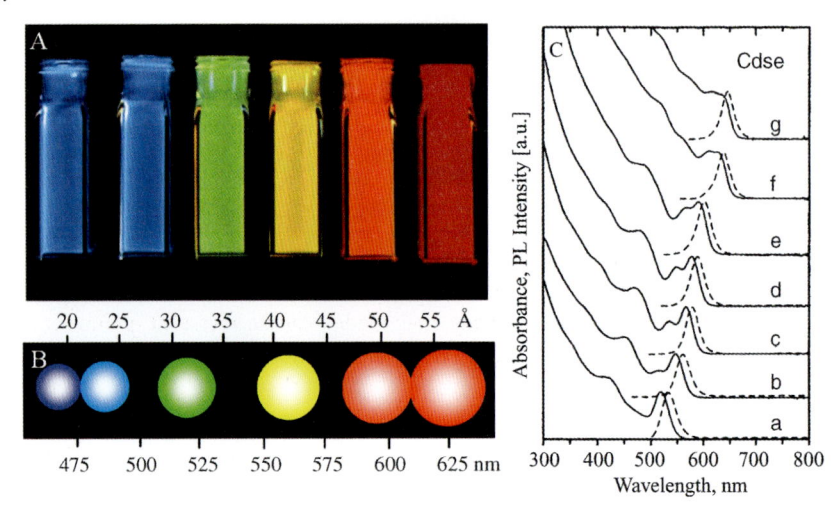

Figure 17.1 (A) Size-dependent photoluminescence color of ZnS-shelled CdSe quantum dots. (B) Schematic presentation of size in Å, color, and photoluminescence spectral maxima of CdSe quantum dots. (C) Size-dependent absorption (solid lines) and photoluminescence (broken lines) spectra of CdSe quantum dots. Reprinted with permission from references [4] (A) and [5] (C); copyright [1997, 2001], American Chemical Society.

Among different quantum dots, cadmium selenide (CdSe) attracted much attention in almost all branches of science and technology, especially in nanoscience, nanotechnology and nanobiotechnology (the "Nano World"), rooted in the size-tunable bandgap and photoluminescence color (Figure 17.1) extending throughout the visible region of the electromagnetic spectrum [3, 4]. Probably, the "Nano World" would not be as exciting and colorful as it is today without CdSe quantum dots.

Although extensive investigations of the optical and electronic properties of CdSe quantum dots lifted their status from a fundamental scientific level to direct applications in lasers [6], light emitting diodes [7], solar cells [8], and bioanalyses and bioimaging [9–11], surface defects (defects in the bandgap) and intrinsic "on" and "off" (blinking) photoluminescence remain unresolved. The surface defects and blinking of quantum dots limit the advancement of their applications towards single-quantum dot logic devices and single-molecule imaging. Efforts to improve the photoluminescence properties of quantum dots by decreasing the surface defects and suppression of blinking are underway. Modified syntheses, post-synthesis surface modifications, and photo- and thermal-activations are promising approaches to control the surface defects and blinking of quantum dots.

This chapter presents an overview of the synthesis, ensemble photoluminescence properties and blinking of single CdSe quantum dots. The stress is on (i) widely accepted methods of synthesis, (ii) the origin of photoluminescence and variations of photoluminescence as functions of surface-coating, surface-passivating molecules, chemical environment, and thermal- and photo-activations, and (iii) photoluminescence blinking and variations of blinking as functions of modified synthesis and post-synthesis modifications.

17.2
Synthesis of CdSe Quantum Dots

In the 1980s, CdSe quantum dots were prepared by top-down techniques such as lithography; however, size variations, crystal defects, poor reproducibility, and poor optical properties of quantum dots made them inadequate for advanced applications. Introduction of bottom-up colloidal synthesis of CdSe quantum dots by Murray *et al.* [3] and its further advancements brought radical changes in the properties of quantum dots and their applications in devices and biology. The colloidal syntheses of CdSe quantum dots are broadly classified into organic-phase synthesis and aqueous-phase synthesis.

17.2.1
Synthesis of CdSe Quantum Dots in Organic Phases

17.2.1.1 Synthesis of CdSe Quantum Dots from Dimethyl Cadmium

In 1993, Murray *et al.* synthesized hydrophobically-capped colloidal CdSe quantum dots by the pyrolysis of organometallic precursors of cadmium and selenium [3]. In this synthesis, dimethyl cadmium [$CdMe_2$, 13.35 mmol in 25 mL trioctylphosphine (TOP)] was reacted at high-temperature with the selenide of TOP (TOPSe, 10 mmol in 15 mL TOP) in the presence of trioctylphosphine oxide (TOPO, Tech. grade). TOPSe was prepared by dissolving Se powder or Se shots in TOP at 150 °C. In a typical synthesis of CdSe quantum dots, TOPO was heated to ~300 °C under vacuum for 20 min followed by injection of a mixture of $CdMe_2$ and the TOPSe in an atmosphere of Ar. The growth of CdSe nanocrystals was carried out at 230–260 °C. The as-synthesized sample contained a size distribution (1.2–11.5 nm) of CdSe quantum dots which were separated by size-selective centrifugation from a mixture of 1-butanol and methanol. One of the main limitations to producing narrow size-distributed or "size-focused" quantum dots by this method was Ostwald ripening, a gradual growth of larger quantum dots at the cost of a gradual dissolution of smaller ones. In 1994, Katari *et al.* lifted the limitation of size-distribution by replacing TOP with tributylphosphine (TBP) and avoiding Ostwald ripening [12]. In the modified synthesis, a precursor solution was prepared by dissolving Se powder (800 mg) in TBP (8 g) followed by the addition of 2 g $CdMe_2$. This mixture was diluted to $1/4$ using TBP and injected into a hot (~350 °C) solution of TOPO (12 g). The injection of CdSe precursors into the hot solution of TOPO resulted in spontaneous nucleation of CdSe nanocrystals and a decrease in temperature. Once the temperature was stabilized, an additional amount (0.4 mL) of the precursor solution was added for the growth of the nucleated nanocrystals. Here, Ostwald ripening was avoided by separating the nucleation and growth processes. All the reagents and the reaction were kept under an Ar atmosphere to avoid fire hazard and surface oxidation of the nanocrystals.

17.2.1.2 Synthesis of CdSe Quantum Dots from Cadmium Sources Other Than Dimethyl Cadmium

With the introduction of the colloidal syntheses of CdSe quantum dots, researchers were interested in cadmium precursors other than the volatile and toxic $CdMe_2$. In 2001, Qu et al. replaced $CdMe_2$ with cadmium oxide (CdO), cadmium acetate [$Cd(AcO)_2$], and cadmium carbonate ($CdCO_3$) and synthesized high-quality CdSe quantum dots [13]. A typical example is the synthesis of CdSe quantum dots from Cd(AcO)$_2$ and trioctylphosphine selenide (POTSe). In this reaction, a suspension of Cd(AcO)$_2$ in TOPO or TOPO–phosphonic acid mixture was heated to 250–360 °C under an Ar atmosphere followed by injection of TOPSe. After nucleation of CdSe nanocrystals, the reaction temperature was decreased to 200–320 °C and nanocrystals with different sizes were grown. CdSe nanocrystals with strong quantum confinement (size <4 nm) were obtained in the presence of phosphonic acids and relatively large (4–25 nm) nanocrystals were formed in the presence of fatty acids. Figure 17.2A shows absorption and photoluminescence spectra of narrow size-dispersed CdSe quantum dots obtained from different cadmium precursors. We synthesized CdSe quantum dots with green photoluminescence by reacting $Cd(AcO)_2$ with TOPSe at a low temperature (75 °C) and in the presence of a mixture of TOP and TOPO [14]. In a typical reaction, a round-bottom flask was charged with a mixture of $Cd(Ac)_2 2H_2O$ (0.262 g, 1 mmol) and TOPO (3.86 g, 10 mmol). This mixture was heated at 75 °C for 30 min under continuous Ar purging followed by injection 0.72 g TOPSe (1 mmol) in four aliquots at 5 min intervals. TOPSe was prepared by dissolving Se shots in TOP at 150 °C for 1 h under an Ar atmosphere. The reaction mixture was vigorously stirred at 75 °C for 5 h. The formation of CdSe quantum dots was identified from a gradual deepening of the yellow color in the reaction mixture. Figure 17.2B shows the optical densities at 400 nm for aliquots of samples withdrawn from the reaction mixture at different time intervals. We noted a gradual increase in the absorbance and photoluminescence intensity with time during the reaction due to an increase in the concentration of CdSe quantum dots. After 5 h, the reaction was quenched by adding 1-butanol and CdSe quantum dots were isolated by precipitation from a mixture of 1-buanol and methanol. The inset of Figure 17.2B shows the absorption and photoluminescence spectra of the as-synthesized CdSe quantum dots. We prepared ZnS shells on CdSe quantum dots from hexamethyldisilathiane and diethyl zinc following a literature method [4].

17.2.2
Synthesis of Water-Soluble Quantum Dots

Although direct synthesis of cadmium chalcogenide quantum dots in the aqueous phase was achieved nearly 30 years ago, the synthesis of size-controlled CdSe quantum dots in the aqueous phase became possible only recently. Rogach et al. successfully synthesized CdSe quantum dots in the aqueous phase from a mixture of cadmium perchlorate (4.7 mmol), sodium hydroselenide (2.2 mmol), and 2-mercaptoethanol or 1-thioglycerol (11.54 mmol) [15]. In a typical synthesis, the above mixture was refluxed in an aqueous NaOH solution (pH = 11.2) in a N_2

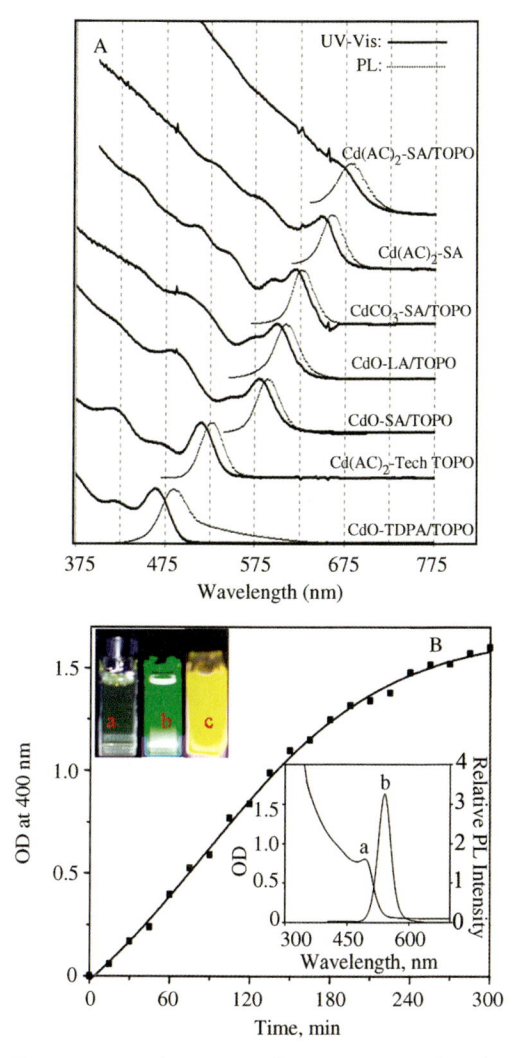

Figure 17.2 (A) Absorption and photoluminescence spectra of CdSe quantum dots prepared from CdO, CdCO$_3$, and Cd(AcO)$_2$ in the presence of different ligands. (B) Increase in the optical density (at 400 nm) of a CdSe quantum dot reaction mixture with time under reaction at 75 °C. Color pictures in the inset of B represent CdSe (a,b) and CdSe–ZnS (c) quantum dot solutions with (b,c) and without (a) UV illumination. Traces in the inset of B are absorption (a) and photoluminescence (b) spectra of CdSe quantum dots prepared at 75 °C. Reprinted with permission from references [13] (A) and [14] (B); copyright [2001, 2005], American Chemical Society.

atmosphere. Despite this method and a few other recent investigations, direct synthesis of CdSe quantum dots in the aqueous phase is still challenging. Thus, conjugation of hydrophilic and amphiphilic shells/molecules on the surface of CdSe quantum dots synthesized in organic phases continues to be attractive. Typical examples of the conversion of CdSe quantum dots from organic-to-aqueous phase involve the exchange of hydrophobic ligands with thioglycolic acid (TGA), hydrophilic dendrimers, silica-shells, amphiphilic polymers, proteins, and sugars [9]. Among these compounds, TGA is widely used in the preparation of water-soluble quantum dots for biological applications; however, low efficiency of ligand exchange and considerable decrease in the photoluminescence quantum efficiency remain.

17.3
Bandgap Structure and Photoluminescence of CdSe Quantum Dots

Size-tunable emission of light in the visible region is the most attractive property of CdSe quantum dots. Strong three-dimensional confinements (the exciton Bohr radius of CdSe is 5.6 nm) of excitons surmounting coulomb interactions, and large surface to volume ratios are the origins of the size-tunable optical properties of quantum dots. Once photoactivated, excitons in a quantum dot cool and relax (intraband relaxation) over $10^{11}\,\text{s}^{-1}$ before inter-band exciton recombination. The inter-band exciton recombination processes include radiative relaxation at the band-edge, phonon-assisted non-radiative relaxations, non-radiative Auger relaxations, and radiative and non-radiative relaxations at the surface defects [16–18]. Among these relaxation processes, the radiative exciton recombination at the band-edge is the origin of the size-tunable photoluminescence color of quantum dots. Photoluminescence of CdSe quantum dots originates from (i) exciton recombination at the band edge; transitions from the lowest unoccupied state, that is, a combination of 5 s orbitals of cadmium, to the highest occupied state, that is, a combination of 4p orbitals of Se and (ii) deactivation of excited electrons (iii) from the surface states. These inter-band radiative relaxations are relatively slow ($<10^9\,\text{s}^{-1}$) and non-exponential. The non-exponentiality originates from carrier-trapping in a distribution of states including shallow surface states. The band-edge states in CdSe QDs ($1S_e 1S_{3/2}$) are degenerate due to asymmetric and crystal-field splitting followed by mixing of carrier exchange perturbations with the angular momentum of the charge carriers [16]. The splitting and mixing provide eightfold degeneracy to the band-edge which is characterized by the total angular momentum (J) values $-2, -1, 0, +1, +2$ for $1S_{3/2}$ and $-1, 0, +1$ for $1S_e$ states. Figure 17.3B shows the degenerate band-edge states and inter-band transitions in CdSe quantum dots. Among the degenerate states, $J = \pm 2$ and $J = 0$ are spin-forbidden states. Therefore, photoactivation of quantum dots close to the band-edge populates ± 1 states. Deactivation of this population takes place via non-radiative relaxation to the forbidden dark exciton states ($J = \pm 2$) followed by radiative or non-radiative exciton recombination. With the characterization of the band-edge structures in CdSe quantum dots, detailed investigations of carrier-relaxation dynamics and photoluminescence became possible. In addition to the above degenerate states, it is necessary to consider surface states (trap states) in the relaxation of photoactivated quantum dots. Surface states are considered on the basis of large surface to volume ratios in quantum dots. In the case of CdSe quantum dots, the surface states are contributed by dangling bonds of Se atoms; for example, $\sim 30\%$ of the surface atoms in a 2.5 nm CdSe

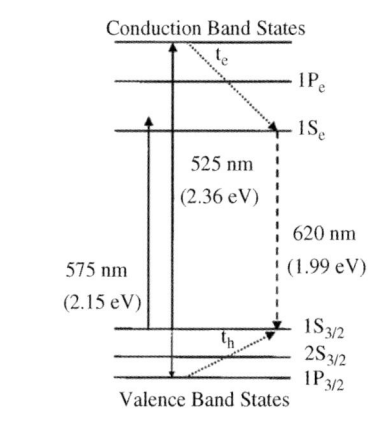

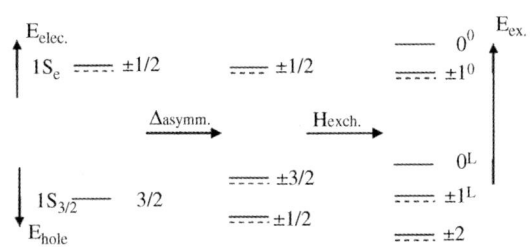

Figure 17.3 Energy level diagram (A) and splitting of $1Se1S_{3/2}$ band-edge states (B) in CdSe quantum dots. Reprinted with permission from references [17] (A) and [16] (B); copyright [1999, 2001], American Chemical Society.

quantum dot are Se atoms. In general, the surface states include shallow traps and deep traps. Carrier relaxations through the deep traps contribute red-shifted deep-trap emission, non-radiative relaxation and low photoluminescence quantum efficiency and large photoluminescence lifetime values.

17.4
Photoluminescence Spectral Shifts

Photoluminescence (and absorption) spectra of semiconductor quantum dots are strongly related to the quantum confinement of excitons (Figure 17.1 and Eq. 17.1). Thus, large shifts in the photoluminescence spectra are attributed to size-variations. For example, in the case of CdSe, the bandgap energy increases from 1.74 (bulk) to >2.75 eV as a result of the quantum confinement of excitons. Other factors such as temperature, pressure, and dielectric environments also contribute to the spectral-shifts. To date, discussions on the spectral shifts of quantum dots have been based on exciton–phonon coupling, confinement energy, surface charges, and surface chemical changes. The spectral shifts in CdSe quantum dots are discussed below with reference to physical and chemical effects.

17.4.1
Physical Effects on Spectral Shifts

The photoluminescence of CdSe quantum dots shifts with pressure [19], temperature [20–22] and applied electric fields [23]. The photoluminescence (and absorption) spectrum of CdSe quantum dots blue shifts with increasing pressure due to an increase in the bulk bandgap energy and strong quantum confinement under high pressure. Interestingly, both blue-shifts and red-shifts in the photoluminescence spectra of CdSe quantum dots are observed independently with increasing temperature. For example, the photoluminescence spectra of CdSe quantum dots continuously blue shift when the temperature is increased from 1.75 to 75 K [20]. As a result of this blue shift, the Stokes shift decreases due to a coupling between exciton and longitudinal optic (LO) phonons followed by tunneling of carriers between localized states on the surface of the quantum dots. On the other hand, the photoluminescence spectra of CdSe quantum dots red shift with increasing temperature >100 K, which is a more general observation [21]. The red shift of the photoluminescence or decreasing bandgap with increasing temperature is due to the dilation of the crystal lattice and lattice–electron interactions, which are bulk properties of CdSe. Thus, the red shift is less related to the quantum confinement effect.

We observed reversible red shifts and blue shifts in the photoluminescence spectra of closely-packed aggregates of CdSe quantum dots [21]. Figure 17.4 shows the photoluminescence spectral shifts during heating and cooling of a solution of CdSe quantum dot aggregates; the photoluminescence spectrum red shifted when the temperature was increased from 298 to 353 K and returned to the original position when re-cooled. In addition to the intrinsic thermal effects on the bulk bandgap energy of CdSe, the red shift is contributed by the relaxation of photoactivated excitons through surface states and interparticle interface states, both of which are thermally populated. Also, carrier tunneling and inter-dot dipole–dipole interactions are involved in the red shift. Figure 17.5 shows various relaxation processes in a photoactivated CdSe quantum dot aggregate. Similar red shifts are observed in the photoluminescence spectra of CdSe quantum dot superstructures in which quantum mechanical tunneling through inter-dot barriers and thermal activation followed by hopping above energy barriers due to exciton-coupling are proposed to account for the red shift.

In addition to the photoluminescence red shifts, broadening of photoluminescence spectra and decrease in the photoluminescence quantum efficiency are reported with increasing temperature. The spectral broadening is due to scattering by coupling of excitons with acoustic and LO phonons [22]. The decrease in the photoluminescence quantum efficiency is due to non-radiative relaxation from the thermally activated state. The Stark effect also produces photoluminescence spectral shifts in CdSe quantum dots [23]. Large red shifts up to 75 meV are reported in the photoluminescence spectra of CdSe quantum dots under an applied electric field of $350\,kV\,cm^{-1}$. Here, the applied electric field decreases or cancels a component in the excited state dipole that is parallel to the applied field; the excited state dipole is contributed by the charge carriers present on the surface of the quantum dots.

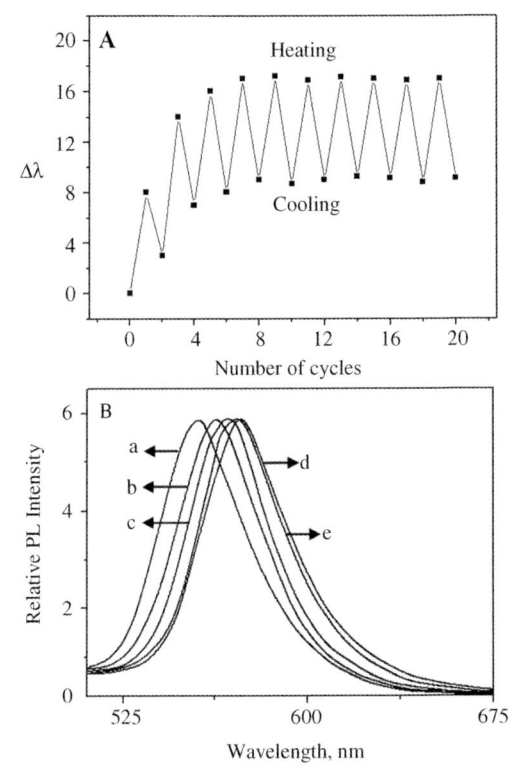

Figure 17.4 (A) Photoluminescence spectral shifts (Δλ) of a solution of CdSe quantum dot aggregates during heating–cooling cycles: photoluminescence spectral maxima were recorded at 298 K during cooling and 353 K during heating. Reversibility of the photoluminescence spectral shift was attained after four heating–cooling cycles. (B) Photoluminescence spectra of a solution of CdSe quantum dot aggregates at (C) 298 K before heating–cooling cycles and (b–e) 353 K after one to four heating cycles. Spectra (b–e) are normalized for intensity with respect to the spectrum before heating (a) [21]. The photoluminescence measurements were carried out by exciting the samples at 450 nm. Reprinted with permission from reference [21]; copyright [2005], American Chemical Society.

17.4.2
Chemical Effects on Spectral Shifts

In addition to the bulk bandgap (1.74 eV), the total energy of an exciton in a CdSe quantum dot is contributed by the kinetic energy of confinement, electron–hole coulomb interactions, self-charging energy of charge carriers and polarization energy. These factors increase the quantum confined bandgap of CdSe by ~1 eV. The contribution of the polarization energy to the total energy is relatively small. For example, $1S_e$–$1S_{3/2}h$ transitions in a size-series of CdSe quantum dots were only slightly (<2 meV) shifted when the solvent was changed from a less polar solvent (hexane) to a more polar solvent (3-iodotoluene) [24]. The slight shift shows that the perturbations of the surface polarization by external dipoles, excluding the Stark

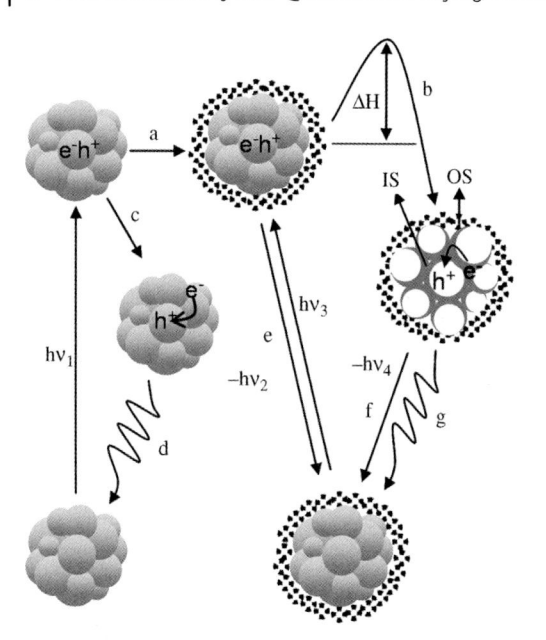

Figure 17.5 Schematic presentation of photoactivation and relaxation processes in a CdSe quantum dot aggregate: (a) surface-passivation of photoexcited quantum dots by solvent molecules or dissolved oxygen, (b) thermal activation followed by the formation of a stabilized state, (c) the formation of deep-trap states, (d) non-radiative relaxation of deep-trapped excitons, (e) radiative relaxation of the surface passivated state, (f) radiative relaxation of the stabilized state, and (g) non-radiative relaxation of the stabilized state. IS and OS are interparticle interface and outer surface, respectively. Reprinted with permission from reference [21]; copyright [2005], American Chemical Society.

effect mentioned in the previous section, alter the bandgaps in quantum dots only slightly and any large shifts in the photoluminescence (and absorption) spectra are mainly contributed by variations of temperature and size.

We observed irreversible blue shifts in the photoluminescence (and absorption) spectra of CdSe quantum dots up to 176 meV (565–523 nm) when photoactivated in organic solvents and in the presence of polymers [25]. Similar large blue shifts are also observed for CdSe quantum dots under photoactivation in different chemical environments such as water, air or oxygen, living cells and polymer films. The large blue shift in the presence of oxygen is attributed to a decrease in the effective quantum confined size due to surface oxidation and the formation of thin layers of $CdSeO_2$ and $CdSeO_3$. The relation between surface oxidation and blue shift was characterized by negligible blue shift in the photoluminescence spectra of quantum dots when photoactivated in vacuum or in an inert atmosphere [26]. Another reason for spectral shifts is dielectric coupling and inter-dot energy transfer in close-packed quantum dots. For example, the photoluminescence spectra of closely-packed quantum dots are red shifted by ~10 nm relative to that of isolated quantum dots [27]. This red shift is a size-distribution effect, that is, dipole–dipole coupling results in Förster resonance energy transfer (FRET) from smaller to larger quantum dots. As a result, the photoluminescence quantum efficiencies of smaller quantum dots decrease and the ensemble photoluminescence spectrum shifts to the red.

17.5
Enhancement of Photoluminescence in CdSe Quantum Dots

Highly luminescent CdSe quantum dots are promising materials for optoelectronic devices and bioimaging. However, the photoluminescence quantum efficiencies of as-synthesized CdSe quantum dots in the classical colloidal syntheses were low (∼10%) [3, 4, 12]. Thus, improvement of the photoluminescence quantum efficiency of CdSe quantum dots has been a subject of great research interest. The low photoluminescence quantum efficiency values of CdSe quantum dots are attributed to non-radiative exciton recombination at the surface states. Different methods for preparing quantum dots with improved photoluminescence quantum efficiencies include (i) fine tuning of quantum dot synthesis by adjusting precursor ratios, ligand composition and nanocrystal growth kinetics [13], (ii) post-synthesis passivation of surface defects by overlaying shells from higher bandgap materials such as ZnS [4], (iii) passivation of surface defects by ligands [28], molecular oxygen [29] and thermal activation [30], and (iv) photoinduced surface-passivation [25, 31]. All these methods contributed considerably to the understanding of the nature of surface states and the involvement of surface states in non-radiative exciton relaxations in CdSe quantum dots. Photoactivation of monolayers and films of core CdSe QDs with low PL quantum efficiency was independently introduced by Cordero et al. [32] and Maenosono et al. [33] in 2000. According to Cordero et al., the photoluminescence quantum efficiency of CdSe quantum dots is improved by the interactions of water molecules adsorbed on the surface. This proposal was revisited by Myung et al. [29]. Although the enhancement of photoluminescence was not reproduced in the presence of water, Myung et al. found considerable enhancement (2–6 times, Figure 17.6A) of photoluminescence when a solution of CdSe quantum dots was saturated with oxygen. In this case, the photoluminescence enhancement is due to the formation of trace amounts of CdO and SeO_2 on the surface of quantum dots and the passivation of surface defects by the oxide layer.

Photoactivation plays an important role in the photoluminescence enhancement of quantum dots. For example, Uematsu et al. [34] found an increase in the photoluminescence quantum efficiency of photoactivated CdSe–ZnS quantum dots in silicon oxide matrices with increasing oxygen composition (Figure 17.6B). This photoluminescence enhancement is contributed by an increase in the number of non-bonding oxygen–hole centers. Wang et al. achieved ∼50 fold enhancement in the photoluminescence quantum efficiency of CdSe quantum dots by photo-assisted etching of the surface defects [35]. All these investigations correlate the origin of the low PL quantum efficiencies of QDs with the non-radiative carrier recombination in the surface defects.

A detailed mechanism of the photoactivated enhancement of photoluminescence in quantum dots was proposed by Jones et al. [31]. They found an increase in the

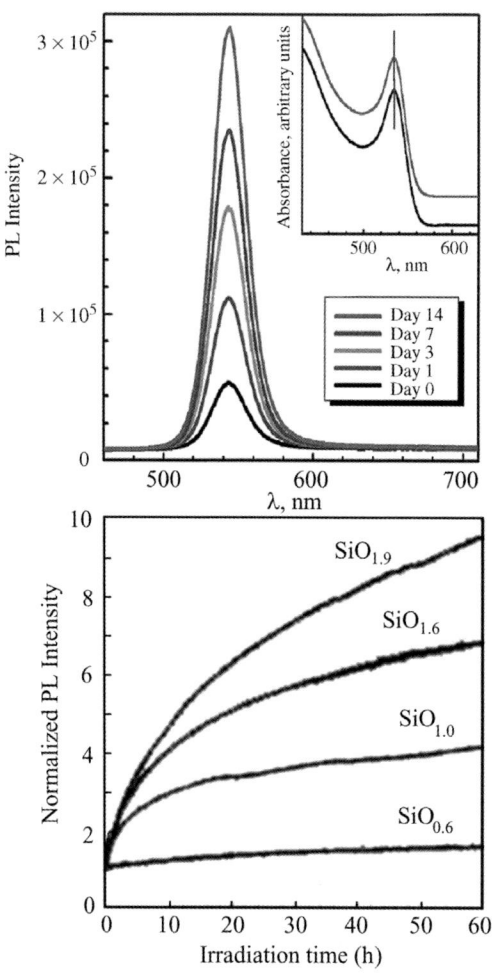

Figure 17.6 (A) Temporal evolution of photoluminescence and UV spectra (B) of CdSe quantum dots dispersed in $CHCl_3$ [29]. (C) The evolution curves of the photoluminescence peak intensity of quantum dot films on four kinds of SiOx substrates [34].
Reprinted with permission from references [29] (A) and [34] (B); copyright [2003], American Chemical Society and copyright [2006], American Institute of Physics.

photoluminescence lifetime and a considerable enhancement in the photoluminescence quantum efficiency of a CdSe–ZnS quantum dot sample under photoactivation. A model involving ligand rearrangement on the surface of photoactivated quantum dots accounts for the enhanced photoluminescence. This model states that the photoluminescence enhancement is due to (i) photoactivation of dark quantum dots into emissive ones and (ii) chemical rearrangement of ligand molecules and

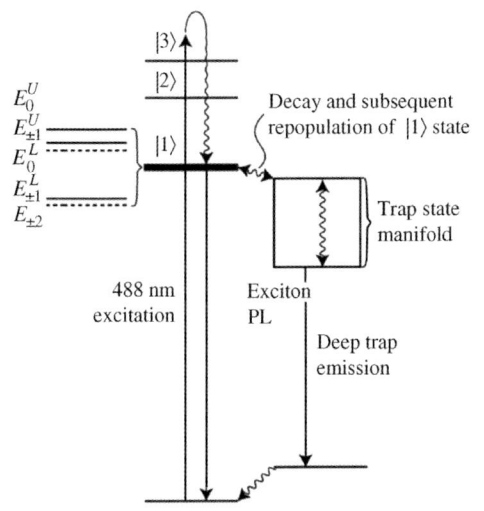

Figure 17.7 Schematic illustration of the decay routes of an exciton generated in CdSe–ZnS quantum dots; Reprinted with permission from reference [31]; copyright [2003], American Chemical Society.

surface states. In other words, photoactivated rearrangement of ligand molecules stabilizes the surface states and increases the probability of trapped charge carriers thermalizing into emissive states (Figure 17.7). More recently, we found that the photoluminescence quantum efficiency of CdSe quantum dots is enhanced (8% to >30%) and considerably stabilized under photoactivation in polar solvents and in the presence of dissolved polymer molecules (Figure 17.8A) [25]. We also found a decreasing photoluminescence quantum efficiency of CdSe quantum dots under continuous photoactivation. The decrease in the photoluminescence quantum efficiency indicates photobleaching of quantum dots or a gradual formation of surface defects, that is, defects are continuously formed on and removed from the surface. Therefore, the stability of the photoactivated photoluminescence depends on the stability of the equilibrium between the photoactivated formation and removal of surface defects (Figure 17.8B). We attributed the enhanced photoluminescence quantum efficiency to a static passivation of the surface defects by polymers and polar molecules.

In general, there are multiple ways to improve the photoluminescence quantum efficiency of as-synthesized CdSe quantum dots: (i) Preparation shells from ZnS and other higher bandgap materials, (ii) surface passivation by oxygen, ligands and polar molecules and (iii) photoactivation in the presence of various surface passivating molecules. It has been possible in recent years to synthesize highly luminescent CdSe quantum dots in a one-pot approach and, therefore, the interest in the surface modifications of quantum dots is focused on providing physical stability, chemical and bioconjugations, reduced cytotoxicity and suppressed blinking.

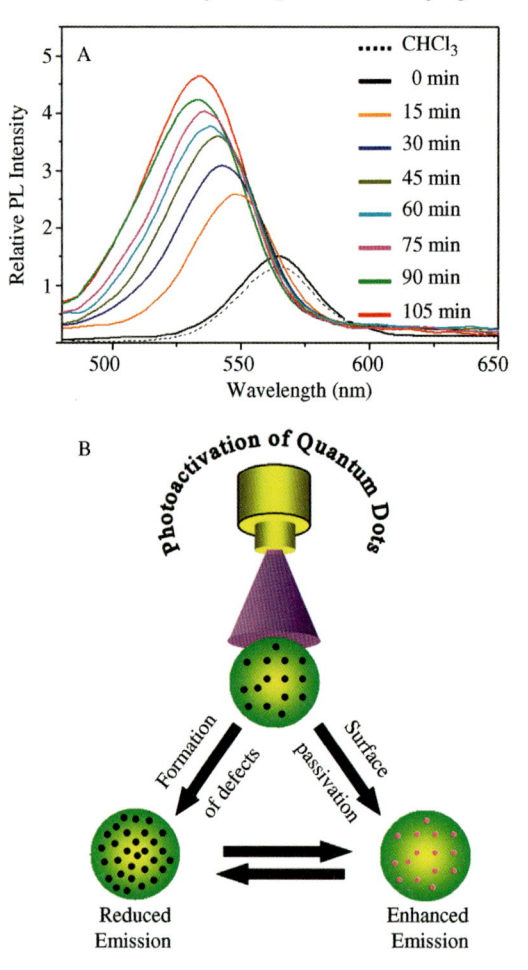

Figure 17.8 (A) Photoluminescence spectra of CdSe quantum dots in CHCl$_3$ in the presence of polybutadiene at different times under photoactivation at 400 nm. The blue shift of the photoluminescence spectra is due to a gradual decrease in quantum dot size. (B) Schematic presentation of the surface passivation and the formation of surface defects in CdSe quantum dots under photoactivation [25]. Reprinted with permission from reference [25]; copyright [2007], American Chemical Society.

17.6
On and Off Luminescence Blinking in Single Quantum Dots

Intermittent on and off photoluminescence (blinking) property of single CdSe and CdSe–ZnS quantum dots was first observed by Nirmal et al. in 1996 [36]. Figure 17.9A shows the typical blinking behavior of a single quantum dot. To date, the origin of the blinking has been correlated with Auger ionization and trapping of photoactivated charge-carriers in defect states. Applications of quantum dots both in technology and biology are limited to a certain extent by this intrinsic blinking. For example, blinking

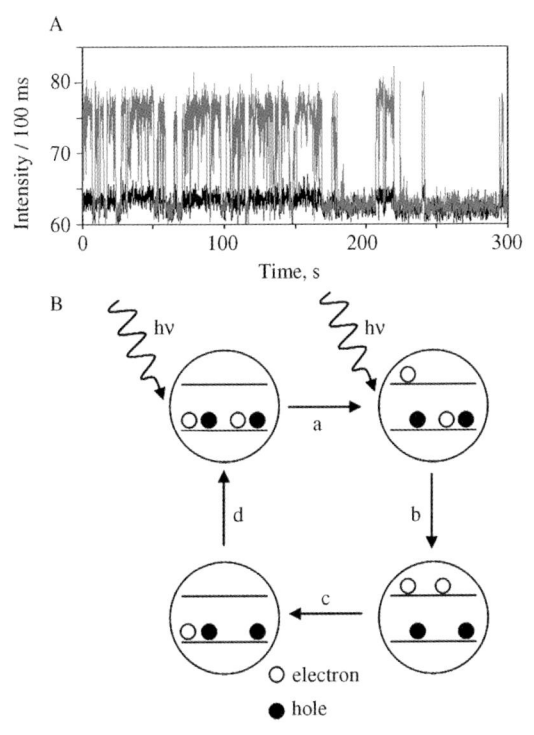

Figure 17.9 (A) Photoluminescence intensity trajectory (gray) of a CdSe–ZnS quantum dot. The high intensity level is the "on" state and the low intensity level is the "off" state. The trace in black is the background intensity. Reprinted with permission from reference [14]; copyright [2005], American Chemical Society. (B) Schematic presentation of Auger ionization in a CdSe quantum dot: (a) photoactivation of a single exciton, (b) photoactivation of a second exciton, (c) electron–electron annihilation and Auger ionization and (d) neutralization of an ionized quantum dot.

of quantum dots is not promising for single-photon logic devices, on demand light emitters, quantum dot switches and continuous imaging and tracking of single molecules. Thus, synthesis of non-blinking quantum dots or complete suppression of blinking by post-synthesis modifications would bring far-reaching changes in the applications of single quantum dots.

Auger ionization and transient trapping of electrons (or holes) in surface states and environment are the origins of blinking [37]. Unlike single molecules, the blinking of a single quantum dot is photoinduced and not spontaneous, which was characterized by an inverse-relation between excitation light-intensity and average "on" times, and excitation light intensity independent "off" times of single CdSe quantum dots [36]. Auger ionization is one of the best models to explain the blinking of quantum dots, which occurs by exciton–exciton annihilation when more than one exciton pairs are photoactivated (Figure 17.9B). Exciton–exciton annihilation creates a high-energy exciton that escapes from the quantum dot and ionizes it. Photoexcitation of an ionized quantum dot always results in non-radiative exciton recombination due to strong coulombic repulsive potentials. Thus, an ionized quantum dot continues to be in the dark state until neutralized. Simply, photoinduced ionization turns "off" a quantum dot and neutralization turns it back "on". On the basis of the Auger ionization model, an increase in the excitation light intensity increases the possibility to turn "off" a quantum dot. On the other hand, neutralization of a photoionized quantum dot is an uncontrolled process. This is the reason why the average "on" time of a quantum dot is excitation light intensity dependent. Thus, blinking can be suppressed if (i) excitation light intensity is kept at a low level, which has practical limitations in single-molecule microscopy and (ii) resonant tunneling of a charge carrier to an outside trap is prevented by removing defects.

17.6.1
Power-Law Statistics of On and Off Time Distributions

Although the Auger ionization model is widely accepted to account for the blinking of quantum dots, exponential decays of "on" and "off" probability densities involving a single trap state are inconsistent with experimental observations of the distributed kinetics; that is, "on" and "off" times are distributed over different (3–5) decades of time (power-law behavior) [38]. A distribution of "off" times over three decades of time for a single CdSe quantum dot is shown in Figure 17.10. The distribution of "off" times indicates the presence of distributed trap states. Thus, trap states with different energies are present on the surface and in the surrounding matrix of a quantum dot. Although an Arrhenius model and a hopping mechanism are assumed to account for the power-law distribution of "off" times, the hopping mechanism is significant based on the lower sensitivity of "off" time distributions to temperature changes. The concepts of the single-exponential behavior and the distributed kinetics have recently been revisited by Tang and Marcus [39] and Frantsuzov and Marcus [40] to account for a cut-off of the power-law behavior at longer time scales. In the latest models, blinking is attributed to hole-trapping and phonon-assisted diffusion-controlled electron transfer processes [39, 40].

17.6.2
Modified Blinking

There are advances in both experimental and theoretical studies to gain an understanding of the origin and distribution of trap states in quantum dots and to prepare non-blinking quantum dots. Electron transfer from a photoexcited quantum dot to an outside trap or an acceptor is a barrier-less process. Thus, preparation of a physical barrier such as an insulating layer on the surface of a quantum dot is insufficient to control the blinking. However, post-synthesis surface modifications suppress the blinking of quantum dots to a certain extent by removing electron traps on the surface. For example, "on" and "off" times in the intensity trajectories of single CdSe quantum dots are considerably increased when overlaid with ZnS shells of different thickness [36]. The increases in the "on" and "off" times show a reduced density of surface defects and a suppression of Auger ionization. Similarly, blinking of single

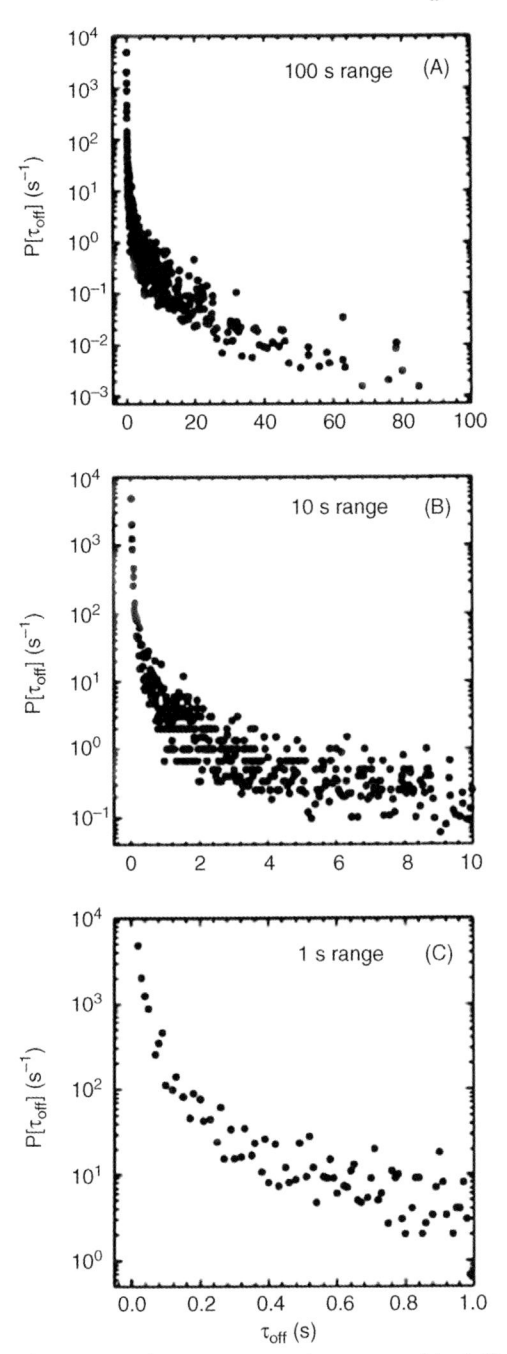

Figure 17.10 Three successive enlargements of the "off" time probability density for a single CdSe quantum dot. Reprinted with permission from reference [38]; copyright [2000], American Institute of Physics.

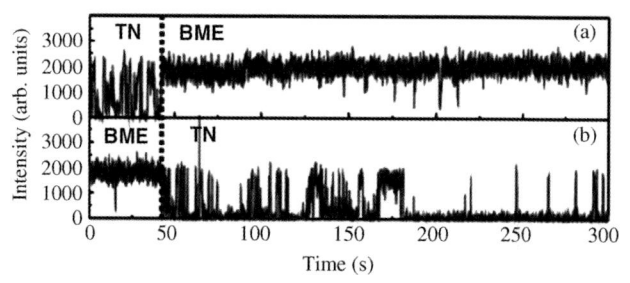

Figure 17.11 (a) A single dot (emission peak at 585 nm) intensity trace as tris-HCl buffer with 140 mM BME was injected at ∼40 s (dotted lines) into the sample, displacing the buffer from the sample. (b) A time trace of a reverse case where BME buffer was washed away using the buffer at ∼40 s. Reprinted with permission from reference [28]; copyright [2004], American Chemical Society.

quantum dots is suppressed in the presence of ligands such as β-mercaptoethanol (Figure 17.11), oligo(phenylene vinylene) and aliphatic amines. In the presence of these ligands, the surface traps are efficiently quenched by electron transfer. Blinking is also considerably affected when the dielectric environment of a quantum dot is changed. Typical examples are suppression of blinking when quantum dots are placed on Ag [41], Au [42] and ITO (indium-tin-oxide) glass [43] surfaces. Recently, we observed a considerable increase in the "on" time of single CdSe–ZnS quantum dots placed on an Ag nanoparticle film [41]. Figure 17.12B shows the intensity trajectories of single CdSe–ZnS quantum dots on an Ag nanoparticle surface and a glass surface. Interestingly, complete "off" levels are absent in the intensity trajectory of a quantum dot placed on the Ag nanoparticle film; instead, stochastic fluctuations of intensity with non-zero intensity levels (pseudo "off" states) are observed. This modified-blinking is associated with reduced average photoluminescence intensity (Figure 17.12C) and lifetime values at ensemble and single dot levels. The reduced photoluminescence intensity and lifetime values are due to non-radiative energy transfer from quantum dots to Ag nanoparticles. However, quantum dots with enhanced and intact photoluminescence intensities were also observed on the Ag nanoparticle film (Figure 17.12D). Figure 17.12A shows the energy transfer and various relaxation processes in a photoactivated quantum dot. The competence of the energy transfer process to compete with carrier-trapping and Auger ionization is the origin of the modified blinking. In other words, the relaxation of excitons in a photoactivated quantum dot that is present on an Ag nanoparticle film is dominated by ultra-fast non-radiative energy transfer to a proximal Ag nanoparticle; the rate of energy transfer is comparable to that of the trapping of charge-carriers.

The suppression of blinking by electron donor molecules, ZnS shells, and noble metal nanoparticles validates the electron trapping mechanism and the distributed kinetics model of blinking. The suppression of blinking and increase in the "on" times are due to (i) the suppression of Auger ionization by removing traps (by electron donor molecules and ZnS shells), and (ii) deactivation of charge carriers via energy transfer to Ag nanoparticles before carrier trapping and Auger ionization.

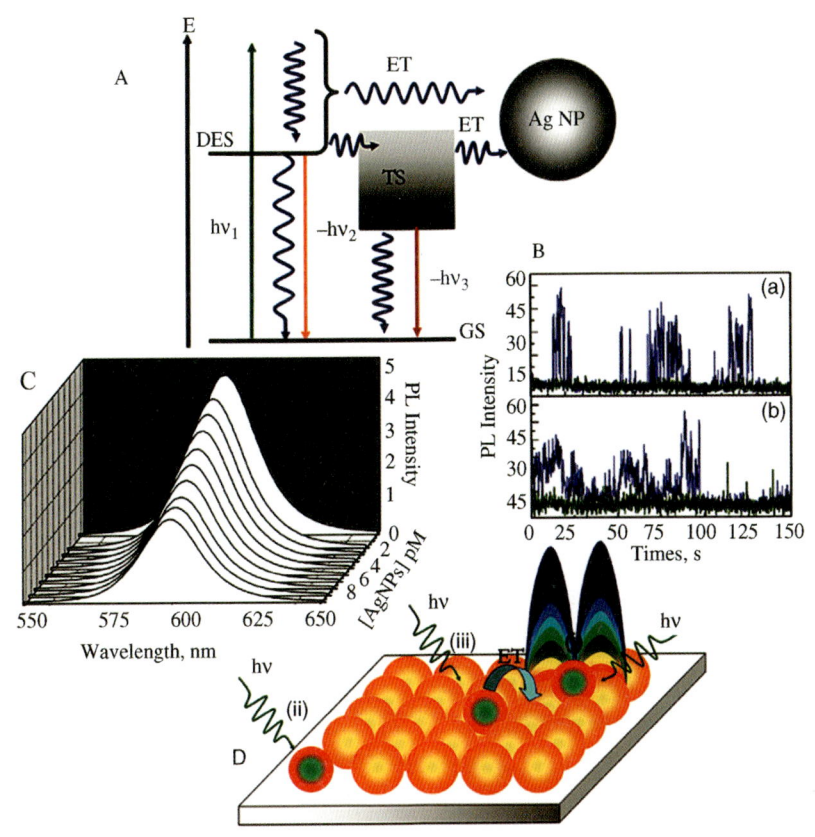

Figure 17.12 (A) Schematic presentation of deactivation and energy transfer processes in a single quantum dot placed on an Ag nanoparticle film. (B) Photoluminescence intensity trajectories of single quantum dots on a glass substrate (a) and on an Ag nanoparticle film (b). The traces in green represent background intensities. (C) Photoluminescence spectra of quantum dot solutions in the presence of different concentrations of Ag nanoparticles. (D) Schematic presentation of (i) a quantum dot without proximal Ag nanoparticles, (ii) a quantum dot involved in energy transfer to a proximal Ag nanoparticle, and (iii) a quantum dot in a strong plasmon field of Ag nanoparticles. Reprinted with permission from reference [41]; copyright [2008], American Chemical Society.

Also, we identified that the "on" and "off" times in the intensity trajectories of single CdSe quantum dots in a sample synthesized at a slow rate at a low temperature (75 °C) are considerably decreased and uniformly distributed [14]. The short-lived "on" states indicate the presence of a high density of surface traps and the short-lived "off" states indicate a narrow distribution of the surface traps. In other words, the rate of ionization was increased by the high density of surface traps, and the narrow distribution of surface traps increased the probability of neutralization. This is an example of deviation from the distributed-kinetics model of blinking. In short, the blinking of quantum dots is due to transient trapping of charge carriers in trap states distributed on the surface and environment and the discrepancies in the blinking in different environments and in the presence of different ligands are due to a difference in the distribution of trap states.

17.7
Conclusions

Strong three-dimensional confinements of excitons provide size-tunable photoluminescence color to CdSe quantum dots. Additionally, brightness and stability of photoluminescence make the CdSe quantum dot a promising material for optoelectronic and electro-optic devices, bioanalyses, and bioimaging. Systematic investigations of various chemical and physical parameters involved in the colloidal synthesis of CdSe quantum dots made their preparation straightforward and improved their optical and electronic properties. The large surface-to-volume ratios and the presence of surface states give temperature-, pressure- and environment-sensitive photoluminescence properties to CdSe quantum dots. However, the intermittent blinking due to Auger ionization makes quantum dots less attractive for various single quantum dot applications. Photoluminescence properties and the stability of quantum dots are considerably improved through advances in the colloidal synthesis and post-synthesis surface modifications. Also, the blinking of single quantum dots is suppressed by preparing shells, tethering electron donor molecules, modified synthesis, and introducing dielectric environments. In addition to the improvement of the optical and electronic properties of CdSe quantum dots, all the above efforts considerably contributed to the general understanding of bandgap structure and exciton relaxation processes in quantum confined systems.

References

1 Brus, L. E. (1984) Electron-electron and electron-hole interactions in small semiconductor crystallites: The size dependence of the lowest excited electronic state. *J. Chem. Phys.*, **80**, 4403–4409.

2 Brus, L. E. (1986) Electronic wave functions in semiconductor clusters: experiment and theory. *J. Phys. Chem.*, **90**, 2555–2560.

3 Murray, C. B., Norris, D. J. and Bawendi, M. G. (1993) Synthesis and characterization of nearly monodisperse CdE (E = sulfur, selenium, tellurium) semiconductor nanocrystallites. *J. Am. Chem. Soc.*, **115**, 8706–8715.

4 Dabbousi, B. O., RodriguezViejo, J., Mikulec, F. V., Heine, J. R., Mattoussi, H., Ober, R., Jensen, K. F. and Bawendi, M. G. (1997) (CdSe)ZnS core-shell quantum dots: Synthesis and characterization of a size series of highly luminescent nanocrystallites. *J. Phys. Chem. B*, **101**, 9463–9475.

5 Talapin, D. V., Rogach, A. L., Kornowski, A., Haase, M. and Weller, H. (2001) Highly luminescent monodisperse CdSe and CdSe/ZnS nanocrystals synthesized in a hexadecylamine-trioctylphosphine oxide-trioctylphospine mixture. *Nano Lett.*, **1**, 207–211.

6 Klimov, V. I., Mikhailovsky, A. A., Xu, S., Malko, A., Hollingsworth, J. A., Leatherdale, C. A., Eisler, H. J. and Bawendi, M. G. (2000) Optical gain and

stimulated emission in nanocrystal quantum dots. *Science*, **290**, 314–317.

7 Schlamp, M. C., Peng, X. G. and Alivisatos, A. P. (1997) Improved efficiencies in light emitting diodes made with CdSe(CdS) core/shell type nanocrystals and a semiconducting polymer. *J. Appl. Phys.*, **82**, 5837–5842.

8 Robel, I., Subramanian, V., Kuno, M. and Kamat, P. V. (2006) Quantum dot solar cells. Harvesting light energy with CdSe nanocrystals molecularly linked to mesoscopic TiO_2 films. *J. Am. Chem. Soc.*, **128**, 2385–2393.

9 Medintz, I. L., Uyeda, H. T., Goldman, E. R. and Mattoussi, H. (2005) Quantum dot bioconjugates for imaging, labelling and sensing. *Nat. Mater.*, **4**, 435–446.

10 Biju, V., Muraleedharan, D., Nakayama, K., Shinohara, Y., Itoh, T., Baba, Y. and Ishikawa, M. (2007) Quantum dot-insect neuropeptide conjugates for fluorescence imaging, transfection, and nucleus targeting of living cells. *Langmuir*, **23**, 10254–10261.

11 Chan, W. C. W. and Nie, S. M. (1998) Quantum dot bioconjugates for ultrasensitive nonisotopic detection. *Science*, **281**, 2016–2018.

12 Katari, J. E. B., Colvin, V. L. and Alivisatos, A. P. (1994) X-Ray photoelectron-spectroscopy of CdSe nanocrystals with applications to studies of the nanocrystal surface. *J. Phys. Chem.*, **98**, 4109–4117.

13 Qu, L. H., Peng, Z. A. and Peng, X. (2001) Alternative routes toward high quality CdSe nanocrystals. *Nano Lett.*, **1**, 333–337.

14 Biju, V., Makita, Y., Nagase, T., Yamaoka, Y., Yokoyama, H., Baba, Y. and Ishikawa, M. (2005) Subsecond luminescence intensity fluctuations of single CdSe quantum dots. *J. Phys. Chem. B*, **109**, 14350–14355.

15 Rogach, A. L., Kornowski, A., Gao, M. Y., Eychmuller, A. and Weller, H. (1999) Synthesis and characterization of a size series of extremely small thiol-stabilized CdSe nanocrystals. *J. Phys. Chem. B*, **103**, 3065–3069.

16 Nirmal, M. and Brus, L. E. (1999) Luminescence photophysics in semiconductor nanocrystals. *Acc. Chem. Res.*, **32**, 407–414.

17 Underwood, D. F., Kippeny, T. and Rosenthal, S. J. (2001) Ultrafast carrier dynamics in CdSe nanocrystals determined by femtosecond fluorescence upconversion spectroscopy. *J. Phys. Chem. B*, **105**, 436–443.

18 Klimov, V. I., McBranch, D. W., Leatherdale, C. A. and Bawendi, M. G. (1999) Electron and hole relaxation pathways in semiconductor quantum dots. *Phys. Rev. B*, **60**, 13740–13749.

19 Kim, B. S., Islam, M. A., Brus, L. E. and Herman, I. P. (2001) Interdot interactions and band gap changes in CdSe nanocrystal arrays at elevated pressure. *J. Appl. Phys.*, **89**, 8127–8140.

20 Nirmal, M., Murray, C. B. and Bawendi, M. G. (1994) Fluorescence-line narrowing in CdSe quantum dots – surface localization of the photogenerated exciton. *Phys. Rev. B*, **50**, 2293–2300.

21 Biju, V., Makita, Y., Sonoda, A., Yokoyama, H., Baba, Y. and Ishikawa, M. (2005) Temperature-sensitive photoluminescence of CdSe quantum dot clusters. *J. Phys. Chem. B*, **109**, 13899–13905.

22 Al Salman, A., Tortschanoff, A., Mohamed, M. B., Tonti, D., van Mourik, F. and Chergui, M. (2007) Temperature effects on the spectral properties of colloidal CdSe nanodots, nanorods, and tetrapods. *Appl. Phys. Lett.*, **90**, 093104.

23 Empedocles, S. A. and Bawendi, M. G. (1997) Quantum-confined stark effect in single CdSe nanocrystallite quantum dots. *Science*, **278**, 2114–2117.

24 Leatherdale, C. A. and Bawendi, M. G. (2001) Observation of solvatochromism in CdSe colloidal quantum dots. *Phys. Rev. B*, **63**, 165315.

25 Biju, V., Kanemoto, R., Matsumoto, Y., Ishii, S., Nakanishi, S., Itoh, T., Baba, Y. and Ishikawa, M. (2007) Photoinduced photoluminescence variations of CdSe quantum dots in polymer solutions. *J. Phys. Chem. C*, **111**, 7924–7932.

26 Nazzal, A. Y., Wang, X. Y., Qu, L. H., Yu, W., Wang, Y. J., Peng, X. G. and Xiao, M. (2004) Environmental effects on photoluminescence of highly luminescent CdSe and CdSe/ZnS core/shell nanocrystals in polymer thin films. *J. Phys. Chem. B*, **108**, 5507–5515.

27 Kanemoto, R., Anas, A., Matsumoto, Y., Ueji, R., Itoh, T., Baba, Y., Nakanishi, S., Ishikawa, M. and Biju, V. (2008) Relations between dewetting of polymer thin films and phase-separation of encompassed quantum dots. *J. Phys. Chem. C*, **112**, 8184–8191.

28 Hohng, S. and Ha, T. (2004) Near-complete suppression of quantum dot blinking in ambient conditions. *J. Am. Chem. Soc.*, **126**, 1324–1325.

29 Myung, N., Bae, Y. and Bard, A. J. (2003) Enhancement of the photoluminescence of CdSe nanocrystals dispersed in $CHCl_3$ by oxygen passivation of surface states. *Nano Lett.*, **3**, 747–749.

30 Wuister, S. F., van Houselt, A., Donega, C. D. M., Vanmaekelbergh, D. and Meijerink, A. (2004) Temperature antiquenching of the luminescence from capped CdSe quantum dots. *Angew. Chem. Int. Ed.*, **43**, 3029–3033.

31 Jones, M., Nedeljkovic, J., Ellingson, R. J., Nozik, A. J. and Rumbles, G. (2003) Photoenhancement of luminescence in colloidal CdSe quantum dot solutions. *J. Phys. Chem. B*, **107**, 11346–11352.

32 Cordero, S. R., Carson, P. J., Estabrook, R. A., Strouse, G. F. and Buratto, S. K. (2000) Photo-activated luminescence of CdSe quantum dot monolayers. *J. Phys. Chem. B*, **104**, 12137–12142.

33 Maenosono, S., Dushkin, C. D., Saita, S. and Yamaguchi, Y. (2000) Optical memory media based on excitation-time dependent luminescence from a thin film of semiconductor nanocrystals. *Jpn. J. Appl. Phys.*, **39**, 4006–4012.

34 Uematsu, T., Maenosono, S. and Yamaguchi, Y. (2006) Photoinduced fluorescence enhancement in CdSe/ZnS quantum dot monolayers: Influence of substrate. *Appl. Phys. Lett.*, **89**, 031910.

35 Wang, Y., Tang, Z. Y., Correa-Duarte, M. A., Pastoriza-Santos, I., Giersig, M., Kotov, N. A. and Liz-Marzan, L. M. (2004) Mechanism of strong luminescence photoactivation of citrate-stabilized water-soluble nanoparticles with CdSe cores. *J. Phys. Chem. B*, **108**, 15461–15469.

36 Nirmal, M., Dabbousi, B. O., Bawendi, M. G., Macklin, J. J., Trautman, J. K., Harris, T. D. and Brus, L. E. (1996) Fluorescence intermittency in single cadmium selenide nanocrystals. *Nature*, **383**, 802–804.

37 Efros, A. L. and Rosen, M. (1997) Random telegraph signal in the photoluminescence intensity of a single quantum dot. *Phys. Rev. Lett.*, **78**, 1110–1113.

38 Kuno, M., Fromm, D. P., Hamann, H. F., Gallagher, A. and Nesbitt, D. J. (2000) Nonexponential "blinking" kinetics of single CdSe quantum dots: A universal power law behavior. *J. Chem. Phys.*, **112**, 3117–3120.

39 Tang, J. and Marcus, R. A. (2005) Diffusion-controlled electron transfer processes and power-law statistics of fluorescence intermittency of nanoparticles. *Phys. Rev. Lett.*, **95**, 107401.

40 Frantsuzov, P. A. and Marcus, R. A. (2005) Explanation of quantum dot blinking without the long-lived trap hypothesis. *Phys. Rev. B*, **72**, 155321.

41 Matsumoto, Y., Kanemoto, R., Itoh, T., Nakanishi, S., Ishikawa, M. and Biju, V. (2008) Photoluminescence quenching and intensity fluctuations of CdSe-ZnS quantum dots on an Ag nanoparticle film. *J. Phys. Chem. C*, **112**, 1345–1350.

42 Ito, Y., Matsuda, K. and Kanemitsu, Y. (2007) Mechanism of photoluminescence enhancement in single semiconductor nanocrystals on metal surfaces. *Phys. Rev. B*, **75**, 033309.

43 Verberk, R., Chon, J. W. M., Gu, M. and Orrit, M. (2005) Environment-dependent blinking of single semiconductor nanocrystals and statistical aging of ensembles. *Physica E*, **26**, 19–23.